ANIMAL SIGNALING AND FUNCTION

ANIMAL SIGNALING AND FUNCTION

AN INTEGRATIVE APPROACH

Edited by

Duncan J. Irschick

Department of Biology and Organismic and Evolutionary Biology Program
University of Massachusetts at Amherst
Amherst, MA, USA

Mark Briffa

Marine Biology and Ecology Research Centre
Plymouth University
Plymouth, UK

Jeffrey Podos

Department of Biology and Organismic and Evolutionary Biology Program
University of Massachusetts at Amherst
Amherst, MA, USA

WILEY Blackwell

Library of Congress Cataloging-in-Publication Data:

Animal signaling and function : an integrative approach / edited by Duncan J. Irschick, Mark Briffa, Jeffrey Podos.
    pages cm
  Includes bibliographical references and index.
  ISBN 978-0-470-54600-0 (Cloth)
  1. Animal communication. I. Irschick, Duncan J., editor. II. Briffa, Mark., editor. III. Podos, Jeffrey, 1967- editor.
  QL776.A538 2015
  591.59–dc23

                                                                    2014028491

10 9 8 7 6 5 4 3 2 1
Printed and bound in Singapore by Markono Print Media Pte Ltd

CONTENTS

CONTRIBUTORS

Michael J. Angilletta Jr., School of Life Sciences, Arizona State University, Tempe, AZ, USA

Jonathan D. Blount, Centre for Ecology and Conservation, School of Biosciences, University of Exeter, Penryn, Cornwall, UK

Gerald Borgia, Department of Biology and Behavior, Ecology, Evolution, and Systematics Program, University of Maryland, College Park, MD, USA

Mark Briffa, Marine Biology and Ecology Research Centre, Plymouth University, Plymouth, UK

Gregory F. Grether, Department of Ecology and Evolutionary Biology, University of California, Los Angeles, CA, USA

Eileen A. Hebets, School of Biological Sciences, University of Nebraska-Lincoln, Lincoln, NE, USA

Justin P. Henningsen, Organismic and Evolutionary Biology Program, University of Massachusetts at Amherst, Amherst, MA, USA

Jerry F. Husak, Department of Biology, University of Massachusetts at Amherst, Amherst, MA, USA

Duncan J. Irschick, Department of Biology and Organismic and Evolutionary Biology Program, University of Massachusetts at Amherst, Amherst, MA, USA

Jason Keagy, Department of Biology and Behavior, Ecology, Evolution, and Systematics Program, University of Maryland, College Park, MD, USA

Darrell J. Kemp, Department of Biological Sciences, Macquarie University, Sydney, NSW, Australia

Josephine M. Orledge, Centre for Ecology and Conservation, School of Biosciences, University of Exeter, Penryn, Cornwall, UK

Sheila Patek, Department of Biology, Duke University, Durham, NC, USA

Jeffrey Podos, Department of Biology and Organismic and Evolutionary Biology Program, University of Massachusetts at Amherst, Amherst, MA, USA

Nick J. Royle, Centre for Ecology and Conservation, School of Biosciences, University of Exeter, Penryn, Cornwall, UK

Bieke Vanhooydonck, Department of Biology, University of Antwerp, Antwerp, Belgium

Dustin J. Wilgers, School of Biological Sciences, University of Nebraska-Lincoln, Lincoln, NE, USA

Robbie S. Wilson, School of Biological Sciences, The University of Queensland, St Lucia, QLD, Australia

1

INTRODUCTION

Duncan J. Irschick,[1] Mark Briffa,[2] and Jeffrey Podos[1]

[1]*Department of Biology, Organismic and Evolutionary Biology Program,
University of Massachusetts at Amherst, Amherst, MA, USA*
[2]*Marine Biology and Ecology Research Centre, Plymouth University, Plymouth, UK*

Animal signals are among nature's most compelling and diverse phenomena. Human cultures have long celebrated the expression of elaborate signals and displays, such as colors, songs, and dances of birds, which impress with their exuberance. Yet equally impressive are subtle modes of communication that had until recently eluded our detection. Some examples include the low-voltage electrical signals emitted and detected by some fishes as they orient, navigate, and communicate (Lissmann, 1958); the emission of pheromone plumes leading moths on a path upwind toward mates (David *et al.*, 1983); the inaudible, ultrasonic echolocation cries of bats (Griffin, 1958); the ultraviolet reflectance structures of many birds, butterflies, and flowers (Sheldon *et al.*, 1999); and the subtle substrate-borne signals that insects like lacewings use to communicate species identity (Wells and Henry, 1992). In many animal groups, signals express structures that are species-specific (e.g., Sueur, 2002) and that are partitioned over time and space (e.g., Luther, 2009). And many animal displays involve the coordination of multiple modalities, perhaps as a way to signal simultaneously to multiple audiences, or alternatively to enhance detectability, discriminability, and memorability. Documenting the diversity and intricacies of natural signaling modes, structures, and strategies is of itself a highly worthwhile endeavor.

Animal Signaling and Function: An Integrative Approach, First Edition.
Edited by Duncan J. Irschick, Mark Briffa, and Jeffrey Podos.
© 2015 John Wiley & Sons, Inc. Published 2015 by John Wiley & Sons, Inc.

Signals also demand our attention because they hold additional conceptual relevance in the fields of animal behavior and evolutionary biology (Andersson, 1994; Berglund *et al.*, 1996; Maynard-Smith and Harper, 2003). Signals and communication behavior turn out to be central to understanding varied processes of fundamental interest such as how animals optimize their social interactions, how animals choose mates, and how new species arise. We define signals as traits that are produced by senders, which transmit information through the environment, and which help receivers decide if and how to respond. Typically, but not always, both sender and receiver benefit via this transfer of information. This definition encompasses the presentation of morphological structures specialized for transmitting information to other individuals (e.g., a colorful anoline lizard dewlap) as well as elaborate displays that require high levels of skill, such as bird song (e.g., Podos and Nowicki, 2004; Byers *et al.*, 2010). The majority of communication occurs within species, and signals thus evolve primarily in the context of social selection (West-Eberhard, 1983). When signals of co-occurring species overlap in structure, they tend to diverge through a process of reproductive character displacement, thus emphasizing interspecific distinctions (e.g., Grant and Grant, 2010). Within species, much communication occurs between the sexes as each vies to maximize reproductive success, typically in circumstances in which the interests of signalers and receivers conflict with one another (Searcy and Nowicki, 2005). The signals that mediate these interactions, and other conflicts of interest, have been the focus of a large body of work in recent decades, with contributions from both modeling and empirical perspectives (e.g., Andersson, 1994; Johnstone 1995; Briffa and Hardy, 2013).

Yet despite years of research, our state of knowledge concerning sexual signals and their evolutionary basis has remained surprisingly unsettled. Some of this can be explained by a lack of certainty about which sexual selection models are most broadly applicable, whether it is possible to identify relevant null models, and the degree to which we should assume that signals convey information that is reliable (e.g., Hunt *et al.*, 2004a, 2004b). Most well-known is the difficulty in reconciling classic Fisherian (runaway) models of sexual selection with those requiring that signals provide reliable indicators of sender attributes (e.g., Maynard-Smith and Harper, 2003; Prum, 2010). From an empirical standpoint, Fisherian models of sexual selection require a genetic association of signal and preference traits, the demonstration of which still remains mostly beyond reach (Prum, 2010). Indicator models, by contrast, require that "high-quality" senders possess "good genes" (Møller and Alatalo, 1999) and are thus desirable as mates (the "sexy son" hypothesis, Zeh, 2004). Yet in practice it is daunting to determine whether a signaler possesses high genetic quality, and therefore most studies attempt to find a more pragmatic proxy. For example, some models of sexual signal evolution assume costs and benefits to the possession of a signal, such as a diminished flight performance as a result of unusually elongated tail feathers (Balmford *et al.*, 1993), or increased energetic or developmental costs (e.g., drumming in wolf-spiders, Kotiaho *et al.*, 1998; vocalization in frogs, Wells and Tiagen, 1989; see Kotiaho, 2001). This integration of physiological and mechanistic methods with more traditional sexual selection theory has been formalized as the functional approach to sexual selection (Lailvaux and Irschick, 2006; Mowles *et al.*, 2010). This approach has gained significant traction over the past decade, with many studies emerging to test theories of sexual selection across a range of behavioral contexts. Our goal in this volume is to bring together a wide variety of papers applying diverse approaches to this topic, ranging across empirical, experimental, and theoretical perspectives. As a result, this work should hold special interest for researchers in three fields: sexual selection, physiological ecology, and functional morphology.

Functional approaches hold the promise of providing insight into several key aspects of sexual selection theory, especially in regard to signal honesty and the handicap hypothesis. The handicap hypothesis is predicated on the notion that we should be able to define individual male quality and relate it to measurements of sexual signal elaboration (e.g., size, color, and shape) as well as to reproductive effort and output. Researchers have devoted much effort toward this end, focusing on quality traits such as condition (Kodric-Brown and Nicoletto, 1993; Jakob *et al.*, 1996; Kotiaho, 1999; Peig and Green, 2010) and levels of parasitism. Yet such measures can be intrinsically problematic (e.g., Jakob *et al.*, 1996; Green, 2000; Peig and Green, 2010). For example, while values of condition may shed some light on an animal's overall health and vigor, simple observations of human or animal sporting events shows that one cannot easily predict human athletic performance based on external appearance (consider the case of the legendary thoroughbred horse Seabiscuit, which outperformed many other larger and more imposing horses in the 1930s and 1940s). On this point, it is important to recognize that no one trait will likely represent a valid measure of quality for all species. But we can ask whether certain kinds of traits offer a more general and satisfying link to our underlying model of individual quality. Over the last decade, and especially within the last few years, functional research has emphasized the utility of measurements of either whole-organism performance capacity (e.g., maximum sprint speed, bite force, locomotor endurance) or physiological variables such as metabolic rate and lactic acid level (e.g., Garland *et al.*, 1990; Briffa *et al.*, 2003; Huyghe *et al.*, 2005; Lappin and Husak, 2005; Wilson *et al.*, 2007; reviewed in Lailvaux and Irschick, 2006; Mowles *et al.*, 2010).

Although the first applications of a functional approach in the study of communication focused on sexual signals, it has now been applied to signals of individual quality that occur in an array of contexts, for example, during agonistic behavior that can occur over resources other than mates (e.g., Briffa *et al.*, 2003; Mowles *et al.*, 2010). Furthermore, the case for a useful interplay between the domains of sexual and non-sexual signals seems increasingly clear from a conceptual viewpoint as well as from a methodological one. As discussed above, the handicap hypothesis is often assumed to be most relevant to the context of sexual signaling, but it also pertains to the question of signal honesty during agonistic encounters as well as signals between prey and predators. Similarly, models of repeated signals are most often assumed to be relevant to animal contests even though it was first suggested in 1997 (Payne and Pagel, 1997) that these models could explain signals in other contexts as well (Mowles and Ord 2012). Thus, the functional approach to the analysis of animal signals is relevant to a wide range of contexts, which are reflected in the chapters of this volume.

The logic of using performance or physiology traits as metrics of individual quality is straightforward. Whereas the role of variables such as condition or parasite levels for dictating the outcome of male fights is unclear, divergence among signalers in performance and physiology seems often far more obvious to us, and perhaps for females choosing mates as well (for female choice, which variables form the basis for it remain far less clear, Wong and Candolin, 2005). For example, for animals that fight by biting one another, the measurement of bite force is likely to be particularly important for determining who will win or lose the fight. Similarly, for animals that fight each other for relatively long time periods, measurements of locomotor endurance or perhaps physiological measurements of lactic acid buildup over time (Schuett and Grober, 2000; Briffa and Elwood, 2001) can inform us which males are well-suited to fight for periods, and which are likely to become exhausted (and why). A second reason for why a functional approach is useful is that performance or physiological traits may offer more holistic overall metrics of male "vigor" because they emerge as a result of many lower-level processes (Arnold, 1983; Bennett and Huey, 1990; Garland and Losos, 1994; Irschick and Garland, 2001). A very fast animal, to illustrate, is one that is likely to be generally healthy across the board, because of running's intense demand on its muscular and skeletal systems, which in turn rely on cellular and metabolic efficiency and capacity. Finally, apart from studies of sexual selection, there is a long and vital tradition of measuring performance and physiological traits in a wide variety of animals and relating variation in them to variation in habitat use, behavior, and morphology (see above references).

In practice, the integration of functional traits into studies of sexual selection can take several forms. First, we can ask whether there is any linkage between performance or physiological traits and the shape, design, or size of sexual signals, a methodology that ultimately tests whether sexual signals are honest. Second, we can ask whether male reproductive success or its correlates, such as dominance, is enhanced by improved performance and physiology, especially in the context of the use of signaling during such encounters. Finally, we can generally examine the evolutionary relationships between sexual signals and functional traits to understand how and why their linkage has arisen.

The chapters in this book showcase the wide variety and utility of functional approaches for enhancing our understanding of signaling evolution, across a range of such contexts. In the second chapter, Royle *et al.* focus their discussion on oxidative stress, and outline how it may serve a causal link between life-history tradeoffs and signal evolution, particularly in taxa under strong sexual selection. Oxidative stress is a price animals pay for using oxygen in its typical reactive form, which in sufficient concentration can as a byproduct cause cells to degrade in structure and function. Selection should thus favor antioxidant defenses, which in turn can compete in life-history development and evolution with investment in elaborate secondary sexual traits. This hypothesis is being supported by multiple emerging lines of evidence. A particularly interesting point emphasized in this chapter is the diversity of ways in which oxidative stress and responses to it can interface with proximate mechanisms that underlie signal expression.

In Chapter 3, Husak *et al*. review a rapidly expanding literature on interrelationships among costs (e.g., energetic, reproductive costs), performance traits, and sexually selected traits. The authors divide their attention between receiver-dependent and receiver-independent costs, and emphasize the interface between these types of capacities and organisms' overall performance capacities. One point of this chapter is that evidence is accumulating for significant function costs in signal evolution. Moreover, in parallel to the discussion of life-history tradeoffs in Chapter 2, Husak *et al*. focus on the idea that animals may evolve "compensatory traits" in response to the negative effects of sexually selected traits.

Chapter 4, by Borgia and Keagy, focuses specifically on the evolution of complex songs, which is an emerging area in which a functional approach is yielding some answers where prior approaches had stalled. The neuroanatomy of bird song has been well-studied, yet the links between the anatomy of the brain and song behavior, as well as these links with social behavior and learning, remain poorly understood. Keagy and Borgia examine how their own work on bowerbirds, a fascinating species in which males construct colorful nests that are designed to attract females, sheds light on the link between social behavior, song, and color signals.

In Chapter 5, Kemp and Grether show how a totally different kind of signal, namely color, offers exciting opportunities to characterize linkages between sexual selection theory and animal function. A main point emphasized by these authors is that colors come in many different forms and vary widely in degrees of phenotypic plasticity, ranging from those that are largely invariant from birth (and thus cannot really be changed) to those that are under considerable environmental influence. It is this latter set of colors, of which the most common form are carotenoid pigments, that have been of particular interest in the realm of sexual selection. Accumulating evidence indicates that such pigments, which are acquired through the consumption of food such as fruit, are limited in nature, and thus the acquisition of them, and their expression in brilliant colors, may be a strong indicator of male quality. Understanding the functional and mechanistic underpinnings of color production, and how animals vary in this trait thus allows us to more clearly understand why different colors have evolved.

Chapter 6, Briffa, evaluates the way in which signals are important for understanding how animals resolve conflicts. His historical approach shows that there has been a steady succession of models aiming to understand how animals resolve fights, especially through the use of signals, which in many cases, are designed to resolve fights without males resorting to violence that could injure either participant. The fact that sexual signals are so strongly linked with functional traits that play a key role during male fights indicates that the resolution of fights may often occur with the use of agonistic signals as advertisements of male quality, and particularly male ability to either persist in the contest or hurt the other opponent. Such examples suggest another key feature of signals that advertise individual quality.

While recent work on communication has perhaps been dominated by sexual signals, signals that advertise quality may also occur in non-reproductive contexts, such as during fights over resources other than mates (see Chapter 1 in Bradbury and Vehrencamp, 2012 for a discussion).

In Chapter 7, Podos and Patek return to acoustic signals, presenting a broad framework for asking how proximate mechanisms of acoustic production can shape signal evolution and divergence. They focus on three interrelated facets of acoustic production: biomechanics, size, and performance, and consider how each constrains and provides opportunities for signal divergence. A proximate focus on acoustic signal production, the authors argue, provides a useful complement to more traditional analyses of signal evolution that adopt optimality-based approaches.

In Chapter 8, Wilson and Angilletta continue the theme of animal contests, this time focusing on the question of the honesty of agonistic signals. The ability to convey false information is characteristic of humans and may even have contributed to the evolution of large brain size. It is therefore a fascinating topic and "bluffing" or exaggerating could clearly be of benefit to any animal involved in a conflict-of-interest situation. After reviewing the underlying theory of honest signals, they focus on how crustaceans have been used as model species to test these ideas. Moreover, they demonstrate the application of functional performance techniques, such as analysis of claw strength, to the analysis of signal honesty. This approach has given many new insights into the question of signal honesty, potentially providing alternative explanations for apparent bluffing during a fight.

Finally, in Chapter 9, Wilgers and Hebets turn to the condition-dependency of animal signals. Although the term *condition* is intuitive and widely used, it is a somewhat difficult concept to define. Nevertheless, signals are often influenced by an individual's health and vigor and may thus be indicative of viability. Many studies have relied on body condition as a proxy for available energy reserves. In this chapter, the authors discuss the advantages of measuring energy reserves directly and explore the potential for genetic correlates of condition to yield new insights about the links between resource allocation and signals. Thus they promote the idea of moving beyond "black box" proxies for condition, such as body size measurements. Such metrics may mean different things for different individuals, species, and taxa. Therefore, the authors explain, we would do better to focus on analyzing the actual proximate mechanisms that may underlie condition and the signals that advertise this state.

This collection thus assembles some of the premier researchers in behavioral ecology and functional morphology, discussing some of the newest ideas to emerge at these fields' interface. It is our hope that this book will generate new ways of thinking about sexual signals, animal function, and performance, and thereby open new avenues for collaborative research and new ways of testing theories both classic and emerging.

REFERENCES

Andersson, M. (1994) *Sexual Selection*, Princeton University Press, Princeton, New Jersey.

Arnold, S.J. (1983) Morphology, performance, and fitness. *American Zoologist*, **23**, 347–361.

Balmford, A., Thomas, A.L., and Jones, I.L. (1993) Aerodynamics and the evolution of long tails in birds. *Nature*, **631**, 628–631.

Bennett, A.F. and Huey, R.B. (1990) Studying the evolution of physiological performance. *Oxford Surveys in Evolutionary Biology*, **7**, 251–284.

Berglund, A., Bisazza, A., and Pilastro, A. (1996) Armaments and ornaments: an evolutionary explanation of traits of dual utility. *Biological Journal of the Linnean Society*, **58**, 385–399.

Bradbury, J.W. and Vehrencamp, S.L. (2012) *Principles of Animal Communication*, 2nd edn, Sinauer, Sunderland, MA.

Briffa, M. and Elwood, R.W. (2001) Decision rules, energy metabolism and vigour of hermit-crab fights. *Proceedings of the Royal Society of London, Series B: Biological Sciences*, **268**, 1841–1848.

Briffa, M., Elwood, R.W., and Russ, J.M. (2003) Analysis of multiple aspects of repeated signal: Power and rate of rapping during shell fights in hermit crabs. *Behavioral Ecology*, **14**, 60–65.

Briffa, M. and Hardy, I.C.W. (2013) Preface, in *Animal Contests* (eds I.C.W. Hardy and M. Briffa), Cambridge University Press, Cambridge, pp. xxi–xxiv.

Byers, J., Hebets, E., and Podos, J. (2010) Female mate choice based upon male motor performance. *Animal Behaviour*, **79**, 771–778.

David, C.T., Kennedy, J.S., and Ludlow, A.R. (1983) Finding of a sex-pheromone source by gypsy moths released in the field. *Nature*, **303**, 804–806.

Garland, T. Jr., Hankins, E., and Huey, R.B. (1990) Locomotor capacity and social dominance in male lizards. *Functional Ecology*, **4**, 243–250.

Garland, T. Jr. and Losos, J.B. (1994) Ecological morphology of locomotor performance in squamate reptiles, in *Ecological Morphology: Integrative Organismal Biology* (eds P.C. Wainwright and S.M. Reilly), University of Chicago Press, Chicago, pp. 240–302.

Grant, B.R. and Grant, P.R. (2010) Songs of Darwin's finches diverge when a new species enters the community. *Proceedings of the National Academy of Sciences of the United States of America*, **107**, 20156–20163.

Green, A.J. (2000) Mass/length residuals: measurements of body condition, or generators of spurious results? *Ecology*, **82**, 1473–1483.

Griffin, D.R. (1958) *Listening in the Dark*, Yale University Press, New Haven, CT.

Hunt, J., Bussiere, L.C., Jennions, M.D., and Brooks, R. (2004a) What is genetic quality? *Trends in Ecology and Evolution*, **19**, 329–333.

Hunt, J., Brooks, R., Jennions, M.D., Smith, M.J., Bentsen, C.L., and Bussiere, L.C. (2004b) High quality male field crickets invest heavily in sexual display but die young. *Nature*, **432**, 1024–1027.

Huyghe, K., Vanhooydonck, B., Scheers, H., Molina-Borja, M., and Van Damme, R. (2005) Morphology, performance and fighting capacity in male lizards, *Gallotia galloti*. *Functional Ecology*, **19**, 800–807.

Irschick, D.J. and Garland, T. Jr. (2001) Integrating function and ecology in studies of adaptation: studies of locomotor capacity as a model system. *Annual Reviews of Ecology and Systematics*, **32**, 367–396.

Jakob, E.M., Marshall, S.D., and Uetz, G.W. (1996) Estimating fitness components: a comparison of body condition indices. *Oikos*, **77**, 61–67.

Johnstone, R.A. (1995) Sexual selection, honest advertisement and the handicap principle: reviewing the evidence. *Biological Review*, **70**, 1–65.

Kodric-Brown, A. and Nicoletto, P. (1993) The relationship between physical condition and social status in pupfish *Cyprinodon pecosensis*. *Animal Behaviour*, **46**, 1234–1236.

Kotiaho, J.S. (1999) Estimating fitness: comparison of body condition indices revisited. *Oikos*, **87**, 399–400.

Kotiaho, J. (2001) Costs of sexual traits: a mismatch between theoretical considerations and empirical evidence. *Biological Reviews*, **76**, 365–376.

Kotiaho, J., Alatalo, R.V., Mappes, J., Nielson, M.G., Parri, S., and Rivero, A. (1998) Energetic costs of size and sexual signaling in a wolf spider. *Proceedings of the Royal Society of London, Series B: Biological Sciences*, **265**, 2203–2209.

Lailvaux, S. and Irschick, D.J. (2006) A functional perspective on sexual selection: insights and future prospects. *Animal Behaviour*, **72**, 263–273.

Lappin, A.K. and Husak, J. (2005) Weapon performance, not size, determines mating success and potential reproductive output in the collared lizard (*Crotaphytus collaris*). *American Naturalist*, **166**, 426–436.

Lissmann, H.W. (1958) On the function and evolution of electric organs in fish. *Journal of Experimental Biology*, **35**, 156–191.

Luther, D. (2009) The influence of the acoustic community on songs of birds in a neotropical rain forest. *Behavioral Ecology*, **20**, 864–871.

Maynard-Smith, J. and Harper, D. (2003) *Animal Signals*, Oxford University Press, Oxford.

Møller, A.P. and Alatalo, R.V. (1999) Good-genes effects in sexual selection. *Proceedings of the Royal Society of London, Series B: Biological Sciences*, **266**, 85–91.

Mowles, S.L., Cotton, P.A., and Briffa, M. (2010) Whole-organism performance capacity predicts resource holding potential in the hermit crab *Pagurus bernhardus*. *Animal Behaviour*, **80**, 277–282.

Mowles, S.L. and Ord, T.J. (2012) Repetitive signals and mate choice: insights from contest theory. *Animal Behaviour*, **84**, 295–304.

Payne, R.J.H. and Pagel, M. (1997) Why do animals repeat costly signals? *Animal Behavior*, **54**, 109–119.

Peig, J. and Green, A.J. (2010) The paradigm of body condition: a critical reappraisal of current methods based on mass and length. *Functional Ecology*, **6**, 1323–1332.

Podos, J. and Nowicki, S. (2004) Performance limits on birdsong, in *Nature's Music: The Vocal Life of Birds* (eds P. Marler and H. Slabbekoorn), Elsevier Academic Press, San Diego, pp. 318–342.

Prum, R.O. (2010) The Lande–Kirkpatrick mechanism is the null model of evolution by intersexual selection: implications for meaning, honesty, and design in intersexual signals. *Evolution*, **64**, 3085–3100.

Schuett, G.W. and Grober, M.S. (2000) Post-fight levels of plasma lactate and corticosterone in male copperheads, *Agkistrodon contortrix* (Serpentes, Viperidae): differences between winners and losers. *Physiology & Behavior*, **71**, 335–341.

Searcy, W.A. and Nowicki, S. (2005) The Evolution of Animal Communication: *Reliability and Deception in Signaling Systems*, Princeton University Press, Princeton, NJ.

Sheldon, B.C., Andersson, S., Griffith, S.C., Ornborg, J., and Sendecka, J. (1999) Ultraviolet colour variation influences blue tit sex ratios. *Nature*, **402**, 874–877.

Sueur, J. (2002) Cicada acoustic communication: potential sound partitioning in a multi-species community from Mexico (Hemiptera : Cicadomorpha : Cicadidae). *Biological Journal of the Linnean Society*, **75**, 379–394.

Wells, M.M. and Henry, C.S. (1992) The role of courtship songs in reproductive isolation among populations of green lacewings of the genus Chrysoperla (Neuroptera, Chrysopidae). *Evolution*, **46**, 31–32.

Wells, K.D. and Tiagen, T.L. (1989) Calling Energetics of Neotropical Treefrog, *Hyla microcephala*. *Behavioral Ecology and Sociobiology*, **25**, 13–22.

West-Eberhard, M.J. (1983) Sexual selection, social competition and speciation. *Quarterly Review of Biology*, **58**, 155–183.

Wilson, R.S., Angilletta, M.J., James, R.S., Navas, C., and Seebacher, F. (2007) Dishonest signals of strength in male slender crayfish (*Cherax dispar*) during agonistic encounters. *The American Naturalist*, **170**, 284–291.

Wong, B.B.W. and Candolin, U. (2005) How is female mate choice affected by male competition? *Biological Reviews*, **80**, 559–571.

Zeh, J.A. (2004) Sexy sons: a dead end for cytoplasmic genes. *Proceedings of the Royal Society of London, Series B: Biological Sciences*, **271**, S306–S309.

2

EARLY LIFE-HISTORY EFFECTS, OXIDATIVE STRESS, AND THE EVOLUTION AND EXPRESSION OF ANIMAL SIGNALS

Nick J. Royle, Josephine M. Orledge, and Jonathan D. Blount

Centre for Ecology and Conservation, College of Life and Environmental Sciences,
University of Exeter, Penryn, Cornwall, UK

INTRODUCTION

The expression of signals involves costs (Maynard Smith and Harper, 2003; Searcy and Nowicki, 2005). These costs are important as they are widely thought to maintain the reliability of signals and therefore their efficacy in communication (Maynard Smith and Harper, 2003). However, relatively little is known about the physiological mechanisms underlying such costs of signaling. Oxidative stress has generally been suggested to play a key mediating role in the evolution of animal life-histories, including signals (e.g., Blount, 2004; Catoni *et al.*, 2008; Costantini, 2008; Dowling and Simmons, 2009; Monaghan *et al.*, 2009). In this chapter, we focus particularly on the role of oxidative stress in mediating the evolution and expression of animal signals. The majority of empirical work in this area has concentrated on sexual signals, in particular carotenoid-based sexually selected traits, but there is increasing evidence that oxidative stress also affects signals expressed during growth and development, not just in adulthood. In fact it is likely that virtually any signal produced by an animal will be affected by oxidative stress in some way. We review the evidence for long-term effects of environmental variation experienced during growth and development on the expression of signals throughout an organism's life, proximately mediated by oxidative stress, and

Animal Signaling and Function: An Integrative Approach, First Edition.
Edited by Duncan J. Irschick, Mark Briffa, and Jeffrey Podos.
© 2015 John Wiley & Sons, Inc. Published 2015 by John Wiley & Sons, Inc.

evaluate the conceptual issues raised and relationships involved. We begin by defining what constitutes a signal in this context.

SIGNALING

What is a Signal?

Maynard Smith and Harper in their 2003 book *Animal Signals* define a signal as "any act or structure which alters the behaviour of other organisms, which evolved because of that effect, and which is effective because the receiver's response has also evolved." A corollary of this is that if the signal alters the behavior of others it follows that it must benefit the receiver to behave in a way that is also beneficial to the signaler, otherwise signalers would not respond. Both sides, in other words, benefit from the exchange. These characteristics distinguish signals from coercion (Maynard Smith and Harper, 2003). A cue, on the other hand, is defined as a feature that can be used by an animal as a guide to future action (Hasson, 1994). This contrasts with signals, which evolve *because* of their effects on others (Maynard Smith and Harper, 2003). So, for example, size may be a cue, but not a signal. However, a behavior that conveys *information* about size can be a signal (Maynard Smith and Harper, 2003).

Honesty of Signals

Signals are not always honest (e.g., mimicry in warningly colored organisms), but to be effective in stimulating the appropriate response from receivers, they must be honest most of the time (Maynard Smith and Harper, 2003). So what maintains the reliability (honesty) of signals? Zahavi (1975) suggested that costs of signaling maintain their honesty. There are, however, two components to signaling costs. There are costs associated with the transmission of information unambiguously between signaler and receiver, which must be paid even when there is no motivation to be dishonest (efficacy costs; Guilford and Dawkins, 1991). The second type of cost is that required to maintain the honesty of the signal (strategic cost; Grafen, 1990a, 1990b). Strategic costs can be divided into "receiver-dependent" and "receiver-independent" costs. The former are costs that arise from the response of receivers to a signal, whereas the latter are costs that are imposed regardless of how receivers respond (Searcy and Nowicki, 2005). Specification of these different costs allows the identification of different types of signals.

Handicaps and Indices

Handicaps (strategic signals) can therefore be defined as signals whose reliability is maintained because the costs of producing the signals are greater than the costs required for efficacy, so that they are costly to produce or have costs associated with the consequences of signal expression (Zahavi, 1975; Grafen, 1990a, 1990b; Adams and Mesterton-Gibbons, 1995). A different form of honest signal, where there is a causal relationship between the intensity of the signal and the quality of the signaler, which cannot be faked, is known as an *index* (Maynard Smith and Harper, 1995). The relationship between the fundamental frequency of a vocalization and body size (e.g., the roar of a red deer stag during the rut in relation to the stag's size) is frequently cited as a good example of an index (e.g., Maynard Smith and Harper, 2003), as the fundamental frequency of the vocalization is primarily determined by the size of the vocal-production apparatus, which is correlated with body size (Searcy and Nowicki, 2005). However, the correlation between body size and vocal-production apparatus is not always that tight (Searcy and Nowicki, 2005). This may, at least in part, be a consequence of trade-offs during growth and development, mediated by oxidative stress. As a result, it is not always clear when a signal is an index as opposed to a handicap (Searcy and Nowicki, 2005). This also means that the honesty of signals can be corrupted (Royle *et al.*, 2002a).

This review is concerned with the effects of resource allocation trade-offs experienced during development (early life-history effects) in maintaining the reliability of signals expressed both during growth and development, and during adulthood. In particular, we emphasize the role of the oxidative status of individuals (oxidative stress) in mediating these effects on signal expression. Consequently, we are primarily concerned with "receiver-independent" signals.

EARLY LIFE-HISTORY EFFECTS AND RESOURCE ALLOCATION TRADE-OFFS

What are Early Life-History Effects?

The early life-history of an organism covers the period from conception to developmental maturity (Henry and Ulijaszek, 1996; Lindström, 1999). Early life-history effects therefore refer to the long-term consequences of perturbations during an individual's development. In general, the earlier in development that these perturbations occur, the stronger are the effects (Lindström, 1999). Environmental conditions affect the early development of individuals through maternal (and paternal) effects, which have downstream effects on growth and the allocation of resources to competing functions (e.g., the development of the immune system) that are both dependent upon, and are determined by, the variation in environmental conditions experienced during ontogeny. Early life-history effects are key drivers of evolutionary processes (Badyaev and Uller, 2009) and can also have important consequences for group (e.g., Linksvayer *et al.*, 2009) and population (e.g., Plaistow and Benton, 2009) dynamics. One of the most fundamental characteristics of an organism to be affected by variation in resource availability during development is the rate of growth.

Costs of Growth

Although it is commonly assumed that higher rates of growth lead to higher fitness, the fact that growth rates are not always maximal illustrates that growth can be costly (Metcalfe and Monaghan, 2001). The costs of rapid growth are varied, and include, for example, reduced investment in protein maintenance (in rats; Samuels and Baracos, 1995), deferred sexual maturation (salmonids; Morgan and Metcalfe, 2001), weight loss during metamorphosis in butterflies (Fischer *et al.*, 2004), reduced lifespan (zebra finches; Birkhead *et al.*, 1999; mice; Ozanne and Hales, 2004), lower competitive ability (swordtails; Royle *et al.*, 2005), impaired locomotor performance (salmonids; Farrell *et al.*, 1997; larval anurans; Arendt, 2003; swordtails; Royle *et al.*, 2006a, 2006b), and increased risk of predation (damselflies; Stoks *et al.*, 2005). These costs can also be paid over a range of time scales, from immediate (e.g., reduced rate of bone ossification in bluegill sunfish; Arendt and Wilson, 2000) to long term (e.g., increased risk of heart disease in humans; Singhal and Lucas, 2004; Singhal *et al.*, 2004). Consequently, for individuals that have experienced significant variation in resource availability during development, there will be an optimal balance between immediate investment in growth and the costs of this growth.

Trade-Offs during Growth and Development

As the examples given above illustrate, costs of compensation are highly diverse in form and widespread across taxonomic groups. So what are the benefits to individuals of growing rapidly, given that there are substantial costs to pay? Birkhead *et al.* (1999) and Blount *et al.* (2003) manipulated nestling diet in zebra finches, *Taeniopygia guttata*, so that individuals fed on a suboptimal nestling diet were relatively stunted at fledging, but largely caught up in size when subsequently put on an improved diet. As a result, in terms of morphology they were virtually indistinguishable from well-fed control birds when re-measured several months later, but they had markedly reduced blood antioxidant defenses (Blount *et al.*, 2003) and suffered a reduced adult lifespan (Birkhead *et al.*, 1999). Compensating zebra finches therefore appear to preferentially allocate resources toward sexual attractiveness at the expense of potential reproductive lifespan.

Selection therefore favors rapid growth and investment in secondary sexual traits, but at a cost. So what are the primary selective forces driving this process? Competition for limited resources during development (e.g., Royle *et al.*, 1999), reduced fitness associated with small size (increased mortality; Metcalfe and Monaghan, 2001), and poorly developed sexual signals (reduced mating success; Blount *et al.*, 2003) are the most likely candidates. For example, when zebra finch nestlings are reared under conditions of higher competition, individuals had faster growth, despite receiving less food than individuals reared under conditions of lower conflict (Royle *et al.*, 2006a). As a consequence, they were less attractive as adults (Royle *et al.*, 2002b), suggesting that there were substantial costs, paid downstream, associated with allocating proportionately more resources to growth at the expense of maintenance and/or development.

Although there is a wealth of information on the functional outcome of such costs, and there is substantial inter-specific variation in how these costs are expressed (see Section "Costs of growth"), relatively little is known about the underlying mechanism(s) involved. Oxidative stress is emerging as a strong candidate as an overarching mechanism to explain variation in such life-history trade-offs, because virtually all activities generate reactive oxygen species (ROS) (Blount, 2004; Catoni *et al.*, 2008; Costantini, 2008; Dowling and Simmons, 2009; Monaghan *et al.*, 2009).

OXIDATIVE STRESS AS A MEDIATOR OF RESOURCE ALLOCATION TRADE-OFFS

What is Oxidative Stress?

Oxygen is a basic necessity for survival. However, in the form of ROS it can cause substantial damage to intracellular macromolecules that, unchecked, can lead to the degeneration of cellular structure and function (Genestra, 2007). ROS are predominantly highly reactive oxygen-derived free radicals (an atom or molecule with a single, unpaired electron), the most prevalent of which are superoxide ions (O_2^-), hydrogen peroxide (H_2O_2), and hydroxide radicals (OH^-) (Cash *et al.*, 2007). Nitric oxide and its derivatives also react with free radicals to form reactive nitrogen species (RNS) (Patel *et al.*, 1999), but will not be considered further in this review. Oxidative stress is the net outcome of a physiological imbalance between the production of and the removal, by antioxidants, of ROS resulting in increased cellular levels of the latter (Kregel and Zhang, 2007) (Figure 2.1).

How are ROS Generated?

A variety of processes and molecules contribute to the generation of ROS. However, the principal source of ROS is provided by mitochondria (Dalle-Donne *et al.*, 2006). Mitochondria fulfill the energy requirements of cells as a result of the synthesis of the high-energy biochemical compound adenosine triphosphate (ATP) through the oxidation of foodstuffs (Campbell and Reece, 2008). The following

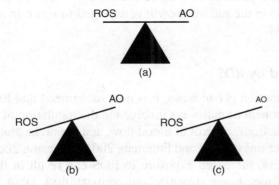

Figure 2.1. Oxidative stress. (a) No oxidative stress – the production of reactive oxygen species (ROS) is counterbalanced by antioxidant (AO) capacity. An imbalance between production of ROS and their removal by antioxidants can arise either as a consequence of, (b) an increase in ROS without an increase in antioxidant activity, or (c) a reduction in the availability of antioxidants for defense.

paragraphs provide an outline of how and why ROS are generated in mitochondria during respiration, based on accounts in Campbell and Reece (2008) and Lane (2003, 2005).

Glucose is broken down in the reactions that take place in the Kreb's cycle, where carbon and oxygen atoms are stripped out and discharged as carbon dioxide. Hydrogen atoms are split into protons and electrons, and the electrons are passed down the respiratory chain by a series of carriers (protein complexes embedded into membranes), with each carrier successively reduced then oxidized by the connecting link in the chain. The energy released by the electrons in this series of successive redox reactions is also used to pump protons across the mitochondrial membrane. In the final step of the process, electrons pass from cytochrome C to oxygen, and are reunited with protons to form water. The diffusion of protons flowing back across the membrane powers the formation of ATP to generate the chemical energy used by the cell.

The speed of respiration is generally controlled by demand for ATP, unless food or oxygen availability is very low. Respiration can also be affected by the balance between oxidation and reduction in the respiratory chain. Electron carriers can either be oxidized or reduced, but not both at once. If the carrier already has an electron, it cannot receive another one until it has passed on the current electron. The speed of respiration is quickest when half of the electron carriers within the chains are oxidized and half are reduced (redox poise). Carriers normally pass on electrons to the next carrier, but if the successive carrier is full, then the electron may be passed to something else instead, such as oxygen, to create a free radical. The probability of this loss of "poise" depends upon a variety of factors, including the number of carriers within the chains (there are normally thousands), how frequently they are replaced, and the balance between the number of electrons entering and leaving the respiratory chain (supply and demand). The consequence of loss of poise is a slowing of respiration and an increase in the amount of free radicals formed. Free radical production is therefore closely tied to metabolic rate, which is why an increase in the rate of growth is expected to result in an increase in the production of ROS.

Damage Caused by ROS

Not all ROS production is bad news. It is now recognized that ROS production is an essential component of pathways involved in the regulation of gene expression, neurotransmitter actions, control of blood flow, and the orchestration of inflammatory signaling reactions (Castro and Freeman, 2001; Genestra, 2007; Seifried et al., 2007). Nevertheless, increased exposure to ROS can result in lipid peroxidation (which alters cell membrane structure and permeability), DNA damage (leading potentially to uncontrolled cell proliferation or accelerated cell death), and protein oxidation (Kregel and Zhang, 2007; Seifried et al., 2007). In addition, detrimental adjustments to cellular signaling can result in cellular senescence, an irreversible state in which a cell remains metabolically active but is incapable of cellular division, or even apoptosis (programmed cell death; Berg et al., 2004).

Antioxidant Defenses

Antioxidants are used to maintain a balanced redox state in cells. They significantly delay, inhibit, or prevent free radical induced tissue damage by impeding the formation of radicals, scavenging them, or by promoting their decomposition (Halliwell and Gutteridge, 2007). Antioxidants are a collection of compounds distinguished by their proficiency to be oxidized more readily than vital cellular components and to upregulate repair systems (Genestra, 2007). They are situated in subcellular compartments, organelles, and extracellular spaces and comprise a wide diversity of molecules including natural fat-soluble antioxidants such as vitamins A and E, carotenoids and ubiquinones, antioxidant enzymes, water-soluble antioxidants such as uric acid and ascorbic acid, and the thiol redox system.

There are three major levels of antioxidant defence (Surai and Sparks, 2001). The first level, which is primarily involved in the prevention and restriction of free radical formation, includes endogenously derived antioxidant enzymes such as superoxide dismutase and glutathione peroxidase. Superoxide dismutase eliminates superoxide radicals by converting them to oxygen and hydrogen peroxide, with the latter eliminated by catalase (Barja, 2004). The second level of antioxidant defense, which consists of fat-soluble (predominantly vitamin E and carotenoids) and water-soluble (predominantly ascorbic acid and glutathione) chain-breaking antioxidants, provides protection to lipid membranes through the prevention and restriction of lipid peroxidation (Niki, 1996). Specific enzymes used in the repair and reconstitution of lipid membranes form the third level of antioxidant defense (Surai, 1999).

Fat-soluble antioxidants (vitamin E and carotenoids) cannot be synthesized *de novo* by animals, so are derived from the diet (Goodwin, 1984; Surai, 1999). Carotenoids (the intensely colored pigments responsible for red to yellow coloration of many plants, insects, bird plumages, and marine animals), for example, are derived from a diet of higher plants and algae, or are secondarily acquired from other animals. The biological actions of carotenoids continue to be debated, but include functions as immunostimulants (e.g., Blount *et al.*, 2003; McGraw and Ardia, 2003; Saks *et al.*, 2003; but see e.g., Hõrak *et al.*, 2007), potentially also as photoprotectants (e.g., McGraw and Hill, 2004), and as antioxidants (e.g., Blount *et al.*, 2002; Alonso-Alvarez *et al.*, 2004; McGraw, 2005; Hõrak *et al.*, 2007; but see Costantini and Møller, 2008; Cohen and McGraw, 2009). We discuss these issues in more detail below.

The eight compounds, α-, β-, δ-, and γ-tocopherol and α-, β-, δ-, and γ-tocotrienols, for which vitamin E is a generic descriptor, are derived from seed oils and the leaves of higher plants (Surai, 2002). Diet composition is therefore important in improving antioxidant protection; however, bioavailability, the rate and extent at which a compound appears in the blood, is also important. Bioavailability of dietary-derived antioxidants is dependent upon a range of factors, including the type and amount of antioxidants consumed, the process by which specific antioxidants are assimilated, and the nutritional status of the individual (Catoni *et al.*, 2008). Antioxidants have been suggested to be a scarce, limiting resource in natural diets (e.g., Olson and Owens, 1998; Grether *et al.*, 2001; Hill *et al.*, 2002; de Ayala *et al.*, 2006), although how general such limitations are has been questioned (Hadfield and Owens, 2006) and constraints in the ability to assimilate ingested antioxidants or to partition them among competing demands *in vivo* are likely to be at least as important in shaping life-history trade-offs in animals (Catoni *et al.*, 2008).

Mechanisms of Action of Dietary Antioxidants

Fat-soluble dietary-derived antioxidants, such as vitamin E and carotenoids, are predominantly located in cell membranes, whereas water-soluble antioxidants, such as vitamin C, are located in the cytoplasm (Surai, 2002). Vitamin C has high free radical scavenging activity, reacting directly with superoxide and hydroxyl radicals and is also important in vitamin E recycling (Surai, 2002). Vitamin E is closely associated with polyunsaturated fatty acids in cell membranes and is considered the most effective chain-breaking antioxidant, scavenging lipid peroxyl radicals and so preventing the propagation of lipid peroxidation (Surai, 2002). Vitamin E is also important in recycling carotenoids (e.g., Mortensen *et al.*, 2001), and has various non-antioxidant properties, including roles in the transcription and expression of some genes and the upregulation of some enzymes (Surai, 2002).

In the light of their role as pigments, carotenoids have received most attention from behavioral and evolutionary ecologists studying animal signals. Carotenoids have a wide array of other properties, including roles as antioxidants and in the recycling of antioxidants (Surai, 2002). Carotenoids are also immunostimulants. This can occur directly, for example, through the regulation of membrane fluidity and particular cell signaling pathways, or indirectly, via their ability to act as antioxidants to mop up free radicals produced as a consequence of immune system upregulation (Chew and Park, 2004). However, because carotenoid radicals have pro-oxidant characteristics, unless there are other antioxidants such as vitamin E to recycle the carotenoids (Packer *et al.*, 1979) the ability of carotenoids to act as antioxidants can be compromised (Catoni *et al.*, 2008). The antioxidant capacity of carotenoids *in vivo* is therefore debatable (e.g., Hartley and Kennedy, 2004; Isaksson *et al.*, 2007; Costantini and Møller, 2008), and likely to be contingent on the presence of other antioxidants (Catoni *et al.*, 2008).

However, the antioxidant functions of carotenoids are context-dependent and tissue-specific (Surai, 2002), and while many studies have reported the antioxidant capacity of carotenoids (or lack of) in blood plasma, few studies have measured the antioxidant effects of carotenoids in specific tissues (but see e.g., Blount *et al.*, 2002; McGraw *et al.*, 2005). Further research is therefore required before the importance of carotenoids as antioxidants, and how this relates to the information content of carotenoid-based signals, can be dismissed. In addition, potential links between non-antioxidant functions of carotenoids as modulators of gene expression, intercellular signaling and enzyme activity, and the information content of signals have been entirely unexplored.

Depending upon their distribution in tissues and their biochemical properties, particular antioxidants may directly affect the expression of a given trait (a "focal" antioxidant effect; e.g., the direct, singular, effects of vitamin C on sperm production; Catoni *et al.*, 2008), and/or have indirect effects on traits, as a result of interactions with focal antioxidants, or effects on the overall oxidative status of individuals. Such effects of "non-focal" antioxidants are therefore dependent upon the presence of focal antioxidants (e.g., Bertrand *et al.*, 2006; Pike *et al.*, 2007).

Antioxidants, Oxidative Stress, and Rapid Growth

Provision of dietary antioxidants has been shown to increase growth rate in a number of different species (e.g., Cucco *et al.*, 2006; de Ayala *et al.*, 2006; O'Brien and Dawson, 2008) and several studies have also shown that increased rates of growth, as predicted (see Section "How are ROS generated?"), are associated with higher oxidative stress (Alonso-Alvarez *et al.*, 2007; Nussey *et al.*, 2009) or production of defensive antioxidant enzymes (e.g., De Block and Stoks, 2008). Dietary antioxidants can also reduce susceptibility to oxidative damage (Lin *et al.*, 2005). A recent study by Hall *et al.* (2010) manipulated antioxidant availability and levels of within-brood competition and then measured oxidative damage in nestling red-winged blackbirds, *Agelaius phoeniceus*, and showed that there is a trade-off between investment in growth and self-maintenance (oxidative stress) in favor of growth (see Box 2.1). Similarly, Ewen *et al.* (2009) showed that maternally derived carotenoids mediated the trade-off between growth and investment in immune response to ectoparasites. Related work on "metabolic programming," mainly in mammals, whereby conditions experienced by individuals during early development have significant long-term effects on body mass and homeostasis (Tamashiro and Moran, 2010), have shown that oxidative stress is an important mediator of these effects. This may occur directly through the modification of gene expression or indirectly through the oxidation of lipids, proteins, DNA, or RNA during key windows in development (Luo *et al.*, 2006). Alterations in placental ROS, for example, modulates fetal growth and development and can lead to permanent changes in adult function, including the probability of development of diabetes, hypertension, and obesity (Dennery, 2010). Studies such as these that incorporate physiological measurements are important to understanding the proximate mechanisms underpinning the costs of maintaining key ecological and evolutionary processes that shape animal life-histories (Harshman and Zera, 2007).

What Sorts of Signals will be Affected by Oxidative Stress?

Given the potential ubiquity of oxidative stress in the evolution of animal life-histories (e.g., Dowling and Simmons, 2009; Monaghan *et al.*, 2009) virtually any signal produced by an animal is likely to be affected by oxidative stress, even if only indirectly; for example, as a byproduct of trade-offs during growth affecting body size, and therefore the ability to signal size (e.g., the roar of a red deer stag). In addition, the central nervous system is very rich in polyunsaturated fatty acids, has a high metabolic oxidative rate, and high concentrations of some pro-oxidants, but relatively low levels of antioxidants, so is prone to oxidative stress (Praticò, 2008). As a result, there may be many aspects of behavior that are affected to some degree by the redox status of individuals. Oxidative stress also has a more direct role in determining the expression of sexually selected signals in adulthood, particularly carotenoid-based sexual signals, via acquisition of dietary antioxidants, for example. Less obviously, perhaps, oxidative stress is an important determinant of signals expressed during growth and development. In the next section, we outline how the oxidative status of individuals involved in signaling between parents and offspring may affect the outcome of the provision of parental investment (PI) in species that feed dependent offspring during post-natal development.

SIGNALS EXPRESSED DURING DEVELOPMENT

In species that provide parental care, offspring are often reared together in broods, so close relations provide the social environment in which growth and develop-ment take place. Parents determine the initial social environment that offspring are born or hatch out into, and often overproduce offspring, which creates a squeeze on resources (Mock and Parker, 1997). Variation in the mating system and the relat-edness of family members leads to conflicts over the provision of costly PI, which is magnified by scarcity of resources (Parker et al., 2002a). This has led to the co-evolution of signaling mechanisms (involving begging behavior of offspring) to resolve these conflicts (Royle et al., 2004). Parents can simultaneously track resources in the environment and increase control over the behavioral dynamics of the social environment their offspring experience via maternal effects.

Maternal Effects and Signaling between Parents and Offspring

Maternal effects occur whenever offspring phenotype is affected by the pheno-type of the mother, and are peculiar in that they may have both environmental and heritable genetic components (Mousseau and Fox, 1998). As a result, they can have important consequences for evolutionary processes (Mousseau and Fox, 1998; Badyaev and Uller, 2009) and constitute a major source of phenotypic vari-ation. Particular maternal and early-life history effects also have functions as sig-nals; mothers can signal information about resource availability in the environment (e.g., Hinde et al., 2009), offspring signal their need and/or quality, mediated by maternal effects, when soliciting PI (e.g., Helfenstein et al., 2008) and sexual sig-nals produced during adulthood can be associated with the provision of maternal resources (e.g., Biard et al., 2009). In this section, we outline how maternal effects can influence the signals exchanged between parents and offspring during the pro-vision of PI and, therefore, the fitness of parents and offspring, using birds as an example.

One of the most obvious, and important, ways in which variation in the mater-nal environment and phenotype can affect variation in offspring phenotype is via the resources allocated to the egg (Royle *et al.*, 1999; Rubolini *et al.*, 2006). Variation in egg size, which is dependent on maternal condition (Christians, 2002), for example, is known to affect offspring growth and development and may have pervasive effects on body size and fecundity in adulthood (reviewed in Christians, 2002; Badyaev, 2008). More recent work has focused not just on the size of eggs but variation in key constituents, such as steroid hormones (e.g., Groothuis *et al.*, 2005) and antioxidants (e.g., Royle *et al.*, 1999; Surai, 2002). Variation in egg constituents can have important effects on offspring fitness, but such effects need to be considered more widely in the context of how conflicts over the provision of PI are resolved within families to appreciate their role in signaling.

Most birds hatch their offspring asynchronously, which imposes a phenotypic handicap on one or more of their offspring (Mock and Parker, 1997). This leads to a structured family where the last hatched "marginal" offspring generally suffer reduced survival and fitness prospects compared to their older "core" nest-mates (Mock and Forbes, 1995). Where creation of a structured family serves a resource tracking and/or insurance function ("parental optimism"; Mock and Forbes, 1995) phenotypic handicapping of certain offspring, either through hatching asynchrony, or more fine-scale tuning of egg components within the clutch such as androgen (e.g., Schwabl, 1993; Groothuis *et al.*, 2005) or antioxidant (e.g., Royle *et al.*, 1999) concentrations, allows parents to stack the deck in their favor in resolving conflicts over PI.

The Resolution of Parent–Offspring Conflict through Signaling

Parent–offspring conflict occurs because the optimal amount of investment that parents are expected to supply for a brood is less than the optimal amount that offspring are expected to demand, with the magnitude of the difference dependent upon the mating system (Trivers, 1974). This outlines the "battleground" of conflict (Godfray, 1991), but establishing how the conflict is resolved (i.e., how much PI is actually provided to offspring) requires an understanding of how the demand and supply mechanisms interact with one another (Mock and Parker, 1997; Parker *et al.*, 2002a). Offspring solicit PI using a costly means of communication (e.g., begging), such that an increase in begging leads to an increased probability of gaining more PI, but at some cost (Mock and Parker, 1997). The way in which the mechanisms of demand and supply interact depends upon the control of resource allocation (Royle *et al.*, 2002a).

If parents actively control resource allocation, honest signaling models of parent–offspring conflict suggest that offspring advertise their "need" to parents, signaled by the intensity of their begging, and parents allocate resources directly in relation to the strength of the signal (Godfray, 1991, 1995). In contrast, models of scramble competition (e.g., Stamps *et al.*, 1978; Macnair and Parker, 1979; Parker *et al.*, 2002b), where offspring control resource allocation, suggest that parents passively feed competing offspring, with the one presenting the largest overall stimulus receiving the food. Here, competitive ability is the signal that parents respond to.

The amount of PI supplied at conflict resolution is predicted to be at the parents' ideal level under honest signaling, but higher than the parents' ideal (and therefore nearer the ideal of the offspring) in scramble competition (Parker *et al.*, 2002a). It is notoriously difficult to distinguish between the models empirically (Royle *et al.*, 2002a) and it is likely that control of resource allocation is dynamic rather than static, varying both with environmental conditions, offspring age (parental control is expected to decrease as offspring age; Royle *et al.*, 2002a) and the costs of signal efficacy. If, for example, parents allocate food actively to their young, variation in competitive ability within the brood will increase the efficacy costs for parents of reading the begging signals if the cost of signaling for offspring varies (e.g., Kilner, 2001) for young with identical levels of "need" (Royle *et al.*, 2004). Begging is honest regardless of who controls resource allocation, in that an increase in solicitation is balanced by an increase in costs to offspring, but when there is scramble competition, begging only relates directly to true need if all offspring are competitively equal, but still correlates with true need as the costs that offspring can afford will increase with the potential rewards of increased solicitation (Rodríquez-Gironés *et al.*, 2001a; Parker *et al.*, 2002a).

The Costs of Signaling between Offspring and Parents

Models for the evolution of begging signals require costs in order to maintain stability (Parker *et al.*, 2002a). Despite this, empirical evidence for substantial costs of begging is lacking, with most studies of the energetic cost of begging indicating that the metabolic costs are low (reviewed by Chappell and Bachman, 2002) and the various studies looking at the trade-off between begging and growth rate proving inconclusive (e.g., Kilner, 2001; Rodríquez-Gironés *et al.*, 2001b; Leonard *et al.*, 2003). This has led to the conclusion that the reliability of begging signals may be primarily maintained by the benefits associated with the signaling outcomes, rather than the cost (e.g., Searcy and Nowicki, 2005).

However, studies of the energetics of begging have mostly utilized measurements of oxygen consumption, which ignores anaerobic components of energy metabolism (Weathers *et al.*, 1997), and studies that use variation in growth rate as a proxy for fitness do not take account of the role of oxidative damage in determining this trade-off. Although yet to be tested empirically, measurements of the costs of begging mediated by oxidative damage should provide costs in a suitable evolutionary currency (Parker *et al.*, 2002a), given the long-term consequences of oxidative stress on animal life-histories (e.g., Monaghan *et al.*, 2009). Moreover, increased provisioning has been shown to increase oxidative stress in parents (Alonso-Alvarez *et al.*, 2004; Wiersma *et al.*, 2004), emphasizing the role of oxidative damage as a proximate cost of reproduction (Salmon *et al.*, 2001). Significant costs associated with responding to signals (i.e., provisioning) also imply significant costs for signalers (i.e., begging nestlings). The failure to quantify these costs empirically in a suitable currency does not necessarily mean that they are insubstantial or do not exist. Studies that have manipulated availability of dietary antioxidants and quantified effects on begging behavior have produced equivocal results, with supplementary antioxidants having no effect on begging behavior in red-winged blackbirds *A. phoeniceus* (Hall *et al.*, 2010), but a positive effect on begging in yellow-legged gulls *Larus michahellis* (Noguera *et al.*, 2010). What are needed are studies that manipulate begging effort and then quantify effects of the manipulation on oxidative damage.

The maintenance of stability of signals using costs assumes that the marginal costs are higher for poor quality signalers than high quality signalers, but few studies have tested this assumption (Kotiaho, 2001). In particular, there is little if any evidence for these differential costs in studies of begging (Searcy and Nowicki, 2005). However, recent evidence indicates that "marginal" offspring have greater costs of oxidative stress than "core" offspring in structured families of red-winged blackbirds, unless dietary availability of antioxidants is sufficiently high to mitigate these costs (Hall *et al.*, 2010). These costs are paid as a consequence of the trade-off between growth and self-maintenance, tipped in favor of growth (see Box 2.1). Consequently, this provides evidence that poor quality signalers ("marginal" offspring) have higher marginal costs of signaling at the same level as higher quality signalers ("core" offspring; Hall *et al.*, 2010).

Box 2.1. Oxidative Stress and Differential Costs of Signaling in Red-Winged Blackbirds

Red-winged blackbirds nest in marshy habitat, and reproductive success is strongly linked to mean temperatures experienced during the chick rearing period; hence the availability of insect prey (Forbes *et al.*, 2001). Despite this environmental unpredictability, clutch size is remarkably consistent between years (3–5 eggs, with a modal clutch size of 4). However, although there is little variation in clutch size, variation in fledging success is considerable among years (Forbes *et al.*, 2001). Chicks are hatched asynchronously, and families are highly structured. Nestling blackbirds have extremely rapid rates of growth (among the highest recorded for any species; Starck and Ricklefs, 1998), which is related to the high rates of extra-pair paternity, and consequently, high intrafamilial conflict over PI (Royle *et al.*, 1999).

Yolk testosterone concentration increases (Lipar *et al.*, 1999), but yolk antioxidant concentration decreases (Royle *et al.*, 2011) with laying order in red-winged blackbirds, which may be a maternal effect to boost marginal chicks' growth and survival, should environmental or developmental conditions allow (Schwabl, 1993; Royle *et al.*, 2001). Increased provision of yolk testosterone leads to an increase in the mass of the musculus complexus, a group of muscles used to bring nestlings to an upright position, which may enhance begging and growth (Lipar and Ketterson, 2000). Neck height is the strongest determinant of feeding success in nestling blackbirds, so generally favors older, core brood chicks (Teather, 1992). However, increased testosterone should increase the competitiveness of marginal chicks hatching from later-laid eggs, if the costs of testosterone (e.g., reduced immune function and increased oxidative stress; von Schantz *et al.*, 1999) can be balanced by sufficient provision of antioxidants (Royle *et al.*, 2001). Scramble competition and other visual displays largely determine success in feeding (Forbes and Glassey, 2000), favoring larger, core brood chicks, suggesting food allocation by parents is primarily passive (Royle *et al.*, 2002a).

(continued)

Marginal offspring suffer considerably higher rates of mortality than core brood chicks during post-hatch growth and development, particularly last-hatched chicks (Hall *et al.*, 2010). However, marginal offspring that do survive to fledging age grow just as rapidly as their older, core siblings, but have greater oxidative damage (as indicated by greater levels of the lipid peroxidation product, malondialdehyde, in their blood plasma). Sibling scramble competition for resources (a signal of quality and need by offspring to parents; see Section "The costs of signaling between offspring and parents") selects for rapid growth. Larger, more competitive, individuals can secure disproportionate shares of parental resources, which enables them to grow faster and make them even more competitive (power begets power; Mock and Parker, 1997). Consequently, in order to be able to compete with higher quality nest-mates, which are better provisioned, marginal offspring must allocate relatively more resources to growth compared to self-maintenance (Hall *et al.*, 2010). Marginal offspring therefore sacrifice oxidative status on the altar of rapid growth. Supplementation of nestlings with dietary antioxidants confirmed the outcome of this trade-off in favor of growth, as individuals that received more antioxidants throughout development allocated these extra resources to increasing growth rate, rather than reducing oxidative damage, irrespective of whether they were core or marginal offspring (Hall *et al.*, 2010).

Carotenoids, Mouth Color, and Begging Signals in Altricial Nestlings

In addition to these effects on the competitiveness of individuals (involved in signaling quality), antioxidants may also have more direct effects on signals during parental provisioning of resources to offspring. For example, Helfenstein *et al.* (2008) recently showed, using great tits *Parus major*, that carotenoid supplementation of parents during egg-laying had a positive effect on begging intensity of nestlings from small broods (i.e., nestlings of poorer quality parents benefited from increased carotenoid availability). This demonstrated that maternal effects can directly, not just indirectly, influence post-hatch offspring signals.

Begging signals comprise a suite of inter-related components including vocalization and visual display. The latter may involve physical components, such as jostling for position and vocalizations, in addition to visual display involving color (Kilner, 1997). Both the inside of the mouth (the palate) and the fleshy flanges that border the mouth of the nestlings of many altricial species of birds are known to be important in communication between offspring and parents during parental provisioning (Kilner, 1997; Saino *et al.*, 2000). Variation in mouth color may affect detectability (e.g., Kilner and Davies, 1998; Wiebe and Slagsvold, 2009) or may be an honest signal of offspring need (Kilner, 1997) or quality (Saino *et al.*, 2000).

Numerous studies have shown, both between and within broods, that parents reward offspring with more intensely colored mouthparts (reviewed in Dugas, 2009) and it has been assumed that this mouth coloration is carotenoid-based. Carotenoids are now known to be important pigmentary components of the fleshy mouth flanges of nestling altricial birds that determine coloration (Thorogood *et al.*, 2008; Dugas, 2009). Supplementation of carotenoids to nestlings increases blood plasma concentrations of carotenoids and alters the coloration of both palate and flanges (Thorogood *et al.*, 2008). In the hihi, *Notiomystis cincta*, a passerine bird endemic to New Zealand, carotenoid supplementation increased the chroma of the yellow coloration of the palate and the flanges of nestlings and also decreased the UV-brightness and UV-chroma of the flanges. The authors suggested that this combination of reduced UV reflectance and a greater saturation of carotenoid coloration is the signal that parents responded to during parental provisioning (Thorogood *et al.*, 2008).

A signaling function for carotenoid-based mouth coloration of nestlings is also supported by a study of house sparrows, *Passer domesticus*, by Dugas (2009). He manipulated flange coloration directly and showed that parents responded by allocating more food to chicks with mouthparts that appeared more carotenoid-rich. He suggested that because carotenoid-based coloration is positively associated with condition, especially in sparrows (e.g., Loiseau *et al.*, 2008) parents were allocating resources based on signals of quality (*sensu* Saino *et al.*, 2000), reasoning that if parents have a degree of control over resource allocation then it is best to allocate resources preferentially to offspring that will provide the highest future returns on that investment (Dugas, 2009).

However, not all studies support a signaling function for mouth coloration in nestlings of altricial birds. Wiebe and Slagsvold (2009) found no evidence that UV reflectance or yellow-color reflectance determined parental allocation of food within the nests of three different cavity-nesting species, although broods with nestlings that had dark-painted mouth flanges received less food, suggesting that flange coloration has evolved for detectability in cavity-nesting species. At present, it is therefore difficult to draw any firm conclusions about the role of carotenoids in signaling offspring quality or need during parental provisioning, especially as, even where a signal function has been supported, the effects are relatively small (Dugas, 2009) and are likely to diminish further as nestlings age and parental control over resource allocation declines (*sensu* Royle *et al.*, 2002a). Nevertheless, given the importance of carotenoids in determining mouth coloration in altricial nestling birds, it is likely that oxidative stress plays a central role in mediating this form of parent–offspring communication during the provision of PI.

SIGNALS EXPRESSED DURING ADULTHOOD

Nutrition during Growth and the Expression of Sexual Signals as an Adult

Variation in resources available during development can have profound effects on the allocation of resources in adulthood, particularly in species with strongly sexually selected traits. In stalk-eyed flies *Cyrtodiopsis dalmanni*, for example, eyespan width in males is a sexually selected trait, with females preferring males with wider eyespans (Wilkinson and Reillo, 1994). Eyespan of males varies with the nutritional environment experienced during growth and development (i.e., it is a condition-dependent trait; Rowe and Houle, 1996), with genetic variation in the degree of response to the amount of resources available (David *et al.*, 2000). High quality males can afford to allocate more resources to the sexual signal than poor quality males, at the same cost (Maynard Smith and Harper, 2003). It was also notable that variation in eyespan width in response to variation in diet was much greater than variation in non-sexual, morphological traits (David *et al.*, 2000).

This bias in the allocation of scarce resources during growth to sexual signals can also be seen in the (compensatory) response of green swordtail fish, *Xiphophorus helleri*, to an improvement in diet during development. Male swordtails virtually cease body growth following maturation, at which point they preferentially invest in growth of the "sword," a secondary sexual trait that is an extension of the caudal fin. Females prefer males with long swords, and, like stalk-eyed flies, male swordtails do not provide any parental care, just "good genes" (e.g., Basolo, 1990). Compensating males catch up completely in body size, sword length, and appearance (Royle *et al.*, 2005) and females find them similarly attractive to males that have not experienced a period of resource depression during growth (Walling *et al.*, 2007). As they do not suffer a cost to longevity (Royle *et al.*, unpublished) compensating males consequently have a similar-sized reproductive window of opportunity to that of males that did not experience a change in resource availability during development (Figure 2.2), but pay costs of reduced competitiveness in access to females (Royle *et al.*, 2005, 2006b) and reduced ability to escape from predators (Royle *et al.*, 2006c).

Exposure to Parasites during Growth and the Expression of Sexual Signals

The expression of sexually selected traits is also sensitive to other environmental variables experienced during growth and development, such as exposure to parasites. Borgia *et al.* (2004) studied adult satin bowerbirds, *Ptilonorhynchus violaceus*, to determine whether male display could provide an indication of parasitic infections experienced during juvenile life history stages. The study found that more attractive males had experienced a lower parasite burden as juveniles but there was no relationship between current adult parasite burden and male attractiveness (Borgia *et al.*, 2004). Consequently, the attractiveness of males was primarily a reflection of their ability to remain healthy and fight off infection during development, rather than a short-term indicator of their quality and current health status.

Carotenoid-Based Sexual Signals and Immunity

Many sexual signals are carotenoid-based. Carotenoid pigments are responsible for the red to yellow coloration of many birds, fish, reptiles, amphibians, and insects. Numerous mate choice studies have demonstrated the preference of females for males possessing signals with greater carotenoid pigmentation (Blount, 2004). The capacity of carotenoids to act as sexual signals is believed to result from their relative rarity in nature, combined with their roles in critical physiological functions such as antioxidant defense and immune function (Saks *et al.*, 2003).

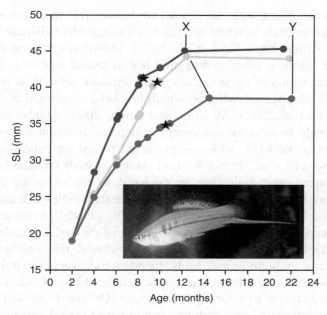

Figure 2.2. Growth trajectories (standard length) of male green swordtail fish that experienced different resource availability during development. The green line shows the average trajectory of males that received *ad lib* resources, whereas the red line shows the average trajectory of males that received a restricted diet throughout development. The yellow line shows males that were released from a restricted diet at 6 months of age and subsequently re-aligned growth (growth compensation). Following maturation (indicated by stars), males compensated for their poor start in life by accelerating growth of both body size and sword length, so that there was no difference in body size or sword length compared to *ad lib* fed males by the time body size reached an asymptote. As a result, the window of reproductive opportunity (difference between points X and Y) was similar for compensating and *ad lib* fed males. Males that remained on a restricted diet, however, had reduced mature body size and a smaller window of reproductive opportunity, putting them at a double disadvantage (see Section "Nutrition during growth and the expression of sexual signals as an adult"). (*See insert for color representation of this figure.*)

The parasite-mediated sexual selection hypothesis suggests that a higher intensity of plumage coloration signals increased genetic resistance to parasites (Hamilton and Zuk, 1982). Parasites of the gastrointestinal tract, such as *Coccidia*, can disrupt epithelial cell permeability and in turn influence the absorption of carotenoids (Baeta *et al.*, 2008). In addition, ectoparasites may directly reduce male coloration. Experiments to induce immune activation using parasite infection (e.g., Hõrak *et al.*, 2004), analysis of infection clearance rates (e.g., Lindström and Lundström, 2000), and heterophil counts (Dufva and Allander, 1995) have provided evidence to demonstrate a link between immune function and carotenoid-based plumage coloration. Other experiments have used dietary supplementation with carotenoids to demonstrate that both sexual signal expression and immune function are carotenoid-dependent (Blount *et al.*, 2003; McGraw and Ardia, 2003; Saks *et al.*, 2003). The trade-off between the use of carotenoids in immune function and their use as sexual signals is based on the premise that activation of the immune system diminishes the level of carotenoids available for signaling, a prediction which has been confirmed in experimental studies (Faivre *et al.*, 2003; McGraw and Ardia, 2003; Alonso-Alvarez *et al.*, 2004). This could be because investment in immune function incurs an increase in the production of ROS, which is predicted to reduce expression of sexual signals (von Schantz *et al.*, 1999). So, for example, a study on male house finches, *Carpodacus mexicanus*, following an epidemic of mycoplasmal conjunctivitis found survivors had significantly redder plumage than males that did not survive (Nolan *et al.*, 1998). In contrast, a study on ring-necked pheasants, *Phasianus colchicus*, found that although supplementation of dietary antioxidants (vitamin E) during early life reduced parasite load in adulthood, this was not reflected in ornamental traits, including carotenoid-dependent traits such as wattles: there was no evidence that sexually selected traits indicated the ability of individuals to resist infection, either currently or during development (Orledge *et al.*, 2012a).

It has been suggested that both carotenoid-based sexual signals and immune function are ultimately determined by nutritional status (Hill, 2000; McGraw, 2005). If immune function and carotenoid-dependent signals are both reliant upon general nutrient uptake, then correlations between such traits may occur in the absence of a causal relationship between the two (Navara and Hill, 2003). There may also be limits to the utilization of carotenoids. A carotenoid supplementation experiment using blackbirds, *Turdus merula*, for example, showed that circulating plasma carotenoid levels reach an asymptote at higher levels of supplementation (Baeta *et al.*, 2008; see also Alonso-Alvarez *et al.*, 2004). Whether such a "carrying capacity" occurs as a result of a limitation on an individual's absorption or transportation capabilities was unresolved (Baeta *et al.*, 2008). In general, interpretation of carotenoid supplementation studies is compromised by a lack of knowledge of natural intake rates of carotenoids (and other antioxidants) in the wild (Monaghan *et al.*, 2009).

Carotenoid-Based Signals as Indicators of Oxidative Stress

Due to the potential role of carotenoids as antioxidants, the intensity of carotenoid-mediated sexual ornaments could signal the oxidative status of individuals (von Schantz *et al.*, 1999; see also Section "Carotenoids, mouth color, and begging signals in altricial nestlings"). There is currently considerable controversy in the ecological literature over the role of carotenoids as antioxidants (e.g., Costantini and Møller, 2008; Isaksson and Andersson, 2008), but synergistic relationships between carotenoids and other antioxidants, and how this may underpin the information content of sexual signals, deserves further study (Catoni *et al.*, 2008). The carotenoid protection theory suggests that because carotenoids do not necessarily always function as antioxidants, and oxidation alters coloration, carotenoid-based sexual traits might actually signal the abundance of non-pigmentary antioxidants (e.g., vitamin E or melatonin) that protect carotenoids from oxidation and so make them available for sexual signaling (Hartley and Kennedy, 2004). This hypothesis has received empirical support in carotenoid and melatonin-supplemented adult zebra finches (Bertrand et al., 2006) and sticklebacks (Pike et al., 2007). Carotenoid-based signals may therefore advertise the oxidative status of individuals, if only indirectly (Monaghan et al., 2009). However, a recent study on ring-necked pheasants by Orledge et al. (2012b) found no support for the carotenoid-protection hypothesis when antioxidants were supplemented during development, rather than in adulthood. In this case, although there were no synergistic early-life effects of antioxidants on carotenoid-based signals (wattles), there were positive effects on other traits important in competition with other males (i.e., growth rate and size of individuals at adulthood), suggesting that oxidative status during development may be an important regulator of competition-dependent traits in general (see Section "Competition-dependent sexual signals").

Testosterone and the Oxidative Handicap Hypothesis

The expression of male sexually selected traits is also affected by androgens such as testosterone (Blas et al., 2006; Buchanan et al., 2003). However, testosterone can be an immune suppressant (Peters, 2000, 2007). An increase in the expression of testosterone-dependent ornamentation may therefore compromise male immune system function (Folstad and Karter, 1992). Recent research indicates that carotenoid pigments and testosterone are two components of a complex integrated physiological mechanism connecting ornamentation and immunity (Peters, 2000, 2007). Lipoprotein molecules bind to and transport carotenoids from the intestine and deliver them to the integument and other sites for incorporation into sexual signals. For example, there is a strong correlation between carotenoid-mediated beak coloration and the accumulation of lipoproteins in the blood of zebra finches (Peters, 2000), and evidence suggests that testosterone aids the formation of pivotal carotenoid transporting lipoproteins by upregulating the production of cholesterol, which is a major constituent of these lipoproteins (McGraw and Parker, 2006).

The synergy between testosterone (immunosuppressant) and carotenoid (immunoenhancing) pathways may account for the ability of males with high levels of testosterone to withstand the associated increase in oxidative stress (Peters, 2000). The oxidative handicap hypothesis suggests that an increase in testosterone leads to elevated metabolic rates, which increase ROS production and oxidative stress. Male red-legged partridges, *Alectoris rufa*, given extra testosterone, had increased carotenoid availability in the blood but reduced expression of carotenoid-mediated sexual signals (Alonso-Alvarez et al., 2008). High testosterone promoted the maintenance of oxidative status at the cost of color (i.e., reduced sexual signal expression; Alonso-Alvarez et al., 2008). Allocation of resources therefore favors self-maintenance over reproduction in this study (cf. Hall *et al.*, 2010). In another work, it has recently been shown that experimentally elevated testosterone increased circulating antioxidant levels and also enhanced ornamentation in male red grouse, *Lagopus lagopus scoticus*; however this was at the cost of increased oxidative damage in circulation (Mougeot *et al.*, 2009). In addition, males that had initially smaller carotenoid-based ornaments (combs) suffered more oxidative damage than those with initially larger ornaments when forced to increase testosterone levels (Mougeot *et al.*, 2009). These studies are consistent with the idea that male sexually selected traits signal the oxidative status of the individual (Alonso-Alvarez *et al.*, 2008; Mougeot *et al.*, 2009).

Early Life-History Effects and the Expression of Carotenoid-Mediated Sexual Signals

Despite the large amount of research effort that has been focused on the expression of carotenoid-mediated sexual signals, virtually all these studies have concentrated on, and manipulated, carotenoid availability in adulthood, and largely ignored the downstream effects of antioxidant availability during growth and development on the expression of sexual signals. This seems surprising, given what is known about the effects of early life exposure to antioxidants on physiological function at adulthood (e.g., Blount *et al.*, 2003) and evidence that the expression of sexual signals is more dependent on resource availability experienced during growth than resource availability as an adult (e.g., Borgia *et al.*, 2004; McGraw, 2005). Exposure to carotenoids during development, for example, determines the ability to assimilate and metabolize carotenoids, independent of current dietary availability, as adults (e.g., Koutsos *et al.*, 2003; Biard *et al.*, 2005, 2007; McGraw, 2005; Isaksson and Andersson, 2008). Similar effects of variation in early nutritional conditions on physiological function and the expression of sexual signals are also well established (e.g., Ohlsson *et al.*, 2002; Blount *et al.*, 2003; Royle *et al.*, 2005). However, a comparative study has taken this one step further by demonstrating a link between egg-based maternal effects and the expression of carotenoid-based sexual signals.

Biard *et al.* (2009) hypothesized that because maternally derived carotenoids can increase the efficiency of carotenoid metabolism post-hatching, egg-yolk carotenoids may be advantageous to offspring in allocating resources to carotenoid-based sexual signals. If these signals are under directional selection, then a positive genetic correlation between yolk carotenoid content and subsequent offspring performance based on the expression of the sexual signal would be expected to accelerate the rate of evolution of carotenoid-based sexual signals (Biard *et al.*, 2009). They therefore predicted a positive relationship between yolk carotenoid concentration and expression of carotenoid-based sexual signals across species. As predicted, yolk carotenoid concentration increased with the extent of carotenoid-based signals, but concentrations of the non-pigmentary antioxidants in yolk, vitamins A and E, were not related to the expression of carotenoid sexual signals, providing support for a direct relationship between the deposition of carotenoids in yolk and sexual signals, rather than the "carotenoid protection" hypothesis for expression of carotenoid-mediated traits (Hartley and Kennedy, 2004; see Section "Carotenoid-based signals as indicators of oxidative stress").

The results of this study are particularly notable as they indicate that maternally derived carotenoids may ultimately mediate the trade-offs that occur during post-hatch development between investment in self-maintenance and reproduction (e.g., Hall *et al.*, 2010), with the reliability of the signals being maintained by the costs to the mother of the initial investment in the egg (Biard *et al.*, 2009). Maternal effects are rarely shown to have effects that are this pervasive (e.g., Lindholm *et al.*, 2006), further highlighting the potential importance of the early life-history effects of variation in dietary antioxidants and oxidative stress on the expression and evolution of sexual signals in birds in particular. Recent work by Walker *et al.* (2013), on supplementing carotenoids to neonatal hihis, *N. cincta*, provides support for this relationship, with supplemented males having more saturated yellow breeding plumage as adults than controls. In contrast, Orledge *et al.* (2012b) found no relationship between antioxidant availability during development (supplements of vitamin E, carotenoids, or a combination of both) and the expression of sexual ornaments in ring-necked pheasants (regardless of whether they were carotenoid-dependent or not).

Early Life-History Effects, Oxidative Stress, and Avian Song

The effects of variation in the environmental conditions experienced during growth and development are not confined to morphometric and carotenoid-based sexual signals however. The early environment also affects characteristics of avian song. In many avian species, females choose mates based on the quality of their song (Searcy and Nowicki, 2005). Experimental manipulation of the environment experienced during early life through imposing nutritional stress or by injection with the stress hormone corticosterone can lead to a reduction in the size of the high vocal center (HVC), a song control nucleus known to be a key determinant of song complexity (Buchanan *et al.*, 2004). Learned features of song development can serve as reliable indicators of male quality, signaling a male's experience of, and developmental response to, stress experienced during early life (Buchanan *et al.*, 2003, 2004; Nowicki *et al.*, 2002; Zann and Cash, 2008). The HVC has a higher susceptibility to stress experienced during development than other areas of the brain (Buchanan *et al.*, 2004), and it has been suggested that the physiological mechanism behind this trade-off in allocation of resources to the HVC may be related to the dynamics of the response of corticosteroid receptors to elevated levels of the stress hormone corticosterone (Buchanan *et al.*, 2004; MacDonald *et al.*, 2006). However, the vertebrate brain also has a particularly high requirement for oxygen, contains a high volume of vulnerable polyunsaturated fatty acids, and contains free radical releasing enzymes (Halliwell and Gutteridge, 2007). This means that the HVC is likely to be particularly susceptible to oxidative damage by free radicals (von Schantz *et al.*, 1999), and suggests that the complexity of avian song, which is a sexually selected trait, is also likely to be determined by resource allocation trade-offs during growth and development that are mediated by oxidative stress.

COMPETITION-DEPENDENT SEXUAL SIGNALS

Early life-history effects are characterized by acquisition of scarce resources and the trade-offs that subsequently occur in the allocation of these resources to traits (van Noordwijk and de Jong, 1986). Greater acquisition of resources reduces the trade-off in resource allocation, but scarcity of resources leads to greater levels of competition, so the competitiveness of individuals determines their resource acquisition ability, which increases competitiveness, which in turn increases resource acquisition ability, and so on, through a process of positive feedback (see Figure 2.3: "power begets power" as Mock and Parker, 1997 put it). As

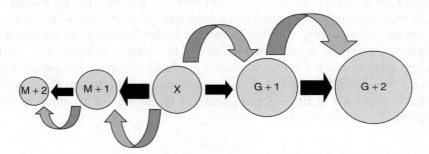

Figure 2.3. Illustration of model of resource allocation during development under competition. Discs represent the amounts of resources acquired, with the size of the disc corresponding to the magnitude of the resource. The black arrows indicate the amount of resources allocated to maintenance, and the gray, curved arrows represent the amount of resources allocated to growth. An individual with a moderate amount of initial resources, indicated by the disc marked X, can subsequently either prioritize allocation of resources to maintenance (M + 1, M + 2) or growth (G + 1, G + 2). If there is significant competition for scarce resources, then allocating resources to maintenance at the expense of growth (i.e., X to M) will result in individuals being out-competed for resources by larger conspecifics, which will reduce the relative amount of resources that can be subsequently acquired (M + 1). Positive feedback between competitiveness and the ability to acquire resources results in a reduction in the amount of resources acquired over time (M + 1 to M + 2 and beyond), which reduces the absolute amount of resources that can be allocated to maintenance as well as growth. Death through starvation is the endpoint, unless there is an increase in overall resource availability (and, therefore, a reduction in competition). Alternatively, allocation of resources to growth at the expense of (a threshold minimum level of) maintenance will lead to increased competitiveness, which feeds back positively on resource acquisition ability (X to G + 1 to G + 2). The greater amount of resources available as a result of increased competitiveness means there will also be an increase in the amount of resources available for maintenance as well as growth. If the level of competition for resources should drop, then extra resources can be allocated toward maintenance (i.e., future investment), rather than growth, in contrast to the response of individuals following an M strategy in the same circumstances (which have to prioritize putting resources toward catching up in size).

the examples provided above have outlined, variation in resource availability during growth and development can affect the expression of signals both during development and in adulthood. Since it is the competitiveness of individuals that determines access to scarce resources, any signal sensitive to early life-history effects will be a "competition-dependent" signal (Orledge et al., 2012b).

What are Competition-Dependent Signals?

The social environment provided by conspecifics is unusual in that the effects on individuals are both environmental and genetic (Moore et al., 1997; Wolf and Brodie, 1998). Indirect genetic effects (IGEs) occur whenever genes expressed in one individual affect the phenotype of another (Wolf and Brodie, 1998) and are of interest because they can result in complex and non-intuitive evolutionary responses to selection (Moore et al., 1997; Wolf, 2003). Models of competition-dependence (Harris et al., 2008; Wolf et al., 2008) show that competition affects the accumulation of resources that contribute to the overall condition of an individual (sensu Rowe and Houle, 1996), so is an important determinant of the expression of condition-dependent traits (Wolf et al., 2008).

Variation in Sexual Signals can be Maintained by Competition-Dependence

The expression of condition-dependent traits has been proposed as a way to resolve the paradox of the lek, whereby variation in sexual signals can be maintained, despite the depleting effect of directional selection imposed by female choice, through mutational input. This is because condition is influenced by a large number of physiological processes, so a large number of loci contribute to variation in trait expression (Rowe and Houle, 1996). Recent models by Harris et al. (2008) and Wolf et al. (2008) show that competitive interactions among conspecifics during development provide an important source of genetically based variation in condition that maintains variation in indicator traits ("competition-dependent" traits or signals).

The unique combination of direct and IGEs acting on individuals via social competition for resources determines the distribution of traits within groups of competing individuals. Consequently, the trait is indicative of relative competitive ability, making such indicator traits a highly reliable indicator of male quality within groups (Wolf et al., 2008). Moreover, heritable variation may be maintained via the "treadmill of social competition" (Dickerson, 1955), where non-transitive competitive success (i.e., no genotype is best across all environments) leads to long-term cyclical variation in allele frequencies despite constant directional selection (Harris et al., 2008). Selection for increased condition produced by mate choice for indicator traits may be countered by the evolution of increased competitiveness in the social environment, resulting in no net change in condition (Wolf et al., 2008).

The potential for social competition to influence traits arises during the development and/or the expression of the trait(s) involved, because individuals often develop in groups with other conspecifics (Wolf *et al.*, 2008). In the context of species that have biparental care, for example, this is likely to be some form of nursery (*sensu* Mock and Parker, 1997), where dependent offspring are in close proximity to one another and receive food and other resources from parents (e.g., nests of altricial species of birds or a small mammal carcass provided by burying beetle parents for their offspring). The resource richness of the environment experienced by an individual being reared in such a nursery will be determined, at least partly, by the competitive ability of that individual relative to that of its conspecifics (Wolf *et al.*, 2008).

Competition-Dependent Signals and Oxidative Stress

Ecological characteristics such as the number of competitors, the relatedness of the individual to its competitors, the mechanism of resource partitioning (e.g., passive allocation of food/scramble competition vs active allocation of food/honest signaling; Parker *et al.*, 2002a, 2002b), and the degree to which competitive dominance relations among genotypes are transitive (Wolf *et al.*, 2008) will determine the nature of the competitive interactions. Increased competition and competitiveness directly (through increasing metabolic rate) and indirectly (through increasing allocation of resources to growth rate at the expense of self-maintenance; e.g., Alonso-Alvarez *et al.*, 2007; Hall *et al.*, 2010) affect the oxidative status of individuals, which proximately mediates the expression of competition-dependent signals (Orledge *et al.*, 2012b). The key point is that relative competitive ability can determine the expression of indicator traits via effects on the oxidative status of individuals, but the exact evolutionary dynamics will depend upon the availability of resources in the environment, the species involved, and the nature of competition (Wolf *et al.*, 2008).

CONCLUSIONS

In this review, we have presented evidence in support of a central role of oxidative stress in the evolution and expression of morphological, carotenoid-dependent, and vocal signals throughout development, from maternal effects, through parent–offspring communication during the provision of PI, to the signaling that occurs as an adult during mate choice. However, despite a wealth of studies highlighting the importance of early life-history effects on the expression of traits during adulthood, and a large body of work on carotenoid-based sexual signals, relatively few studies have made the connection between early life-history effects and the expression of oxidative stress-mediated sexual signals, such as carotenoid-based traits (notable exceptions include a study by Morales *et al.*, 2009, who showed that the red-bill spot of Larid gulls, which is carotenoid-based, is used in both sexual signaling and signaling during parent–offspring interactions, and the work on pheasants by Orledge *et al.*, 2012a, 2012b and on hihis by Walker *et al.*, 2013). Instead, most studies have concentrated on, and manipulated, antioxidant availability in adulthood and ignored downstream/programming effects of antioxidant availability during growth and development on the expression of sexual signals.

We suggest that the key to understanding these important relationships, and the evolution and expression of (sexual) signals in general, is the relative competitive ability of individuals (see Figures 2.3 and 2.4). Variation in the environment

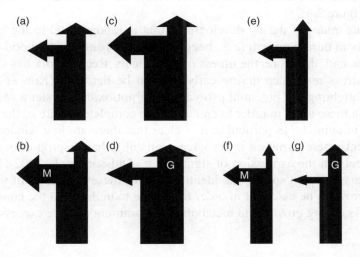

Figure 2.4. Resource allocation strategies under varying levels of resource acquisition. In (a) and (b) demand > supply; in (c) and (d) demand ≈ supply; in (e)–(g) demand ≫ supply. The width of the base represents the total amount of resources acquired. Allocations to growth are donated by the vertical arrows and to maintenance by horizontal arrows, with the width of the arrow in both cases indicating the relative amount of resources allocated to each function. Under moderate levels of competition (a and b) resources should be preferentially allocated to growth, with a minimal (threshold) amount allocated to maintenance (a), as this strategy increases competitiveness and future resource acquisition ability (see Figure 2.3). In these circumstances, growth is favored over maintenance (i.e., a is a better strategy than b). However, if competition is low (c and d), then the extra resources acquired would be better allocated to maintenance (c) rather than growth (d), as selection will favor allocation of resources to components of fitness that affect expression of "future" traits (e.g., sexual signals). In addition, it is expected that selection will act against allocating extra resources to growth under conditions of relaxed competition, as the costs of oxidative stress will not be offset by any gains in resource acquisition ability. When resources are severely restricted and competition is severe (e–g), individuals would be expected to allocate the minimal (threshold) amount of resources to maintenance and the rest to growth. This strategy increases the probability that individuals will be able to compensate for their poor start in life if resource availability improves. In contrast, putting more than the minimal level of resources into maintenance (f) may lead to reduced competitiveness and therefore reduced survival prospects (see Figure 2.3). Likewise, putting less than the threshold minimum levels of resources into maintenance (g), will, by definition, result in the rapid accumulation of damage and, ultimately, death.

experienced by individuals during development is common due to differences in resource availability. Competition for resources leads to trade-offs in favor of traits that increase competitive ability, but a cost of increased competition is an increase in oxidative stress, unless relative competitive ability is such that sufficient antioxidants can be obtained to mitigate these costs and maintain competitive status (i.e., there is a positive feedback between the costs and the maintenance of competitiveness; see Figure 2.4).

It is this trade-off during development that is fundamental to the expression of signals both during growth (e.g., begging behavior) and in adulthood (e.g., sexual signals), and, therefore, the fitness of individuals. Recent work has shown that oxidative stress resistance during early life can be heritable (Kim *et al.*, 2010, 2013), highlighting the potential evolvability of antioxidant systems. More work is needed in these areas in order to create a more complete picture of the evolution of signals in animals. In particular, it is clear that there are key windows during development where variation in resource availability has disproportionate downstream effects on the expression of signals (e.g., Ohlsson *et al.*, 2002), but this is likely to vary between species, so identification of these "windows of vulnerability" may prove to be essential in order to be able to understand the consequences of trade-offs during growth and metabolic programming that are expressed during adulthood.

ACKNOWLEDGMENTS

Some of the ideas expressed in this chapter were developed while NJR was in receipt of a postdoctoral fellowship from NERC. JDB was supported by a Royal Society University Research Fellowship. JMO was supported by a NERC CASE studentship awarded to NJR and JDB.

REFERENCES

Adams, E.S. and Mesterton-Gibbons, M. (1995) The cost of threat displays and the stability of deceptive communication. *Journal of Theoretical Biology*, **175**, 405–421.

Alonso-Alvarez, C., Bertrand, S., Devevey, G., Prost, J., Faivre, B., and Sorci, G. (2004) Increased susceptibility to oxidative stress as a proximate cost of reproduction. *Ecology Letters*, **7**, 363–368.

Alonso-Alvarez, C., Bertrand, S., Faivre, B., and Sorci, G. (2007) Increased susceptibility to oxidative damage as a cost of accelerated somatic growth in zebra finches. *Functional Ecology*, **21**, 873–879.

Alonso-Alvarez, C., Perez-Rodriquez, L., Mateo, R., Chastel, O., and Vinuela, J. (2008) The oxidation handicap hypothesis and the carotenoid allocation trade-off. *Journal of Evolutionary Biology*, **21**, 1769–1797.

Arendt, J.D. (2003) Reduced burst speed is a cost of rapid growth in anuran tadpoles: problems of autocorrelation and inferences about growth rates. *Functional Ecology*, **17**, 328–334.

Arendt, J.D. and Wilson, D.S. (2000) Population differences in the onset of cranial ossification in pumpkinseed sunfish (*Lepomis gibbosus*): a potential cost of rapid growth. *Canadian Journal of Fish and Aquatic Science*, **57**, 351–356.

Badyaev, A.V. (2008) Maternal effects as generators of evolutionary change. *Annals of the New York Academy of Sciences*, **1133**, 151–161.

Badyaev, A.V. and Uller, T. (2009) Parental effects in ecology and evolution: mechanisms, processes and implications. *Philosophical Transactions of the Royal Society*, **364**, 1169–1177.

Baeta, R. *et al.* (2008) Carotenoid trade-offs between parasitic resistance and sexual display: an experimental study in the blackbird (*Turdus merula*). *Proceedings of the Royal Society B: Biological Sciences*, **275**, 427–434.

Barja, G. (2004) Free radicals and aging. *Trends in Neurosciences*, **27**, 595–600.

Basolo, A. (1990) Female preference predates the evolution of the sword in swordtail fish. *Science*, **250**, 808–810.

Berg, D.M. *et al.* (2004) Redox imbalance. *Cell and Tissue Research*, **318**, 201–213.

Bertrand, S., Faivre, B., and Sorci, G. (2006) Do carotenoid-based sexual traits signal the availability of non-pigmentary antioxidants? *Journal of Experimental Biology*, **209**, 4414–4419.

Biard, C., Surai, P.F., and Møller, A.P. (2005) Effects of carotenoid availability during laying on reproduction in the blue tit. *Oecologia*, **144**, 32–44.

Biard, C., Surai, P.F., and Møller, A.P. (2007) An analysis of pre- and post-hatching maternal effects mediated by antioxidants in the blue tit. *Journal of Experimental Biology*, **20**, 326–339.

Biard, C., Gil, D., Karadas, F., Saino, N., Spottiswoode, C.N., Surai, P.F., and Møller, A.P. (2009) Maternal effects mediated by antioxidants and the evolution of carotenoid-based signals in birds. *American Naturalist*, **174**, 696–708.

Birkhead, T.R., Fletcher, F., and Pellatt, E.J. (1999) Nestling diet, secondary sexual traits and fitness in the zebra finch. *Proceedings of the Royal Society B: Biological Sciences*, **266**, 385–390.

Blas, J., Perez-Rodriquez, L., Bortolotti, G.R., Vinuela, J., and Marchant, T.A. (2006) Testosterone increases bioavailability of carotenoids: insights into the honesty of sexual signaling. *Proceedings of the National Academy of Sciences of the United States of America*, **103**, 18633–18637.

Blount, J.D. (2004) Carotenoids and life-history evolution in animals. *Archives of Biochemistry and Biophysics*, **430**, 10–15.

Blount, J.D. *et al.* (2002) Carotenoids and egg quality in the lesser black-backed gull (*Larus fuscus*): a supplemental feeding study of maternal effects. *Proceedings of the Royal Society B: Biological Sciences*, **269**, 29–36.

Blount, J.D., Metcalfe, N.B., Arnold, K.E., Surai, P.F., Devevey, G.L., and Monaghan, P. (2003) Neonatal nutrition, adult antioxidant defences and sexual attractiveness in the zebra finch. *Proceedings of the Royal Society B: Biological Sciences*, **270**, 1691–1696.

Borgia, G. *et al.* (2004) Juvenile infection and male display: testing the bright male hypothesis across individual life histories. *Behavioral Ecology*, **15**, 722–728.

Buchanan, K.L., Spencer, K.A., Goldsmith, A.R., and Catchpole, C.K. (2003) Song is an honest signal of past developmental stress in the European starling (*Sturnus vulgaris*). *Proceedings of the Royal Society B: Biological Sciences*, **270**, 1149–1156.

Buchanan, K.L. *et al.* (2004) Developmental stress selectively affects the song control nucleus HVC in the zebra finch. *Proceedings of the Royal Society B: Biological Sciences*, **271**, 2381–2386.

Campbell, N.B. and Reece, J.B. (2008) *Biology*, 8th edn, Pearson/Benjamin Cummings, San Francisco.

Cash, T.P. *et al.* (2007) Reactive oxygen species and cellular oxygen sensing. *Free Radical Biology and Medicine*, **43**, 1219–1225.

Castro, L. and Freeman, B. (2001) Reactive oxygen species in human health and disease. *Nutrition*, **17**, 161–165.

Catoni, C., Peters, A., and Schaefer, M. (2008) Life history trade-offs are influenced by the diversity, availability and interactions of dietary antioxidants. *Animal Behaviour*, **76**, 1107–1119.

Chappell, M.A. and Bachman, G.C. (2002) Energetic costs of begging behaviour, in *The Evolution of Begging: Competition, Cooperation and Communication* (eds J. Wright and M.L. Leonard), Kluwer, Dordrecht, pp. 143–162.

Chew, B.P. and Park, J.S. (2004) Carotenoid action on the immune response. *Journal of Nutrition*, **134**, 257S–262S.

Christians, J.K. (2002) Avian egg size: variation within species and inflexibility within individuals. *Biological Reviews*, **77**, 1–26.

Cohen, A.A. and McGraw, K.J. (2009) No simple measures for antioxidant status in birds: complexity in inter- and intraspecific correlations among circulating antioxidant types. *Functional Ecology*, **23**, 310–320.

Costantini, D. (2008) Oxidative stress in ecology and evolution: lessons from avian studies. *Ecology Letters*, **11**, 1238–1251.

Costantini, D. and Møller, A.P. (2008) Carotenoids are minor antioxidants for birds. *Functional Ecology*, **22**, 367–370.

Cucco, M., Guasco, B., Malacarne, G., and Ottonelli, R. (2006) Effects of β-carotene supplementation on chick growth, immune status and behaviour in the grey partridge, *Perdix perdix*. *Behavioural Processes*, **73**, 325–332.

Dalle-Donne, I.R. *et al.* (2006) Biomarkers of oxidative damage in human disease. *Clinical Chemistry*, **52**, 601–623.

David, P., Bjorksten, T., Fowler, K., and Pomiankowski, A. (2000) A condition-dependent signalling of genetic variation in stalk-eyed flies. *Nature*, **406**, 186–188.

de Ayala, R.M., Martinelli, R., and Saino, N. (2006) Vitamin E supplementation enhances growth and condition of nestling barn swallows (*Hirundo rustica*). *Behavioral Ecology and Sociobiology*, **60**, 619–630.

De Block, M. and Stoks, R. (2008) Compensatory growth and oxidative stress in a damselfly. *Proceedings of the Royal Society B: Biological Sciences*, **275**, 781–785.

Dennery, P.A. (2010) Oxidative stress in development: nature or nurture? *Free Radical Biology & Medicine*, **49**, 1147–1151.

Dickerson, G.E. (1955) Genetic slippage in response to selection for multiple objectives. *Cold Spring Harbor Symposium of Quantitative Biology*, **20**, 213–224.

Dowling, D.K. and Simmons, L.W. (2009) Reactive oxygen species as universal constraints in life-history evolution. *Proceedings of the Royal Society B: Biological Sciences*, **276**, 1737–1745.

Dufva, R. and Allander, K. (1995) Intraspecific variation in plumage coloration reflects immune-response in great tit (*Parus major*) males. *Functional Ecology*, **9**, 785–789.

Dugas, M.B. (2009) House sparrows, *Passer domesticus*, parents preferentially feed nestlings with mouth colours that appear carotenoid-rich. *Animal Behaviour*, **78**, 767–772.

Ewen, J.G., Thorogood, R., Brekke, P., Cassey, P., Karadas, F., and Armstrong, D.P. (2009) Maternally invested carotenoids compensate costly ectoparasitism in the hihi. *Proceedings of the National Academy of Sciences of the United States of America*, **106**, 12798–12802.

Faivre, B., Gregoire, A., Preault, M., Cezilly, F., and Sorci, G. (2003) Immune activation rapidly mirrored in a secondary sexual trait. *Science*, **300**, 29–31.

Farrell, A.P. *et al.* (1997) Growth enhanced transgenic salmon can be inferior swimmers. *Canadian Journal of Zoology*, **75**, 335–337.

Fischer, K., Zeilstra, I., Hetz, S.K., and Fiedler, K. (2004) Physiological costs of growing fast: does accelerated growth reduce pay-off in adult fitness? *Evolutionary Ecology*, **18**, 343–353.

Folstad, I. and Karter, A.J. (1992) Parasites, bright males and the immunocompetence handicap. *American Naturalist*, **139**, 603–622.

Forbes, S. and Glassey, B. (2000) Asymmetric sibling rivalry and nestling growth in red-winged blackbirds. *Behavioral Ecology and Sociobiology*, **48**, 413–417.

Forbes, S., Glassey, B., Thornton, S., and Earle, L. (2001) The secondary adjustment of clutch size in red-winged blackbirds (*Agelaius phoeniceus*). *Behavioral Ecology and Sociobiology*, **50**, 37–44.

Genestra, M. (2007) Oxyl radicals, redox-sensitive signalling cascades and antioxidants. *Cellular Signalling*, **19**, 1807–1819.

Godfray, H.C.J. (1991) Signalling of need by offspring to their parents. *Nature*, **352**, 328–330.

Godfray, H.C.J. (1995) Signalling of need between parents and young: parent–offspring conflict and sibling rivalry. *American Naturalist*, **146**, 1–24.

Goodwin, T.W. (1984) *The Biochemistry of the Carotenoids*, Chapman and Hall, London.

Grafen, A. (1990a) Biological signals as handicaps. *Journal of Theoretical Biology*, **144**, 517–546.

Grafen, A. (1990b) Sexual selection unhandicapped by the Fisher process. *Journal of Theoretical Biology*, **144**, 475–518.

Grether, G.F., Hudon, J., and Endler, J.A. (2001) Carotenoid scarcity, synthetic pteridine pigments and the evolution of sexual coloration in guppies (*Poecilia reticulata*). *Proceedings of the Royal Society B: Biological Sciences*, **268**, 1245–1253.

Groothuis, T.G.G., Muller, W., von Engelhardt, N., Carere, C., and Eising, C. (2005) Maternal hormones as a tool to adjust offspring phenotype in avian species. *Neuroscience and Biobehavioral Reviews*, **29**, 329–352.

Guilford, T. and Dawkins, M.S. (1991) Receiver psychology and the evolution of animal signals. *Animal Behaviour*, **42**, 1–14.

Hadfield, J.D. and Owens, I.P.F. (2006) Strong environmental determination of a carotenoid-based trait is not mediated by carotenoid availability. *Journal of Evolutionary Biology*, **19**, 1104–1114.

Hall, M.E., Blount, J.D., Forbes, S., and Royle, N.J. (2010) Does oxidative stress mediate the trade-off between growth and self-maintenance in structured families? *Functional Ecology*, **24**, 365–373.

Halliwell, B. and Gutteridge, J.M.C. (2007) *Free Radicals in Biology and Medicine*, 4th edn, Oxford University Press, Oxford.

Hamilton, W.D. and Zuk, M. (1982) Heritable true fitness and bright birds: a role for parasites. *Science*, **218**, 384–387.

Harris, W.E., McKane, A.J., and Wolf, J.B. (2008) The maintenance of heritable variation through social competition. *Evolution*, **62**, 337–347.

Hartley, R.C. and Kennedy, M.W. (2004) Are carotenoids a red herring in sexual display? *Trends in Ecology & Evolution*, **19**, 353–354.

Harshman, L.G. and Zera, A.J. (2007) The cost of reproduction: the devil in the details. *Trends in Ecology & Evolution*, **22**, 80–86.

Hasson, O. (1994) Cheating signals. *Journal of Theoretical Biology*, **167**, 223–238.

Helfenstein, F., Berthouly, A., Tanner, M., Karadas, F., and Richner, H. (2008) Nestling begging intensity and parental effort in relation to prelaying carotenoid availability. *Behavioral Ecology*, **19**, 108–115.

Henry, C.J.K. and Ulijaszek, S.J. (1996) *Long-Term Consequences of Early Environment: Growth, Development and the Lifespan Developmental Perspective*, Cambridge University Press, UK.

Hill, G.E. (2000) Energetic constraints on expression of carotenoid-based plumage coloration. *Journal of Avian Biology*, **31**, 559–566.

Hill, G.E., Inouye, C.Y., and Montgomerie, R. (2002) Dietary carotenoids predict plumage coloration in wild house finches. *Proceedings of the Royal Society B: Biological Sciences*, **269**, 1119–1124.

Hinde, C.A., Buchanan, K.L., and Kilner, R.M. (2009) Prenatal environmental effects match offspring begging to parental provisioning. *Proceedings of the Royal Society B: Biological Sciences*, **276**, 2787–2794.

Hõrak, P., Surai, P.F., Ots, I., and Møller, A.P. (2004) Fat-soluble antioxidants in brood-rearing great tits *Parus major*: relations to health and appearance. *Journal of Avian Biology*, **35**, 63–70.

Hõrak, P., Saks, L., Zilmer, M., Karu, U., and Zilmer, K. (2007) Do dietary carotenoids alleviate the cost of immune activation? An experiment with greenfinches. *American Naturalist*, **170**, 625–635.

Isaksson, C. and Andersson, S. (2008) Oxidative stress does not influence carotenoid mobilization and plumage pigmentation. *Proceedings of the Royal Society B: Biological Sciences*, **275**, 309–314.

Isaksson, C., McLaughlin, P., Monaghan, P. and Andersson, S. (2007) Carotenoid pigmentation does not reflect total non-enzymatic antioxidant activity in plasma of adult and nestling great tits, Parus major. *Funct Ecol*, **21**: 1123–1129. doi: 10.1111/j.1365-2435.2007.01317.x

Kilner, R.M. (1997) Mouth colour is a reliable signal of need in begging canary nestlings. *Proceedings of the Royal Society B: Biological Sciences*, **264**, 963–968.

Kilner, R.M. (2001) A growth cost of begging in captive canary chicks. *Proceedings of the National Academy of Sciences of the United States of America*, **98**, 11394–11398.

Kilner, R. and Davies, N.B. (1998) Nestling mouth colour: ecological correlates of a begging signal. *Animal Behaviour*, **56**, 705–712.

Kim, S.-Y., Noguera, J.C., Morales, J., and Velando, A. (2010) Heritability of resistance to oxidative stress in early life. *Journal of Evolutionary Biology*, **23**, 769–775.

Kim, S.-Y., Noguera, J.C., Tato, A., and Velando, A. (2013) Vitamins, stress and growth: the availability of antioxidants in early life influences the expression of cryptic genetic variation. *Journal of Evolutionary Biology*, **26**, 1341–1352.

Kotiaho, J. (2001) Costs of sexual traits: a mismatch between theoretical considerations and empirical evidence. *Biological Reviews*, **76**, 365–376.

Koutsos, E.A., Clifford, A.J., Calvert, C.C., and Klasing, K.C. (2003) Maternal carotenoid status modifies the incorporation of dietary carotenoids into immune tissues of growing chickens (*Gallus gallus domesticus*). *Journal of Nutrition*, **133**, 1132–1138.

Kregel, K.C. and Zhang, H.J. (2007) An integrated view of oxidative stress in aging: basic mechanisms, functional effects, and pathological considerations. *American Journal of Physiology. Regulatory, Integrative and Comparative Physiology*, **292**, R18–R36.

Lane, N. (2003) A unifying view of ageing and disease: the double-agent theory. *Journal of Theoretical Biology*, **225**, 531–540.

Lane, N. (2005) *Power, Sex, Suicide: Mitochondria and the Meaning of Life*, Oxford University Press, Oxford.

Leonard, M.L., Horn, A.G., and Porter, J. (2003) Does begging affect growth in nestling tree swallows, *Tachycineta bicolor*? *Behavioural Ecology and Sociobiology*, **54**, 573–577.

Lin, Y.-F., Tsai, H.-L., Lee, Y.-C., and Chang, S.-J. (2005) Maternal vitamin E supplementation affects the antioxidant capability and oxidative status of hatching chicks. *Journal of Nutrition*, **135**, 2457–2461.

Lindholm, A.K., Hunt, J., and Brooks, R. (2006) Where do all the maternal effects go? Variation in offspring body size through ontogeny in the live-bearing fish *Poecilia parae*. *Biology Letters*, **2**, 586–589.

Lindström, J. (1999) Early development and fitness in birds and mammals. *Trends in Ecology & Evolution*, **14**, 343–348.

Lindström, K. and Lundström, J. (2000) Male greenfinches (*Carduelis chloris*) with brighter ornaments have higher virus infection clearance rates. *Behavioural Ecology and Sociobiology*, **48**, 44–51.

Linksvayer, T.A., Fondrk, M.K., and Page, R.E. (2009) Honey bee social regulatory networks are shaped by colony-level selection. *American Naturalist*, **173**, E99–E107.

Lipar, J.L. and Ketterson, E.D. (2000) Maternally derived yolk testosterone enhances the development of the hatching muscle in the red-winged blackbird *Agelaius phoeniceus*. *Proceedings of the Royal Society B: Biological Sciences*, **267**, 2005–2010.

Lipar, J.L., Ketterson, E.D., and Nolan, V. (1999) Intra-clutch variation in testosterone content of red-winged blackbird eggs. *Auk*, **116**, 231–235.

Loiseau, C., Fellous, S., Haussy, C., Chastel, O., and Sorci, G. (2008) Condition-dependent effects of corticosterone on a carotenoid-based begging signal in house sparrows. *Hormones and Behavior*, **53**, 266–273.

Luo, Z.C., Fraser, W.D., Julien, P., Deal, C.L., Audibert, F., Smith, G.N., Xiong, X., and Walker, M. (2006) Tracing the origins of "fetal origins" of adult diseases: programming by oxidative stress? *Medical Hypotheses*, **66**, 38–44.

MacDonald, I.F. *et al.* (2006) Early nutritional stress impairs development of a song-control brain region in both male and female juvenile song sparrows (*Melospiza melodia*) at the onset of learning. *Proceedings of the Royal Society B: Biological Sciences*, **273**, 2559–2564.

Macnair, M.R. and Parker, G.A. (1979) Models of parent-offspring conflict. III. Intra-brood conflict. *Animal Behaviour*, **27**, 1202–1209.

Maynard Smith, J. and Harper, D.G.C. (1995) Animal signals: models and terminology. *Journal of Theoretical Biology*, **177**, 305–311.

Maynard Smith, J. and Harper, D.G.C. (2003) *Animal Signals*, Oxford University Press, Oxford.

McGraw, K.J., Adkins-Regan, E. and Parker, R.S. (2005). Maternally derived carotenoid pigments affect offspring survival, sex ratio, and sexual attractiveness in a colorful songbird. *Naturwissenschaften* **92**, 375-380.

McGraw, K.J. and Ardia, D.R. (2003) Carotenoids, immunocompetence, and the information content of sexual colors: an experimental test. *American Naturalist*, **162**, 704–712.

McGraw, K.J. and Hill, G.E. (2004) Plumage color as a dynamic trait: carotenoid pigmentation of male house finches (*Carpodacus mexicanus*) fades during the breeding season. *Canadian Journal of Zoology*, **82**, 734–738.

McGraw, K.J. and Parker, R.S. (2006) A novel lipoprotein-mediated mechanism controlling sexual attractiveness in a colorful songbird. *Physiology and Behavior*, **87**, 103–108.

McGraw, K.J. (2005) Interspecific variation in dietary carotenoid assimilation in birds: links to phylogeny and color ornamentation. *Comparative Biochemistry and Physiology. B*, **142**, 245–250.

Metcalfe, N.B. and Monaghan, P. (2001) Compensation for a bad start: grow now, pay later? *Trends in Ecology and Evolution*, **16**, 254–260.

Mock, D.W. and Forbes, L.S. (1995) The evolution of parental optimism. *Trends in Ecology and Evolution*, **10**, 130–134.

Mock, D.W. and Parker, G.A. (1997) *The Evolution of Sibling Rivalry*, Oxford University Press, Oxford.

Monaghan, P., Metcalfe, N.B., and Torres, R. (2009) Oxidative stress as a mediator of life history trade-offs: mechanisms, measurements and interpretation. *Ecology Letters*, **12**, 75–92.

Moore, A.J., Brodie, E.D., and Wolf, J.B. (1997) Interacting phenotypes and the evolutionary process. 1. Direct and indirect genetic effects of social interactions. *Evolution*, **51**, 1352–1362.

Morales, J., Alonso-Alvarez, C., Perez, C., Torres, R., Serafino, E., and Velando, A. (2009) Families on the spot: sexual signals influence parent–offspring interactions. *Proceedings of the Royal Society B: Biological Sciences*, **276**, 2477–2483.

Morgan, I.J. and Metcalfe, N.B. (2001) Deferred costs of catch-up growth after autumnal food shortage in juvenile salmon. *Proceedings of the Royal Society B: Biological Sciences*, **268**, 295–301.

Mortensen, A., Skibsted, L.H., and Truscott, T.G. (2001) The interaction of dietary carotenoids with radical species. *Archives of Biochemistry and Biophysics*, **385**, 13–19.

Mougeot, F., Martínez-Padilla, J., Webster, L.M.I., Blount, J.D., Pérez-Rodríguez, L., and Piertney, S.B. (2009) Honest sexual signalling mediated by parasite and testosterone effects on oxidative balance. *Proceedings of the Royal Society B: Biological Sciences*, **276**, 1093–1100.

Mousseau, T.A. and Fox, C.W. (1998) *Maternal Effects as Adaptations*, Oxford University Press, New York.

Navara, K.J. and Hill, G.E. (2003) Dietary carotenoid pigments and immune function in a songbird with extensive carotenoid-based plumage coloration. *Behavioral Ecology*, **14**, 909–916.

Niki, E. (1996) α-Tocopherol, in *Handbook of Antioxidants* (eds E. Cadenas and L. Packer), Marcel Dekker, New York, pp. 3–25.

Nolan, P.M., Hill, G.E., and Stoehr, A.M. (1998) Sex, size, and plumage redness predict house finch survival in an epidemic. *Proceedings of the Royal Society B: Biological Sciences*, **265**, 961–965.

Noguera, J.C., Morales, J., Pérez, C., and Velando, A. (2010) On the oxidative cost of begging: antioxidants enhance vocalizations in gull chicks. *Behavioral Ecology*, **21**, 479–484.

Nowicki, S., Searcy, W.A., and Peters, S. (2002) Brain development, song learning and mate choice in birds: a review and experimental test of the "nutritional stress hypothesis". *Journal of Comparative Physiology A*, **188**, 1003–1014.

Nussey, D.H., Pemberton, J.M., Pilkington, J.G., and Blount, J.D. (2009) Life history correlates of oxidative damage in a free-living mammal population. *Functional Ecology*, **23**, 809–817.

O'Brien, E.L. and Dawson, R.D. (2008) Parasite-mediated growth patterns and nutritional constraints in a cavity-nesting bird. *Journal of Animal Ecology*, **77**, 127–134.

Ohlsson, T. *et al.* (2002) Pheasant sexual ornaments reflect nutritional conditions during early growth. *Proceedings of the Royal Society B: Biological Sciences*, **269**, 21–27.

Olson, V.A. and Owens, I.P.F. (1998) Costly sexual signals: are carotenoids rare, risky or required? *Trends in Ecology & Evolution*, **13**, 510–514.

Orledge, J.M., Blount, J.D., Hoodless, A.N., and Royle, N.J. (2012a) Antioxidant supplementation during early development reduces parasite load but does not affect sexual ornament expression in adult ring-necked pheasants. *Functional Ecology*, **26**, 688–700.

Orledge, J.M., Blount, J.D., Hoodless, A.N., Pike, T.W., and Royle, N.J. (2012b) Synergistic effects of supplementation of dietary antioxidants during growth on adult phenotype in ring-necked pheasants, *Phasianus colchicus*. *Functional Ecology*, **26**, 254–264.

Ozanne, S.E. and Hales, C.N. (2004) Lifespan: catch-up growth and obesity in male mice. *Nature*, **427**, 411–412.

Packer, J.E., Slater, T.F., and Willson, R.L. (1979) Direct observation of a free-radical interaction between vitamin E and vitamin C. *Nature*, **278**, 737–738.

Parker, G.A., Royle, N.J., and Hartley, I.R. (2002a) Intrafamilial conflict and parental investment: a synthesis. *Philosophical Transactions of the Royal Society B*, **357**, 295–307.

Parker, G.A., Royle, N.J., and Hartley, I.R. (2002b) Begging scrambles with unequal chicks: interactions between need and competitive ability. *Ecology Letters*, **5**, 206–215.

Patel, R.P. *et al.* (1999) Biological aspects of reactive nitrogen species. *Biochimica et Biophysica Acta, Bioenergetics*, **1411**, 385–400.

Peters, A. (2000) Testosterone treatment is immunosuppressive in superb fairy-wrens, yet free-living males with high testosterone are more immunocompetent. *Proceedings of the Royal Society B: Biological Sciences*, **267**, 883–889.

Peters, A. (2007) Testosterone and carotenoids: an integrated view of trade-offs between immunity and sexual signaling. *BioEssays*, **29**, 427–430.

Pike, T.W., Blount, J.D., Lindström, J., and Metcalfe, N.B. (2007) Availability of non-carotenoid antioxidants affects the expression of a carotenoid-based sexual ornament. *Biology Letters*, **3**, 353–356.

Plaistow, S.J. and Benton, T.G. (2009) The influence of context-dependent maternal effects on population dynamics: an experimental test. *Philosophical Transactions of the Royal Society B*, **364**, 1049–1058.

Praticò, D. (2008) Oxidative stress hypothesis in Alzheimer's disease: a reappraisal. *Trends in Pharmacological Sciences*, **29**, 609–615.

Rodríquez-Gironés, M.A., Enquist, M., and Lachmann, M. (2001a) Role of begging and sibling competition in foraging strategies of nestlings. *Animal Behaviour*, **61**, 733–745.

Rodríquez-Gironés, M.A., Zuniga, J.M., and Redondo, T. (2001b) Effects of begging on growth rates of nestling chicks. *Behavioral Ecology*, **12**, 269–274.

Rowe, L. and Houle, L. (1996) The lek paradox and the capture of genetic variation by condition dependent traits. *Proceedings of the Royal Society B: Biological Sciences*, **263**, 1415–1421.

Royle, N.J., Hartley, I.R., Owens, I.P.F., and Parker, G.A. (1999) Sibling competition and the evolution of growth rates in birds. *Proceedings of the Royal Society B: Biological Sciences*, **266**, 923–932.

Royle, N.J., Surai, P.F., and Hartley, I.R. (2001) Maternally derived androgens and antioxidants in bird eggs: complementary but opposing effects? *Behavioral Ecology*, **12**, 381–385.

Royle, N.J., Hartley, I.R., and Parker, G.A. (2002a) Sexual conflict reduces offspring fitness in zebra finches. *Nature*, **416**, 733–736.

Royle, N.J., Hartley, I.R., and Parker, G.A. (2002b) Begging for control: when are offspring solicitation behaviours honest? *Trends in Ecology & Evolution*, **17**, 434–440.

Royle, N.J., Hartley, I.R., and Parker, G.A. (2004) Parental investment and family dynamics: interactions between theory and empirical tests. *Population Ecology*, **46**, 231–241.

Royle, N.J., Lindström, J., and Metcalfe, N.B. (2005) A poor start in life negatively affects dominance status in adulthood independent of body size on green swordtails *Xiphophorus helleri*. *Proceedings of the Royal Society B: Biological Sciences*, **272**, 1917–1922.

Royle, N.J., Hartley, I.R., and Parker, G.A. (2006a) Consequences of biparental care for begging and growth in zebra finches, *Taeniopygia guttata*. *Animal Behaviour*, **72**, 123–130.

Royle, N.J., Lindström, J., and Metcalfe, N.B. (2006b) Effect of growth compensation on subsequent physical fitness in green swordtails *Xiphophorus helleri*. *Biology Letters*, **2**, 39–42.

Royle, N.J., Metcalfe, N.B., and Lindström, J. (2006c) Sexual selection, growth compensation and fast-start swimming performance in green swordtails *Xiphophorus helleri*. *Functional Ecology*, **20**, 662–669.

Royle, N.J., Hall, M.E., Blount, J.D., and Forbes, S. (2011) Patterns of egg yolk antioxidant co-variation in an avian brood parasite–host system. *Behavioral Ecology and Sociobiology*, **65**, 313–323.

Rubolini, D., Romano, M., Bonisoli Alquati, A., and Saino, N. (2006) Early maternal, genetic and environmental components of antioxidant protection, morphology and immunity of yellow-legged gull (*Larus michahellis*) chicks. *Journal of Evolutionary Biology*, **19**, 1571–1584.

Saks, L. *et al.* (2003) Carotenoid-based plumage coloration of male greenfinches reflects health and immunocompetence. *Oecologia*, **134**, 301–307.

Saino, N., Ninni, P., Calza, S., Martinelli, R., De Bernardi, F., and Møller, A.P. (2000) Better red than dead: carotenoid-based mouth coloration reveals infection in barn swallow nestlings. *Proceedings of the Royal Society B: Biological Sciences*, **267**, 757–761.

Salmon, A.B., Marx, D.B., and Harshman, L.G. (2001) A cost of reproduction in *Drosophila melanogaster*: stress susceptibility. *Evolution*, **55**, 1600–1608.

Samuels, S.E. and Baracos, V.E. (1995) Tissue protein turnover is altered during catch-up growth following *Escherichia coli* infection in weanling rats. *Journal of Nutrition*, **125**, 520–530.

Schwabl, H. (1993) Yolk is a source of maternal testosterone for developing birds. *Proceedings of the National Academy of Sciences of the United States of America*, **90**, 11446–11450.

Searcy, W.A. and Nowicki, S. (2005) *The Evolution of Animal Communication*, Princeton University Press, Princeton.

Seifried, H.E. *et al.* (2007) A review of the interaction among dietary antioxidants and reactive oxygen species. *Journal of Nutritional Biochemistry*, **18**, 567–579.

Singhal, A. and Lucas, A. (2004) Early origins of cardiovascular disease: is there a unifying hypothesis? *Lancet*, **363**, 1642–1645.

Singhal, A. *et al.* (2004) Is slower early growth beneficial for long-term cardiovascular health? *Circulation*, **109**, 1108–1113.

Stamps, J.A., Metcalf, R.A., and Krishnan, V.V. (1978) A genetic analysis of parent–offspring conflict. *Behavioural Ecology and Sociobiology*, **3**, 369–392.

Starck, J.M., and Ricklefs, R.E. (eds.) Avian Growth and Development: Evolution Within the Altricial-Precocial Spectrum, pp. 381–423. Oxford University Press: New York.

Stoks, R., de Block, M., van de Meutter, F., and Johansson, F. (2005) Predation cost of rapid growth: behavioural coupling and physiological decoupling. *Journal of Animal Ecology*, **74**, 708–715.

Surai, P.F. (1999) Vitamin E in avian reproduction. *Poultry and Avian Biology Reviews*, **10**, 1–60.

Surai, P.F. (2002) *Natural Antioxidants in Avian Nutrition and Reproduction*, Nottingham University Press, Nottingham.

Surai, P.F. and Sparks, N.H.C. (2001) Comparative evaluation of the effect of two maternal diets on fatty acids, vitamin E and carotenoids in the chick embryo. *British Poultry Science*, **42**, 252–259.

Tamashiro, K.L.K. and Moran, T.H. (2010) Perinatal environment and its influences on metabolic programming of offspring. *Physiology & Behavior*, **100**, 560–566.

Teather, K.L. (1992) An experimental study of competition for food resources between male and female nestlings of the red-winged blackbird. *Behavioural Ecology and Sociobiology*, **31**, 81–87.

Thorogood, R., Kilner, R.M., and Ewen, J.G. (2008) Spectral mouth colour of nestlings changes with carotenoid availability. *Functional Ecology*, **22**, 1044–1051.

Trivers, R.L. (1974) Parent–offspring conflict. *American Zoologist*, **14**, 249–264.

van Noordwijk, A.J. and de Jong, G. (1986) Acquisition and allocation of resources: their influence on variation in life history tactics. *American Naturalist*, **128**, 137–142.

von Schantz, T., Bensch, S., Grahn, M., Hasselquist, D., and Wittzell, H. (1999) Good genes, oxidative stress and condition-dependent sexual signals. *Proceedings of the Royal Society B: Biological Sciences*, **266**, 1–12.

Walker, L.K., Stevens, M., Karadas, F., Kilner, R.M., and Ewen, J.G. (2013) A window on the past: male ornamental plumage reveals the quality of their early-life environment. *Proceedings of the Royal Society B: Biological Sciences*, **280**, 20122852.

Walling, C.A., Royle, N.J., Metcalfe, N.B., and Lindström, J. (2007) Early nutritional conditions, growth trajectories and mate choice: does compensatory growth lead to an accelerated decline in adult sexual attractiveness? *Behavioral Ecology and Sociobiology*, **61**, 1007–1014.

Weathers, W.W., Hodum, P.J., and Anderson, D.J. (1997) Is the energy cost of begging by nestling passerines surprisingly low? *Auk*, **114**, 133.

Wiebe, K.L. and Slagsvold, T. (2009) Mouth coloration in nestling birds: increasing detection or signaling quality? *Animal Behaviour*, **78**, 1413–1420.

Wiersma, P., Selman, C., Speakman, J.R., and Verhulst, S. (2004) Birds sacrifice oxidative protection for reproduction. *Proceedings of the Royal Society B: Biological Sciences*, **271** (suppl. 5), S360–S363.

Wilkinson, G.S. and Reillo, P.R. (1994) Female choice response to artificial selection on an exaggerated male trait in a stalk-eyed fly. *Proceedings of the Royal Society B: Biological Sciences*, **255**, 1–6.

Wolf, J.B. and Brodie, E.D. (1998) The coadaptation of parental and offspring characters. *Evolution*, **52**, 299–308.

Wolf, J.B. (2003) Genetic architecture and evolutionary constraint when the environment contains genes. *Proceedings of the National Academy of Sciences of the United States of America*, **100**, 4655–4660.

Wolf, J.B., Harris, W.E., and Royle, N.J. (2008) The capture of heritable variation for genetic quality through social competition. *Genetica*, **134**, 89–97.

Zahavi, A. (1975) Mate selection – a selection for a handicap. *Journal of Theoretical Biology*, **53**, 205–214.

Zann, R. and Cash, E. (2008) Developmental stress impairs song complexity but not learning accuracy in non-domesticated zebra finches (*Taeniopygia guttata*). *Behavioural Ecology and Sociobiology*, **62**, 391–400.

3

A PERFORMANCE-BASED APPROACH TO STUDYING COSTS OF RELIABLE SIGNALS

Jerry F. Husak,[1] Justin P. Henningsen,[2] Bieke Vanhooydonck,[4] and Duncan J. Irschick[2,3]

[1] Department of Biology, University of St. Thomas, St. Paul, MN, USA
[2] Organismic and Evolutionary Biology Program, University of Massachusetts at Amherst, Amherst, MA, USA
[3] Department of Biology, University of Massachusetts at Amherst, Amherst, MA, USA
[4] Department of Biology, University of Antwerp, Antwerp, Belgium

INTRODUCTION

When conspecifics interact, the interests of each individual can conflict. Males and females, for example, often differ in what each may gain (or lose) from the other when decisions are made regarding mating (e.g., Darwin, 1871; Trivers, 1972; Emlen and Oring, 1977; Andersson, 1994; Arnqvist and Rowe, 2005). In many animal species, males may maximize their reproductive success by mating with the greatest number of females possible, whereas a female may mate with a single, high-quality male to maximize her reproductive success. Therefore, selection may favor the use of signals to advertise and assess potential mates. A more extreme example of conflicting interests is when two rival individuals each require a limited resource, resulting in a vigorous contest to determine which individual can access it. Because physical interactions can be costly to both winner and loser, it can be beneficial to have a means of reliably assessing a priori which contestant is most likely to win the interaction (reviewed in Andersson, 1994; Berglund et al., 1996; Maynard Smith and Harper, 2003; Taylor and Elwood, 2003; Searcy and Nowicki,

Animal Signaling and Function: An Integrative Approach, First Edition.
Edited by Duncan J. Irschick, Mark Briffa, and Jeffrey Podos.
© 2015 John Wiley & Sons, Inc. Published 2015 by John Wiley & Sons, Inc.

2005; see also Chapter 6 this volume). Although this pattern of rival interactions has been well-documented, the mechanisms that maintain the reliability of signals that are used for assessment remain under intense debate.

When interests of interacting individuals conflict, there are two general explanations for signal reliability: indices and handicaps. Either a signal cannot be faked due to physical constraints (called *indices*) or there are costs associated with the signal (called *handicaps*). Indices are signals that are physically constrained by the trait(s) that they advertise so that deception is impossible (Taylor *et al.*, 2000; Maynard Smith and Harper, 2003). For example, gaping displays of some lizards reveal the size of jaw muscles that are biomechanically linked to bite-force capacity (Lappin *et al.*, 2006), a functional trait important during male–male interactions (Huyghe *et al.*, 2005; Husak *et al.*, 2006, 2009b; Henningsen and Irschick, 2012). In our view, it is important that an index present an incorruptible link between signal intensity and some morphological, physiological, or biomechanical trait, meaning that this latter trait should directly constrain the signal. Genetic correlations or condition-dependent correlations between functional traits and a signal (e.g., Lailvaux *et al.*, 2005; Vanhooydonck *et al.*, 2007; and references therein), on the other hand, are not necessarily incorruptible over evolutionary time (e.g., Roff, 1997; Hau, 2007; Ketterson *et al.*, 2009). Determining whether and how signal intensity and performance are mechanistically linked can help reveal what maintains reliability in the signaling system. We focus our discussion on handicaps, which signal some aspect of sender quality but whose reliability is maintained by the presence of costs to the sender (Zahavi, 1975). Costs of signals can be categorized as either receiver-independent or receiver-dependent costs, based on whether the cost is imposed upon the sender by the intended receiver (Vehrencamp, 2000; Searcy and Nowicki, 2005). Receiver-independent costs are not dependent on the response of the *intended* receiver to the signal, but instead result from the development, production, or maintenance of the signal. Receiver-dependent costs depend on the response of receivers and entail either increased vulnerability to, or increased retaliation from, *intended* receivers. These are explored in more detail in this chapter.

How does one determine the cost of a signal? This question has remained controversial since Zahavi's initial hypothesis (Zahavi, 1975; Zahavi and Zahavi, 1997) and despite subsequent refinements (e.g., Grafen, 1990; Johnstone, 1995; Getty, 1998, 2006). In general, signal reliability can evolve when the fitness cost-to-benefit ratio of a signal is lower for a signaler that is of high quality compared to lower-quality signalers. More specifically, Grafen (1990) suggested that greater intensity signals have higher costs, but the marginal cost of a signal decreases with increasing signaler quality. These two key points have driven empirical research on signal costs in important ways. Many studies acknowledged the first point (greater intensity signal, greater cost) and test for a negative relationship between signal intensity and viability (survival or some presumed proxy; discussed later). The second point (higher quality, lower marginal costs) has been interpreted by many authors to mean that tests of the handicap principle should include quantifying whether signalers of high quality have lower costs and can produce higher intensity signals than signalers of low quality who cannot "afford" to produce greater intensity signals (see Getty, 1998, 2006). However, when considering the costs and benefits of sexually selected signals, fitness should be considered a multiplicative, and not an additive function of viability and fecundity (Getty, 1998, 2006). Thus, for a given signal intensity, a male with greater viability accrues proportionally greater benefits because of the differential survival (and thus greater fecundity over that longer lifetime) that exists between high- and low-quality males. The prediction, then, is that high-quality males will more efficiently convert signaling effort into fitness ("increasing-efficiency criterion," Getty, 2006, p. 86; see also Johnstone, 1997). This interpretation of signal costs and benefits has several theoretical ramifications for empirical research. First, one may find that high-quality males have lower marginal costs than low-quality males, but this is not a necessary prediction for handicap signals. Second, high-quality males may in some instances have higher marginal costs than low-quality males, but this can still result in a reliable signal where high-quality males have a greater optimal signal intensity than low-quality males (see Box 2 in Getty, 2006). Third, from an empirical standpoint, viability costs of signals may be high and detectable (as some studies have shown) or negligible (as many studies have shown). The salient point is that empirical studies designed to test signal reliability should consider cost–benefit trade-offs relative to both signal intensity and signaler "quality" for a complete understanding of how reliability is maintained (Getty, 2006; Murai et al., 2009). However, one striking pattern in the signaling literature is that few studies consider this interpretation, and instead focus on absolute costs of a signal or, in very rare cases, costs relative to male quality (as emphasized by Getty, 1998; Kotiaho, 2001; Getty, 2006; Murai et al., 2009). Because most studies use the older interpretation in their approach, we necessarily discuss them, but our aim is not to criticize. Indeed, multiplicative models of signal costs and benefits help explain why quantifying costs of sexually selected traits has proven difficult and is often characterized by equivocal results (Kotiaho, 2001; Cotton et al., 2004; Oufiero and Garland, 2007; Fowler-Finn and Hebets, 2011).

We propose that quantifying whole-organism performance traits that impact individual fitness and are linked to the signal under investigation can be a useful approach when studying signal costs. A performance trait is generally defined as an organism's ability to accomplish an ecologically relevant task, and typically such performance traits have an intuitive link to viability and/or fecundity (Bennett and Huey, 1990; Irschick and Garland, 2001; Lailvaux and Irschick, 2006; Irschick et al., 2007; Husak et al., 2009a; see also Chapter 1). For the sake of simplicity, and to match previous treatments of performance traits, we do not include here the display or execution of signals or displays, though this is an interesting topic of debate (Husak et al., 2009a). Theory and empirical studies show that selection is strong on whole-organism performance traits (Figure 3.1a; Bartholomew, 1958; Huey and Stevenson, 1979; Arnold, 1983; Pough, 1989; Irschick and Garland, 2001; Irschick et al., 2008). Alternatively, female choice for signal intensity or structures used as signals during rival assessment may result in selection operating directly on morphology (e.g., size or color of a structure), which in turn may produce phenotypic changes in morphological traits underlying performance due to the negative effects

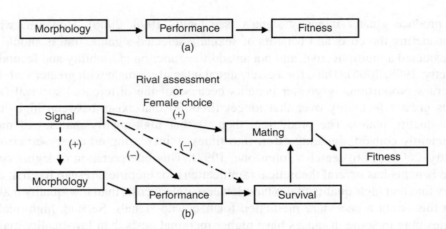

Figure 3.1. (a) Schematic representation of Arnold's (1983) paradigm. Selection operates on performance traits, which are constrained by underlying morphological traits. Ultimately, the response to selection changes the morphological traits, but indirectly via selection on the functional manifestation of morphology: performance. (b) Schematic representation of how female choice on a signal or use of a signal for rival assessment (i.e., direct selection on morphology, not performance) may result in decreased performance and survival (dashed-dotted line). The evolution of compensation (i.e., increased performance to alleviate costs of the signal) results in a correlated indirect influence (dashed line) of sexual selection on the morphological trait(s) underlying the performance trait. Since multiple morphological traits underlie any given performance trait, the indirect morphological response(s) to sexual selection (i.e., the dashed line) on the signal will depend on how those same morphological traits impact fitness via their effects on other performance traits, as well as a multitude of cost–benefit trade-offs, the nature of functional and genetic correlations among traits, and the nature and intensity of selection on the performance trait. (Modified from Husak and Swallow (2011).)

of the signal on performance (Figure 3.1b; Garland and Losos, 1994; Swallow *et al.*, 2009; Husak and Swallow, 2011). Understanding the underlying functional characteristics of a signal can reveal important details about its potential costs. Signals may be structural features of an organism (e.g., exaggerated appendages, color patches, horns) and have either a direct impact on performance (e.g., tail length may affect flight performance), no impact on performance (e.g., color patches will not make an individual run more slowly), or indirect impacts on performance over evolutionary time (e.g., increased conspicuousness from a color patch may result in selection for better escape ability; see discussion below). As an example, the elongated tail streamers of barn swallows (*Hirundo rustica*) have been hypothesized to increase drag during flight (Møller, 1996; Møller *et al.*, 1998; but see Evans, 1998; Buchanan and Evans, 2000; Park *et al.*, 2001). Signals may also be dynamic movements (e.g., aerial displays, lateral compressions, songs) and share morphological, physiological, or biomechanical proximate mediators with performance traits (Vanhooydonck *et al.*, 2007). For example, the duration of lateral compressions during threat displays by male side-blotched lizards (*Uta stansburiana*) is determined by endurance capacity (Brandt, 2003). In the case of side-blotched lizards, males that perform long bouts of lateral compression subsequently have reduced endurance, which may impact survival. Finally, signals may combine structural features and dynamic movements (e.g., waving a structure) and have functional consequences of both types of signal.

Our goal in this chapter is to explore how studying performance traits can increase our understanding of signal costs and, ultimately, signal reliability. We highlight relevant examples across diverse taxa to describe the benefits and limitations of studying physiological and performance traits to understand the costs of animal signals. We first define the different categories of costs that signals may impose on those that bear them, providing examples from a performance-based perspective. We follow with a discussion of how compensatory traits are relevant to signal costs, as well as their importance during signal evolution. Finally, we end with prospects for future research that will increase our understanding of how costs may influence signal reliability.

RECEIVER-INDEPENDENT COSTS

Receiver-independent costs are not enforced by intended conspecific receivers but result from heterospecifics (predators or prey), or by processes internal to the signaler. Here, we discuss three classes of receiver-independent costs: developmental costs, production costs, and maintenance costs. In many cases, these three types of costs may be non-mutually exclusive and a signal may have multiple types of costs.

Developmental Costs

Developmental costs are incurred as the signal develops and before the signal is displayed and may be accrued during investment in display structures, such as the antlers of deer (Andersson, 1986). Dynamic behavioral displays, such as bird song, may also incur developmental costs when the necessary neural circuitry or musculature for motor control is developed to support the complex behavior (Nowicki *et al.*, 1998, 2002). The concept of developmental costs is intuitively appealing, as organisms have finite resources to invest in developing the various aspects of their phenotype, including signals (see also Badyaev, 2004). However, the performance and growth trade-offs involved in increased energy allocation to signals remains poorly studied (Petrie, 1994; Royle *et al.*, 2006; Chapter 2). Significant investment in signals may have multiple phenotypic consequences. First, overall growth may be inhibited directly by redirecting energy into structures required for signaling. Second, overall growth may be indirectly inhibited via "pleiotropic" effects of the hormonal environment required to develop a signal. Third, the energy used to invest in a signal may not be taken away from body size growth *per se*, but instead from investment in other specific body regions.

Because many performance traits scale positively with body size, reduced growth may result in decreased performance (see Dial *et al.*, 2008). Smaller individuals with lower absolute performance capacity may suffer higher mortality, depending on the magnitude of the reduction in performance and how strongly selection operates on that particular performance trait (Irschick *et al.*, 2008). Furthermore, there may be compensatory growth in individuals that have invested heavily in larger signals to catch up in body size, and compensatory growth can decrease performance. For example, in three-spined sticklebacks (*Gasterosteus aculeatus*) from stream habitats, compensatory growth that occurred after food restriction resulted in slower escape speeds (Álvarez and Metcalfe, 2007). Such a growth rate–speed trade-off may be widespread in fishes (Billerbeck *et al.*, 2001; Álvarez and Metcalfe, 2005), as well as other vertebrates (Arendt, 2003). This topic certainly deserves further study in relation to how investment in signals specifically may play a role in this trade-off (Nijhout and Emlen, 1998; Bize *et al.*, 2006; Royle *et al.*, 2006).

Investment in a signal may also take away investment from specific body regions or tissues, such as muscle mass or fat stores. The concept of trade-offs among body regions during development has been a fruitful topic of research for many years (Stearns, 1992; Roff, 2001), and it remains relevant for animal signals, as signal quality can have profound effects on fitness, and significant proportional investment in a structural signal may take resources away from investment in other areas (Tomkins *et al.*, 2005; Lailvaux *et al.*, 2010; but see Marcus and McCune, 1999). Only a few studies have investigated this issue (e.g., Moczek and Nijhout, 2004; Fry, 2006), including in dung beetles (genus *Onthophagus*) in which horn production is associated with decreased allocation to adjacent structures (Emlen, 2001; Emlen *et al.*, 2012). Horns that develop on the head reduce the size of eyes or antennae, depending where on the head horns are produced; horns that develop on the thorax reduce wing size. Similarly, Kawano (1995) found that male rhinoceros beetles with relatively large horns had smaller wings than hornless females and males with smaller horns. Thus, instead of reduced absolute performance, energy shunted away from specific regions of the body for the benefit of a higher quality signal may reduce relative performance (but see McCullough *et al.*, 2012; McCullough and Tobalske, 2013). Clearly, in the case of dung beetle horns, the functional cost to developing horns on the thorax (smaller wings and presumably reduced flight performance) is predictable a priori, but this may not always be the case.

Other functional traits may also be subject to trade-offs. Male fiddler crabs (genus *Uca*) use their major front claws, or chelae, in social displays and for agonistic encounters with other males. The males use the claws to grapple with one another, and males with larger claws generally win more interactions, often with little physical aggression (Hyatt and Salmon, 1979; Jennions and Backwell, 1996; Morrell *et al.*, 2005; Reaney *et al.*, 2008). Levinton and Allen (2005) examined the ontogenetic decrease in claw closing force relative to body size in these animals, and found that as crabs grow and relative closing force decreases, claw closing speed increases. This functional trade-off occurs because claws that produce stronger closing forces have slower closing speeds. In Darwin's finches, beak depth predicts both bite-force capacity and singing performance. Because of a trade-off between force generation and the speed at which the jaws can be moved, birds with deep beaks are able to crack hard seeds, but are poor singers (Podos, 2001; Herrel *et al.*, 2009). Biomechanical and physiological trade-offs are common in many systems (e.g., the sprinting–endurance trade-off: Vanhooydonck *et al.*, 2001; Van Damme *et al.*, 2002; Blake, 2004), and studies testing for signal costs would benefit by considering such trade-offs. There may also be functional trade-offs that are not due to such shared mechanisms. For example, selection for greater bite-force capacity in male lizards may lead to the evolution of exaggerated male head size, but this may reduce sprint speed capacity and acceleration via changes in weight distribution in the body (Aerts *et al.*, 2003; Huyghe *et al.*, 2005; Cameron *et al.*, 2013).

Production Costs

In his original formulation of the handicap principle of reliable signaling, Zahavi (1975) suggested that one test of a male's quality may be risky signaling behavior, such as when birds or frogs vocalize from exposed locations. This represents an example of production costs, which are incurred while giving the signal to a receiver. Selection for conspicuousness has likely played a key role in the evolution of the displays of many animals (e.g., Endler, 1983; Ryan *et al.*, 1982; Kwiatkowski, 2003; Leal and Fleishman, 2004) but may also increase risk of predation for the signaler. Though this idea is frequently cited, compelling empirical examples remain more elusive than one would predict if costs are a major driver of signal reliability. However, in some cases, population-level variation in signal intensity is inversely related to predation levels (e.g., Endler, 1983, 1992; Zuk and Kolluru, 1998; Vanhooydonck *et al.*, 2009). Though it is known that predators may exploit signals to locate prey, it is unclear how the probability of a signaler becoming prey is affected by the interaction between signal intensity and signaler quality (Kotiaho, 2001).

Another popular line of research investigates the energetic cost of signals. Some frogs, birds, spiders, and insects face a substantial increase in metabolic rate when signaling (Table 2 in Kotiaho, 2001). For example, calling frogs increase metabolic rate up to 25-fold above resting levels (Taigen and Wells, 1985). However, to satisfy the criteria of the handicap principle, a metabolic cost must lead to fitness costs by reducing survival or future reproductive output. Hunt *et al.* (2004) provided intriguing experimental evidence of the survival cost of an energetically expensive signaling behavior. By manipulating protein content in the diet of male field crickets (*Teleogryllus commodus*), the authors created high- and low-quality groups of males. The high-quality males had reduced lifespan, but called more over the course of their lifetimes than longer-living, low-quality males. The apparent trade-off between calling effort, body mass, and longevity suggest that, in field crickets, increased energy expenditure may impose viability costs.

A performance-based approach may be useful to explore the production costs of signals that involve dynamic movements and are directly linked to performance traits. Evidence for this is strong for male displays and endurance capacity, the latter of which is a measure of aerobic capacity. For example, endurance capacity is tightly linked to threat display rate in side-blotched lizards (*U. stansburiana*; Brandt, 2003), as well as courtship call rate in decorated crickets (*Gryllodes sigillatus*; Ketola *et al.*, 2009). In both of these cases, displaying reduces endurance capacity available for escaping predators, foraging, or searching for mates. This was elegantly shown by Brandt (2003) with side-blotched lizards in two experiments. First, lizards were forced to run on a treadmill, which resulted in reduced display ability; second, lizards were induced to display at a mirror, which resulted in reduced endurance capacity measured on a treadmill. Although Ketola *et al.* (2009) did not experimentally test whether calling resulted in decreased endurance in crickets, the causal relationship likely exists, given the strong correlation between the two. However, despite the relationship between calling and endurance, Ketola *et al.* (2009) did not find a relationship between basal metabolic rate and calling rate. Since performance is an integrated manifestation of multiple morphological and physiological traits, one may be more likely to find a relationship between signal intensity and performance traits than between signaling intensity and any one lower-level physiological trait.

Maintenance Costs

Signals may impose costs that are not directly related to signal development or production. Maintenance costs result from an individual bearing the signal after it is developed, but when it is not in use. Clearly structural signals may have significant maintenance costs, since they are always present after development and must be carried when not in use. Even dynamic signals, such as threat displays and vocalizations, require having the necessary architecture to produce those signals. For example, male birds must maintain specialized neural circuitry developed for song production (Nowicki *et al.*, 2002). In the case of deer antlers, males must not only allocate energetic resources to develop the antlers, they must then bear the burden of carrying them around when not in use (Andersson, 1986).

One of the most well-investigated maintenance costs of signals is that of decreased locomotor performance (Oufiero and Garland, 2007). Diminished locomotor performance is likely to have negative impacts on the ability of animals to effectively escape predators, forage, and interact with conspecifics (Irschick and Garland, 2001; Husak and Fox, 2006). Signals reduce locomotor performance of males in several species, yet this relationship is absent in other species where decrements are expected. In the swordtail fish *Xiphophorus montezumae*, males with intact swords had higher oxygen consumption than those with swords experimentally removed (Basolo and Alcaraz, 2003). Gonopodium length was negatively associated with the rapid "C-start" escape response in *Gambusia* fishes (Domenici and Blake, 1997; Langerhans *et al.*, 2005). However, the extravagant tail morphology of male guppies (*Poecilia reticulata*) was unrelated to swimming speed (Nicoletto, 1991). Similarly, there was no relationship between sword length and swimming endurance in the swordtail fish *Xiphophorus nigrensis* (Ryan, 1988; see also Oufiero *et al.*, 2014). Surprisingly, male *Xiphophorus helleri* with longer swords were found to have higher "C-start" performance than those with shorter swords (Royle *et al.*, 2006). Similarly equivocal results have been found in other taxa (Allen and Levinton, 2007; Wilson *et al.*, 2007, 2009; Chapter 8).

A study of aerodynamic models of bird tail shape revealed limits of structural ornamentation (not color) that were dependent on both body size and natural history of the bird species. The models showed that some portions of morphological tail space are unattainable for some species due to biomechanical limitations of flight (Evans, 2004). In particular, body size may constrain what types of tail ornamentation can be used as signals by birds that turn during flight. Very small birds can aerodynamically afford only small or seasonal tail ornamentation. Streamers or forked tails can evolve in larger birds because they do not raise the cost of flight as much as elongated inner feathers. Only quite large birds should be able to overcome the aerodynamic costs of elongated pintails or wedge-shaped tails. Experimental manipulation of tail length and measurement of flight performance in Anna's hummingbirds (*Calypte anna*) revealed that performance detriments of tail ornaments are more complex. Metabolic costs for birds with experimentally elongated tails (an ornament) were highest at higher flight speeds, and flight speed was reduced in accordance with metabolic costs (Clark and Dudley, 2009).

RECEIVER-DEPENDENT COSTS

Receiver-dependent costs are imposed upon a signaler by conspecific rivals at the time the signal is given and are not due to developing or maintaining the signal. Thus, they are similar to production costs discussed above, but in the case of receiver-dependent costs, the intended receiver is the one imposing the cost instead of, for example, a predator. Here, we discuss two classes of receiver-dependent costs that are defined by the nature of how the cost is incurred from the receiver. First, vulnerability costs occur while transmitting the signal to a receiver and result from the signal making the sender more vulnerable to physical attack by the receiver. Second, retaliation costs are those incurred by a receiver retaliating against the signal given by a sender. Retaliation costs are not due to the signal making the sender more vulnerable but instead are related to the signal increasing the probability of a receiver physically attacking the signaler after assessing the content of the sender's signal. Although these two types of costs appear similar, they are distinguished by what their production induces from a receiver. Thus, the two types have different predictions: signals that have retaliation costs increase the probability of attack by a given receiver, whereas signals with vulnerability costs increase injury during each attack (Searcy *et al.*, 2008). However, we note that it may be difficult to empirically distinguish these two types of costs. In each case, it is important to note that costs are not necessarily imposed every time the signal is given. Instead, the cost is only imposed if and when a receiver attacks after a signal is given. Key to the evolution of signals that have vulnerability or retaliation costs is the prediction that the signals are reliable indicators of fighting ability, and this is where a functional approach can be useful when studying receiver-dependent costs.

"Fighting Ability" and Receiver-Dependent Costs

The evolution of signals with receiver-dependent costs relies heavily on individuals' fighting ability, which is often referred to as an individual's *resource-holding potential*, or *RHP* (Parker, 1974; Hammerstein, 1981), and much of the available theoretical literature concerning rival contests centers on RHP (Zahavi, 1975; Hurd, 1997; Grafen, 1990; Maynard Smith and Harper, 2003). One important component of fighting ability is body size, with larger individuals typically winning more interactions (Maynard Smith and Parker, 1976; Archer, 1988; Arnott and Elwood, 2009). As theory predicts that contests should last longer and escalate more when asymmetries in body size are small (Maynard Smith and Parker, 1976; Arnott and Elwood, 2009), it is important to consider other asymmetries in RHP. What might determine contest outcome when rivals are very closely matched in body size?

Intrinsic fighting ability, all else equal and barring injuries, is an attribute that changes little between interactions. Ideally, one can quantify aspects of fighting ability for an individual, independent of other individuals or "dominance" status. For example, an individual with fighting ability X may be dominant over individual A with fighting ability Y, but may be subordinate to individual B with fighting ability Z. Whether dominant or subordinate, individual A has fighting ability Y and individual B has fighting ability Z. But how does one measure fighting ability? Our view is that fighting ability, or RHP, can be measured by quantifying functional traits that are important during fights (see also Lailvaux and Irschick, 2006; Briffa and Sneddon, 2007; Husak *et al.*, 2009b; Henningsen and Irschick, 2012). This will allow consideration of individual variation in specific traits that represent components of an individual's phenotype (and genotype), instead of a nebulous "dominance" trait, so that we can elucidate phenotypic (and genotypic) correlations underlying signal reliability.

Several performance traits have been proposed as metrics of fighting ability. Maximal bite-force performance in lizards is a strong predictor of dominance during dyadic interactions between males (Lailvaux *et al.*, 2004; Huyghe *et al.*, 2005; Husak *et al.*, 2006; Henningsen and Irschick, 2012), because biting during fights can result in serious wounds to the recipient of those bites (e.g., Lappin and Husak, 2005). Because of the potential for serious injury from fights, the incidence of biting is rare for many lizard species, and interactions mostly consist of stereotyped displays (Carpenter and Ferguson, 1977; Carpenter, 1978). In green anole lizards (*Anolis carolinensis*), the size of the extendable throatfan (called a *dewlap*) predicts bite-force capacity (Vanhooydonck *et al.*, 2005a), even after the allometric effects of body size have been statistically removed (Irschick *et al.*, 2006). The same is true in highly territorial Caribbean anole lizard species (Figure 3.2; Vanhooydonck *et al.*, 2005b; Lailvaux and Irschick, 2007). Similarly, in collared lizards (*Crotaphytus collaris*), gaping displays (an index) reveal jaw musculature, as well as patches of UV-reflective skin at the corner of the mouth, that both predict bite-force capacity (Lappin *et al.*, 2006). The former is visible when a lizard is viewed front-on, whereas the latter are visible from the front or the side (see Figure 1 in Lappin *et al.*,

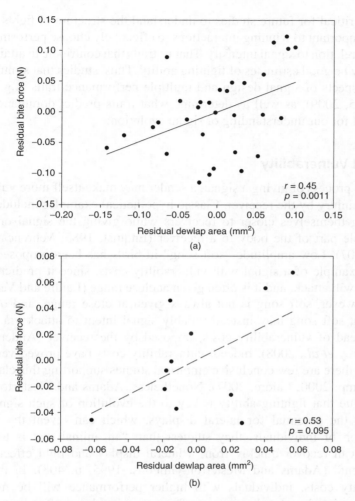

Figure 3.2. Anole dewlaps as a signal of bite-force capacity. The figures show size-corrected (residual) bite-force capacity plotted against size-corrected dewlap area for (a) *Anolis lineatopus* and (b) *Anolis grahami*, two territorial Caribbean anole lizards. (Data modified from Vanhooydonck *et al.* (2005b).)

2006). Signals of fighting ability have also been quantified in other taxa. The size of fiddler crab major chelae (i.e., the crushing claw) predicts the force produced when closed (Sneddon *et al.*, 2000; Lailvaux *et al.*, 2009; see also Dennenmoser and Christy, 2013; Chapter 8). Much theory concerning the evolution of agonistic signals and rival assessment relies on RHP being measured as endurance capacity or "stamina" (Robson and Miles, 2000; Perry *et al.*, 2004; reviewed in Chapter 3 this volume; but see Garland *et al.*, 1990; Huyghe *et al.*, 2005). Other performance traits that have been proposed to be metrics of fighting ability (and act as predictors of dominance) include jumping ability (Lailvaux *et al.*, 2004) and pulling force (Lailvaux *et al.*, 2005), because these traits are important to the way individuals fight.

It will be critical for future studies to understand the structure of fights, and what traits are important to winning interactions to effectively choose performance traits to study in relation to signal intensity. That is, traits that contribute to attaining dominance may be good estimates of fighting ability. Thus, studies that simultaneously quantify aspects of signal design and multiple performance traits (e.g., Lailvaux *et al.*, 2005, 2009), as well as determine what traits predict dominance, will be very useful for our understanding of signal evolution.

Increased Vulnerability

During the process of giving a signal, a sender may make itself more vulnerable to the risk of injury by the receiver. Classic hypothetical examples include signalers positioning themselves closer to receivers while giving the signal or revealing a vulnerable part of the body to a receiver (Enquist, 1985; Vehrencamp, 2000; Laidre, 2007). Low amplitude "soft song" in birds has been proposed by some to be an example of a signal with vulnerability costs, since it predicts when an individual will attack, and it is often given at close range (Laidre and Vehrencamp, 2008). However, soft song is not always given at close range, and others have argued that soft song may instead reliably signal intent to attack via retaliation costs, instead of vulnerability costs, imposed by the receiver (Anderson *et al.*, 2007; Searcy *et al.*, 2008). Indeed, vulnerability costs have proved very difficult to test and there are few conclusive empirical studies supporting this class of costs (Vehrencamp, 2000; Laidre, 2007). Nonetheless, Adams and Mesterton-Gibbons (1995) argue that fighting ability is key to the evolution of such signals. When discussing the potential for lateral displays, which can reveal the vulnerable flank region of individuals, they suggest that, "an animal that is truly strong and agile can recover quickly from a lateral display and fight effectively with its opponent" (Adams and Mesterton-Gibbons, 1995, p. 408). In the case of vulnerability costs, individuals with higher performance will be more likely to give the signal and either recover from the signal or suffer lower injury if the receiver attacks; that is, they win. On the other hand, poor performers will be more susceptible to the vulnerability exposed by giving the signal. Thus, functional measures of fighting ability would be useful in tests of vulnerability costs and which individuals are able to bear them. We predict that individuals of poor fighting ability will be less likely to give the signal and/or will lose more interactions, whereas individuals with high fighting ability will give the signal more and/or win more interactions. Signal intensity should also be positively correlated with functional traits relevant to fighting ability. We further predict that if vulnerability handicaps are associated with fighting ability, then increased performance capacities associated with fighting should be associated with riskier signals, an increased number of risky signals, or a longer duration of risky signals. One difficulty with studying vulnerability costs is determining that giving a signal actually increases vulnerability to attack, but this can be determined with behavioral observations of agonistic interactions and the consequences of giving signals.

Increased Retaliation

In some cases, signals are not costly to produce and do not make the sender more vulnerable to attack. Instead, theory predicts that individuals know their own fighting ability (or motivation to fight) and give a signal according to that self assessment (Hurd and Enquist, 1998; Taylor and Elwood, 2003; Arnott and Elwood, 2009). The signal may be discrete options (i.e., signal X if weak, Y if strong), or it may be along a continuum of intensity such that increased intensity reflects increased fighting ability. The evolutionarily stable strategy is to attack only when a weaker signal or equal-strength signals are given (Hurd, 1997). This allows for tests of bluffers and maintains signal reliability. In this scenario, "weak" individuals that signal "strong" will have to fight "strong" individuals more often than will honest "weak" individuals. Such a scenario leads to the prediction that retaliation costs are especially high for weak individuals. Such theoretical considerations have been applied to explain the evolution of so-called status-signaling badges, which appear to have little to no receiver-independent costs (reviewed in Whiting *et al.*, 2003; Searcy and Nowicki, 2005). Despite the intuitive appeal of this model, results are equivocal, with some studies providing support (e.g., Rohwer, 1977; Møller, 1987) and others not (e.g., Fugle and Rothstein, 1987; Gonzalez *et al.*, 2002). A functional approach can test critical hypotheses of these so-called "social-control" explanations of aggressive signals by providing a means to quantify which individuals are "strong" and which "weak," in a relevant manner (i.e., have high RHP or fighting ability). Instead of casting predictions in terms of dominants and subordinates, one can use performance measures that are relevant to how individuals fight. Although studies of retaliation costs that take a performance-based approach are lacking, existing theoretical predictions should be easily modified to include such data and provide a promising means of testing social-control hypotheses.

COMPENSATORY TRAITS

Although many or most signals are almost certainly under the influence of sexual selection, either through male–male competition or female choice, natural selection works on the entire integrated phenotype of individuals (Lande and Arnold, 1983; Arnold, 1983; Cornwallis and Uller, 2010). Some evidence suggests that natural selection can place "upper limits" on the elaboration and exaggeration of signals (reviewed in Andersson, 1994; Kotiaho, 2001). However, natural selection will not likely do so only by selecting against those individuals with larger or more exaggerated signals. Instead, there may also be concomitant selection on traits that alleviate the negative effects of sexually selected traits, including signals; that is, there may be correlated selection for compensatory traits (Kirkpatrick, 1987; Møller, 1996; Jennions *et al.*, 2001; Tomkins *et al.*, 2005; Oufiero and Garland, 2007; Swallow *et al.*, 2009; Figure 3.1b). Compensation may come in the form of novel or modified structures, physiology, behavior, or performance (Oufiero and Garland, 2007; Fowler-Finn and Hebets, 2011; Husak and Swallow, 2011). Without considering compensatory traits, one may obtain spurious correlations between signal intensity and performance. Indeed, ignoring potential compensatory traits may mask the negative fitness effects of a signal or create the illusion that a signal enhances performance (Oufiero and Garland, 2007).

To illustrate this point, Oufiero and Garland (2007) conducted a simulation study of 57 species of swordtail fishes (genus *Xiphophorus*). In their simulation, they made sword length (a signal) positively related to body length, as well as swimming endurance and the mass of the heart's ventricle (a compensatory trait). When path analysis was used to examine the relationships among these variables, the model that excluded ventricle mass resulted in a path model indicating positive relationships between body length, sword length, and swimming endurance, thus indicating that sword length may be beneficial for endurance (Figure 3.3b). However, when ventricle mass, the compensatory trait, was also included, the path model clearly showed a negative influence of sword length on swimming endurance, as well as the positive compensatory relationship between ventricle mass and endurance (Figure 3.3a). In this hypothetical example, ignoring a compensatory trait led to a contradictory conclusion. Even though the initial data put into the path model was simulated such that body length predicted endurance (with some "noise"), the final path model with all four variables ended up with body length being a negative predictor of endurance, which was likely due to multicollinearity of variables, as the authors pointed out.

The presence of compensatory traits can obfuscate detection of current absolute costs of sexual signals (Møller, 1996; Oufiero and Garland, 2007; Husak

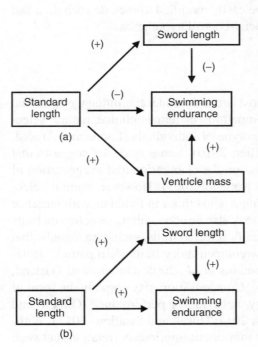

Figure 3.3. Schematic diagram showing the potentially obscuring effects of compensatory traits when studying the effects of signals on performance. The schematics are modified versions of path models from simulated data on 57 swordfish species presented in Oufiero and Garland (2007). Standard length is body size, sword length is the length of the modified caudal fin in fishes of the genus *Xiphophorus*, swimming endurance is aerobic capacity while swimming, and ventricle mass is a measure of heart size. Note how the relationship between the signal (sword length) and performance (swimming endurance) changes when the compensatory trait (ventricle mass) is included (a) compared to when it is left out (b). See text and Oufiero and Garland (2007) for more details.

and Swallow, 2011). There are many examples of compensatory traits, though most come from studies of how the costs of tail ornamentation in birds may be offset with changes in wing size. Among bird species, elongated tail ornaments increase drag, which may hinder flight performance (Thomas, 1993). To apparently alleviate the aerodynamic costs of elongated tails, males of numerous species have evolved larger wings than females, and within males, individuals with larger tail ornaments have larger wings. These findings come from a large taxonomic sample of birds, including barn swallows (*H. rustica*; Møller *et al.*, 1995), scarlet-tufted malachite sunbirds (*Nectarinia johnstoni*; Evans and Hatchwell, 1992; Evans and Thomas, 1992), long-tailed widowbirds (*Euplectes progne*; Craig, 1989), and Jackson's widowbirds (*Euplectes jacksoni*; Andersson, 1992). Husak *et al.* (2011a) found a similar scenario for males within four species of stalk-eyed flies dimorphic for eye span. The elongated eyestalks of male stalk-eyed flies are favored during male–male competition (Burkhardt and de la Motte, 1985; Panhuis and Wilkinson, 1999) and female choice (Burkhardt and de la Motte, 1988; Wilkinson and Reillo, 1994; Wilkinson *et al.*, 1998), but the exaggerated eyestalks result in a much larger moment of inertia for their bodies than females; thus, males require larger torques to turn the body while flying, potentially severely hindering male flight performance (Ribak and Swallow, 2007; see also Worthington and Swallow, 2010). However, contrary to expectations, males of dimorphic species show only minor reductions in flight performance (Swallow *et al.*, 2000) or perform as well as, or better than, females during free-flying turning behavior (Ribak and Swallow, 2007). Among males in the dimorphic species, there was a positive relationship between residual eye span and residual wing area, but this relationship was not found in females (Husak *et al.*, 2011a), which do not have sexual selection acting on eye span. On the other hand, there was no relationship between residual eye span and residual wing area in three monomorphic species studied, and in which there is no mating advantage for exaggerated eyestalks (Wilkinson *et al.*, 1998; Panhuis and Wilkinson, 1999). Because dimorphism and monomorphism have both evolved independently multiple times (Baker and Wilkinson, 2001), it is likely that compensation has also evolved independently multiple times (Husak *et al.*, 2011a). Indeed, the different genera of stalk-eyed flies have apparently modified their wings in different ways to compensate for exaggerated eyestalks (Husak *et al.*, 2013). In all of these examples, the potential reduction in flight performance is apparently offset by increased wing size, which increases power generation during flight.

In addition to intra-specific studies of ornamentation and compensation in the flight apparatus, comparative studies across taxa reveal broad-scale patterns of compensatory trait evolution. In sexually dimorphic widowbirds and bishops (genus *Euplectes*), the elongated tail ornament used as a signal during female mate choice may be up to four times the body length of a male in some species (e.g., long-tailed widowbirds; Andersson and Andersson, 1994) and are predicted to reduce flight performance (Balmford *et al.*, 1993). When looking across 13 species of *Euplectes*, sexual dimorphism in tail length was positively correlated with dimorphism in wing length (Andersson and Andersson, 1994), suggesting that males compensate for elaborate tail ornaments. In a larger phylogenetically based, comparative analysis, tail length dimorphism was correlated with wing length dimorphism across 57 bird species in 13 families (Balmford *et al.*, 1994). A comparative study of 10 stalk-eyed fly species revealed several key findings supporting the hypothesis that wing size and shape have coevolved with eye span to compensate for the exaggerated ornament (Ribak *et al.*, 2009). Among males, but not females, residual eye span was correlated with residual wing length, indicating that when males of a species have relatively long eyestalks, they also have relatively long wings. Further, sexual dimorphism in eye span of stalk-eyed flies was found to be positively evolutionarily correlated with sexual dimorphism in wing length, wing area, and wing shape and negatively evolutionarily correlated with dimorphism in wing loading. However, we caution that the significance of evolving compensatory traits deserves further theoretical and empirical work. Compensatory traits, although reducing the performance decrement of a signal, may lead to viability and/or fecundity losses in other realms. In stalk-eyed flies, for example, the increased wing length of species dimorphic in eye span has the effect of reducing wing-beat frequencies during flight (Husak *et al.*, 2011b), which has unknown fitness consequences.

By simultaneously considering ornament size and the magnitude of compensatory traits, one can conduct more specific tests for the costs of exaggerated ornaments. A recent model proposed by Husak *et al.* (2011a), and expanded by Husak and Swallow (2011) used such an approach (Figure 3.4), where individuals in a population are considered in "compensation space" relative to each other. Although their model focused on male structural ornaments, the same general principles may apply to many types of signals. When looking at relative signal size or intensity plotted against the relative magnitude of some compensatory trait(s), the best-fit regression line (linear or curvilinear) represents the average compensatory ability in the population. Those above the line have larger signals (or those of greater intensity) than their compensatory ability on average, and are "under-compensating" for their signal, whereas those below the line have compensatory mechanisms greater than required for their relative signal magnitude, and are "over-compensating" (Figure 3.4). One can predict absolute costs or relative costs of signals among individuals, based on where they are in "compensation space." It is important to note that the regression line in Figure 3.4 does not represent a biomechanical or physical "optimum" compensatory ability that is calculated to provide the necessary amount of compensation to overcome a cost, but instead represents average compensation for a sex within a population, given the selective environment. This allows a way to compare relative compensatory abilities among individuals within a population to determine which should have higher costs (see studies that examine the fitness of individuals relative to their position in "morphological space"; Kingsolver *et al.*, 2001; Blows, 2007). On average, the "under-compensating" individuals should have viability costs, whereas the "over-compensating" individuals should have viability advantages (Figure 3.4). Further, the more positive the

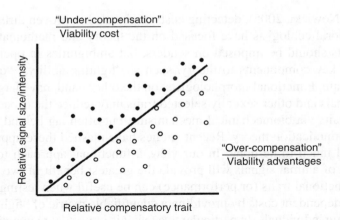

Figure 3.4. Model for examining costs of signals in "compensation space." Individuals above the population average ("under-compensators") should have higher viability costs than those below the line ("over-compensators"). This scenario makes it possible for two individuals to have nearly identical relative signal magnitude, yet be predicted to have very different costs based on how they compensate for their signal. (Modified from Husak and Swallow (2011).)

residual for an individual (i.e., the higher above the line), the higher the viability costs should be.

The predictions for viability relative to signal size or intensity are clear from the model shown in Figure 3.4, but the effects of compensatory traits on fecundity are more complicated. In many if not most cases, an increase in relative signal magnitude will increase the probability of successfully mating with a potential mate (Andersson, 1994). However, the multiplicative nature of fitness (Getty, 1998, 2006) may allow individuals with smaller/less intense relative signals, but "over-compensating" traits, to have similar net fitness due to their predicted higher survival and greater longevity. Husak and Swallow (2011) provide several hypothetical examples of this scenario, and we refer readers there. Of course, specifics will depend on the mating system and life history of the species (e.g., Lailvaux et al., 2010; Lailvaux and Kasumovic, 2011; Emlen et al., 2012), but the predictions represent a model from which investigators can begin to examine ornament costs and life history trade-offs. Further, investigators should explicitly test whether morphological traits, and the performance of that morphology, are compensatory traits, and not traits that are under direct sexual selection, as is known for some locomotor performance traits (reviewed in Lailvaux and Irschick, 2006; Husak and Fox, 2008).

CONCLUSIONS

Even though costs of animal signals have played a prominent role in animal communication theory and empirical research (Maynard Smith and Harper, 1995, 2003;

Searcy and Nowicki, 2005), detecting costs of signals has proven difficult. In general, behavioral ecologists have focused on the theory and mathematical models of how costs should be imposed on senders, but ambiguities in operational definitions for key components to theory, such as "fighting ability," "quality," and "RHP" remain. Functional morphologists, on the other hand, have largely focused on how signals and other sexually selected traits may reduce the capacities of performance traits via biomechanical mechanisms, but often lag behind advances in animal communication theory. Recent studies have meshed these approaches, but there is still much to be done. In our view, a functional approach to investigating the cost of animal signals will provide the greatest benefit in two ways. First, studying functional traits (or performance) can be useful when testing predictions of receiver-dependent costs by providing a quantifiable metric of "fighting ability" for interacting individuals. In particular, we feel it is important to quantify "fighting ability," "strength," and "quality" as non-circular measures, independent of dominance or number of fights won, and performance measures that are important during contest resolution will be useful for this.

Second, a functional approach may allow better tests of predictions for receiver-independent costs by considering compensatory traits, which may obfuscate the detection of signal costs. It is possible that compensatory mechanisms evolve in parallel with signals, making it important to consider the compensatory ability of individuals in order to detect production or maintenance costs. Manipulating signal properties to quantify costs is feasible but compensation can also potentially complicate interpretation of manipulation studies as well (Husak and Swallow, 2011). Manipulation decouples a signal from the trait being signaled and thus creates phenotypes that do not exist within the current range of natural variation and can test whether there are costs to high-intensity signals despite the existence of compensatory mechanisms. However, one must be sure to manipulate beyond the ability of individuals to compensate for their original signal intensity (Husak and Swallow, 2011). Not only will such manipulations test for costs of signals, but they will also help to elucidate the role of compensatory traits in signal evolution. Thus, it is imperative that investigators measure aspects of signal design in addition to multiple morphological and performance traits. Very few studies have done this so far, but those that have, show complex interactions among signal production, reproductive effort, and performance (Lailvaux *et al.*, 2010; Husak *et al.*, 2011a). Whereas we strive for generality in signaling theory, empirical tests of signal costs must consider the importance of context when studying signals, their meanings, and their costs. What is true for a vocal advertisement to potential mates is likely to be quite different than elongated feathers preferred by females or weapons brandished by males.

ACKNOWLEDGMENTS

We thank J. Swallow, S. Lailvaux, C. Oufiero, J. Podos, N. Royle, and an anonymous reviewer for comments that greatly improved the quality of the manuscript.

REFERENCES

Adams, E.S. and Mesterton-Gibbons, M. (1995) The cost of threat displays and the stability of deceptive communication. *Journal of Theoretical Biology*, **175**, 405–421.

Aerts, P., Van Damme, R., Daout, K., and Vanhooydonck, B. (2003) Bipedalism in lizards: whole-body modelling reveals a possible spandrel. *Philosophical Transactions of the Royal Society of London B*, **358**, 1525–1533.

Allen, B.J. and Levinton, J.S. (2007) The costs of bearing a sexually selected ornamental weapon in a fiddler crab. *Functional Ecology*, **21**, 154–161.

Álvarez, D. and Metcalfe, N.B. (2007) The trade-off between catch-up growth and escape speed: variation between habitats in the cost of compensation. *Oikos*, **116**, 1144–1151.

Álvarez, D. and Metcalfe, N.B. (2005) Catch-up growth and swimming performance in sticklebacks: seasonal changes in the cost of compensation. *Canadian Journal of Fisheries and Aquatic Sciences*, **62**, 2169–2173.

Anderson, R.C., Nowicki, S., and Searcy, W.A. (2007) Soft song in song sparrows: response of males and females to an enigmatic signal. *Behavioral Ecology and Sociobiology*, **61**, 1267–1274.

Andersson, M. (1986) Evolution of condition-dependent sex ornaments and mating preferences: sexual selection based on viability differences. *Evolution*, **40**, 804–816.

Andersson, M. (1994) *Sexual Selection*, Princeton University Press, Princeton, NJ.

Andersson, S. (1992) Lek mating and sexual selection in Jackson's Widowbird (*Euplectes jacksoni*). Ph.D. dissertation. University of Göteborg, Göteborg, Sweden.

Andersson, S. and Andersson, M. (1994) Tail ornamentation, size dimorphism and wing length in the genus *Euplectes* (Ploceinae). *Auk*, **111**, 80–86.

Archer, J. (1988) *The Behavioural Biology of Aggression*, Cambridge University Press, Cambridge.

Arendt, J.D. (2003) Reduced burst speed is a cost of rapid growth in anuran tadpoles: problems of autocorrelation and inferences about growth rates. *Functional Ecology*, **17**, 328–334.

Arnold, S.J. (1983) Morphology, performance, and fitness. *American Zoologist*, **23**, 347–361.

Arnott, G. and Elwood, R.W. (2009) Assessment of fighting ability in animal contests. *Animal Behaviour*, **77**, 991–1004.

Arnqvist, G. and Rowe, L. (2005) *Sexual Conflict*, Princeton University Press, Princeton, NJ.

Badyaev, A.V. (2004) Developmental perspective on the evolution of sexual ornaments. *Evolutionary Ecology Research*, **6**, 975–991.

Baker, R.H. and Wilkinson, G.S. (2001) Phylogenetic analysis of eye stalk allometry and sexual dimorphism in stalk-eyed flies (Diopsidae). *Evolution*, **55**, 1373–1385.

Balmford, A., Thomas, A.L.R., and Jones, I.L. (1993) Aerodynamics and the evolution of long tails in birds. *Nature*, **361**, 628–631.

Balmford, A., Jones, I.L., and Thomas, A.L.R. (1994) How to compensate for costly sexually selected tails: the origin of sexually dimorphic wings in long-tailed birds. *Evolution*, **48**, 1062–1070.

Bartholomew, G.A. (1958) The role of physiology in the distribution of vertebrates, in *Zoogeography* (ed. C.L. Hubbs), Washington, DC, American Association for the Advancement of Science, pp. 81–95.

Basolo, A.L. and Alcaraz, G. (2003) The turn of the sword: length increases male swimming costs in swordtails. *Proceedings of the Royal Society of London, Series B: Biological Sciences*, **270**, 1631–1636.

Bennett, A.F. and Huey, R.B. (1990) Studying the evolution of physiological performance. *Oxford Surveys of Evolutionary Biology*, **7**, 251–284.

Berglund, A., Bisazza, A., and Pilastro, A. (1996) Armaments and ornaments: an evolutionary explanation of traits of dual utility. *Biological Journal of the Linnean Society*, **58**, 385–399.

Billerbeck, J.M., Lankford, T.E., and Conover, D.O. (2001) Evolution of intrinsic growth and energy acquisition rates. I. Tradeoffs with swimming performance in *Menidia menidia*. *Evolution*, **35**, 1863–1872.

Bize, P., Metcalfe, N.B., and Roulin, A. (2006) Catch-up growth strategies differ between body structures: interactions between age and structure-specific growth in wild nestling Alpine swifts. *Functional Ecology*, **20**, 857–864.

Blake, R.W. (2004) Fish functional design and swimming performance. *Journal of Fish Biology*, **65**, 1193–1222.

Blows, M. (2007) A tale of two matrices: multivariate approaches in evolutionary biology. *Journal of Evolutionary Biology*, **20**, 1–8.

Brandt, Y. (2003) Lizard threat display handicaps endurance. *Proceedings of the Royal Society of London, Series B: Biological Sciences*, **270**, 1061–1068.

Briffa, M. and Sneddon, L.U. (2007) Physiological constraints on contest behaviour. *Functional Ecology*, **21**, 627–637.

Buchanan, K.L. and Evans, M.R. (2000) The effect of tail streamer length on aerodynamic performance in the barn swallow. *Behavioral Ecology*, **11**, 228–238.

Burkhardt, D. and de la Motte, I. (1985) Selective pressures, variability and sexual dimorphism in stalk-eyed flies (Diopsidae). *Naturwissenschaften*, **72**, 204–206.

Burkhardt, D. and de la Motte, I. (1988) Big 'antlers' are favored: female choice in stalk-eyed flies (Diptera, Insecta), field collected harems and laboratory experiments. *Journal of Comparative Physiology A*, **162**, 649–652.

Cameron, S.F., Wynn, M.L., and Wilson, R.S. (2013) Sex-specific trade-offs and compensatory mechanisms: bite force and sprint speed pose conflicting demands on the design of geckos (*Hemidactylus frenatus*). *Journal of Experimental Biology*, **216**, 3781–3789.

Carpenter, C.C. (1978) Ritualistic social behaviors of iguanid lizards, in *Neurology and Behavior of Lizards: an Interdisciplinary Colloquium* (eds N. Greenberg and P.D. MacLean), National Institute of Mental Health, Rockville, MD, pp. 253–267.

Carpenter, C.C. and Ferguson, G.W. (1977) Variation and evolution of stereotyped behavior in reptiles, in *Biology of the Reptilia*, vol. **7** (eds C. Gans and D.W. Tinkle), Academic Press, London, pp. 335–554.

Clark, C.J. and Dudley, R. (2009) Flight costs of long, sexually selected tails in hummingbirds. *Proceedings of the Royal Society of London, Series B: Biological Sciences*, **276**, 2109–2115.

Cornwallis, C.K. and Uller, T. (2010) Towards an evolutionary ecology of sexual traits. *Trends in Ecology and Evolution*, **25**, 145–152.

Cotton, S., Fowler, K., and Pomiankowski, A. (2004) Do sexual ornaments demonstrate heightened condition-dependent expression as predicted by the handicap hypothesis? *Proceedings of the Royal Society of London, Series B: Biological Sciences*, **271**, 771–783.

Craig, A.J.F.K. (1989) Tail length and sexual selection in the polygynous Longtailed Widow (*Euplectes progne*): a cautionary tale. *African Journal of Science*, **85**, 523–524.

Darwin, C. (1871) *The Descent of Man and Selection in Relation to Sex*, Murray, London.

Dennenmoser, S. and Christy, J.H. (2013) The design of a beautiful weapon: compensation for opposing sexual selection on a trait with two functions. *Evolution*, **67**, 1181–1188.

Dial, K.P., Green, E., and Irschick, D.J. (2008) Allometry of behavior. *Trends in Ecology and Evolution*, **23**, 394–401.

Domenici, P. and Blake, R.W. (1997) The kinematics and performance of fish fast-start swimming. *Journal of Experimental Biology*, **200**, 1165–1178.

Emlen, D.J. (2001) Costs and the diversification of exaggerated animal structures. *Science*, **291**, 1534–1536.

Emlen, D.J., Warren, I.A., Johns, A., Dworkin, I., and Corley-Lavine, L. (2012) A mechanism of extreme growth and reliable signaling in sexually selected ornaments and weapons. *Science*, **337**, 860–864.

Emlen, S.T. and Oring, L.W. (1977) Ecology, sexual selection, and the evolution of mating systems. *Science*, **197**, 215–223.

Endler, J.A. (1983) Natural and sexual selection on color patterns in poeciliid fishes. *Environmental Biology of Fishes*, **9**, 173–190.

Endler, J.A. (1992) Signals, signal conditions, and the direction of evolution. *American Naturalist*, **139**, S125–S153.

Enquist, M. (1985) Communication during aggressive interactions with particular reference to variation in choice of behaviour. *Animal Behaviour*, **33**, 1152–1161.

Evans, M.R. (1998) Selection on swallow tail streamers. *Nature*, **394**, 233–234.

Evans, M.R. (2004) Limits on the evolution of tail ornamentation in birds. *American Naturalist*, **163**, 341–357.

Evans, M.R. and Hatchwell, B.J. (1992) An experimental study of male adornment in the Scarlet-Tufted Malachite Sunbird: II. The role of the elongated tail in mate choice and experimental evidence for a handicap. *Behavioral Ecology and Sociobiology*, **29**, 421–427.

Evans, M.R. and Thomas, A.L.R. (1992) The aerodynamical and mechanical effects of elongated tails in the scarlet-tufted malachite sunbird: measuring the cost of a handicap. *Animal Behaviour*, **43**, 337–347.

Fowler-Finn, K.D. and Hebets, E.A. (2011) More ornamented males exhibit increased predation risk and antipredatory escapes, but not greater mortality. *Ethology*, **117**, 102–114.

Fry, C.L. (2006) Juvenile hormone mediates a trade-off between primary and secondary sexual traits in stalk-eyed flies. *Evolution & Development*, **8**, 191–201.

Fugle, G.N. and Rothstein, S.I. (1987) Experiments on the control of deceptive signals of status in white-crowned sparrows. *Auk*, **104**, 188–197.

Garland, T. Jr. and Losos, J.B. (1994) Ecological morphology of locomotor performance in squamate reptiles, in *Ecological Morphology: Integrative Organismal Biology* (eds P.C. Wainwright and S.M. Reilly), University of Chicago Press, Chicago, pp. 240–302.

Garland, T. Jr., Hankins, E., and Huey, R.B. (1990) Locomotor capacity and social dominance in male lizards. *Functional Ecology*, **4**, 243–250.

Getty, T. (1998) Handicap signalling: when fecundity and viability do not add up. *Animal Behaviour*, **56**, 127–130.

Getty, T. (2006) Sexually selected signals are not similar to sports handicaps. *Trends in Ecology and Evolution*, **21**, 83–88.

Gonzalez, G., Sorci, G., Smith, L.C., and De Lope, F. (2002) Social control and physiological cost of cheating in status signalling male house sparrows (*Passer domesticus*). *Ethology*, **108**, 289–302.

Grafen, A. (1990) Biological signals as handicaps. *Journal of Theoretical Biology*, **144**, 517–546.

Hammerstein, P. (1981) The role of asymmetries in animal contests. *Animal Behaviour*, **29**, 193–205.

Hau, M. (2007) Regulation of male traits by testosterone: implications for the evolution of vertebrate life histories. *BioEssays*, **29**, 133–144.

Henningsen, J.P. and Irschick, D.J. (2012) An experimental test of the effect of signal size and performance capacity on dominance in green anole lizards. *Functional Ecology*, **26**, 3–10.

Herrel, A., Podos, J., Vanhooydonck, B., and Hendry, A.P. (2009) Force–velocity trade-off in Darwin's finch jaw function: a biomechanical basis for ecological speciation? *Functional Ecology*, **23**, 119–125.

Huey, R.B. and Stevenson, R.D. (1979) Integrating thermal physiology and ecology of ectotherms: a discussion of approaches. *American Zoologist*, **19**, 357–366.

Hunt, J., Brooks, R., Jennions, M.D., Smith, M.J., Bentsen, C.L., and Bussiere, L.F. (2004) High-quality male field crickets invest heavily in sexual display but die young. *Nature*, **432**, 1024–1027.

Hurd, P.L. (1997) Is signalling of fighting ability costlier for weaker individuals? *Journal of Theoretical Biology*, **184**, 83–88.

Hurd, P.L. and Enquist, M. (1998) Conventional signalling in aggressive interactions: the importance of temporal structure. *Journal of Theoretical Biology*, **192**, 197–211.

Husak, J.F. and Fox, S.F. (2006) Field use of sprint speed by collared lizards (*Crotaphytus collaris*): compensation and sexual selection. *Evolution*, **60**, 1888–1895.

Husak, J.F. and Fox, S.F. (2008) Sexual selection on locomotor performance. *Evolutionary Ecology Research*, **10**, 213–228.

Husak, J.F. and Swallow, J.G. (2011) Compensatory traits and the evolution of male ornaments. *Behaviour*, **148**, 1–29.

Husak, J.F., Lappin, A.K., Fox, S.F., and Lemos-Espinal, J.A. (2006) Bite-force performance predicts dominance in male Venerable Collared Lizards (*Crotaphytus antiquus*). *Copeia*, **2006**, 301–306.

Husak, J.F., McCormick, S.B., Irschick, D.J., and Moore, I.T. (2009a) Hormonal regulation of whole-animal performance: implications for selection. *Integrative and Comparative Biology*, **49**, 349–353.

Husak, J.F., Lappin, A.K., and Van Den Bussche, R.A. (2009b) The fitness advantage of a high performance weapon. *Biological Journal of the Linnean Society*, **96**, 840–845.

Husak, J.F., Ribak, G., Wilkinson, G.S., and Swallow, J.G. (2011a) Compensation for exaggerated eyestalks in stalk-eyed flies (Diopsidae). *Functional Ecology*, **25**, 608–616.

Husak, J.F., Ribak, G., Wilkinson, G.S., and Swallow, J.G. (2011b) Sexual dimorphism in wing beat frequency in relation to eye span in stalk-eyed flies (Diopsidae). *Biological Journal of the Linnean Society*, **104**, 670–679.

Husak, J.F., Ribak, G., Baker, R.H., Wilkinson, G.S., and Swallow, J.G. (2013) Effects of ornamentation and phylogeny on wing shape evolution in stalk-eyed flies (Diopsidae). *Journal of Evolutionary Biology*, **26**, 1281–1293.

Huyghe, K., Vanhooydonck, B., Scheers, H., Molina-Borja, M., and Van Damme, R. (2005) Morphology, performance and fighting capacity in male lizards, *Gallotia galloti*. *Functional Ecology*, **19**, 800–807.

Hyatt, G. and Salmon, M. (1979) Combat in the fiddler crabs *Uca pugilator* and *Uca pugnax*: a quantitative analysis. *Behaviour*, **65**, 182–211.

Irschick, D.J. and Garland, T. (2001) Integrating function and ecology in studies of adaptation: investigations of locomotor capacity as a model system. *Annual Review of Ecology and Systematics*, **32**, 367–396.

Irschick, D.J., Ramos, M., Buckley, C., Elstrott, J., Carlisle, E., Lailvaux, S.P., Bloch, N., Herrel, A., and Vanhooydonck, B. (2006) Are morphology-performance relationships invariant across different seasons? A test with the green anole lizard (*Anolis carolinensis*). *Oikos*, **114**, 49–59.

Irschick, D.J., Herrel, A., Vanhooydonck, B., and Van Damme, R. (2007) A functional approach to sexual selection. *Functional Ecology*, **21**, 621–626.

Irschick, D.J., Meyers, J.J., Husak, J.F., and Le Galliard, J.F. (2008) How does selection operate on whole-organism functional performance capacities? A review and synthesis. *Evolutionary Ecology Research*, **10**, 177–196.

Jennions, M.D. and Backwell, P.R.Y. (1996) Residency and size affect fight duration and outcome in the fiddler crab *Uca annulipes*. *Biological Journal of the Linnaean Society*, **57**, 293–306.

Jennions, M.D., Møller, A.P., and Petrie, M. (2001) Sexually selected traits and adult survival: a meta-analysis. *Quarterly Review of Biology*, **76**, 3–36.

Johnstone, R.A. (1995) Sexual selection, honest advertisement and the handicap principle: reviewing the evidence. *Biological Reviews*, **70**, 1–65.

Johnstone, R.A. (1997) The evolution of animal signals, in *Behavioural Ecology: an Evolutionary Approach*, 4th edn (eds J.R. Krebs and N.B. Davies), Blackwell Science, Oxford, pp. 155–178.

Kawano, K. (1995) Horn and wing allometry and male dimorphism in giant rhinoceros beetles (Coleoptera: Scarabaeidae) of tropical Asia and Africa. *Annals of the Entomological Society of America*, **88**, 92–99.

Ketola, T., Kortet, R., and Kotiaho, J.S. (2009) Endurance in exercise is associated with courtship call rate in decorated crickets, *Gryllodes sigillatus*. *Evolutionary Ecology Research*, **11**, 1131–1139.

Ketterson, E.D., Atwell, J.W., and McGlothlin, J.W. (2009) Phenotypic integration and independence: hormones, performance, and response to environmental change. *Integrative and Comparative Biology*, **49**, 365–379.

Kingsolver, J.G., Hoekstra, H.E., Hoekstra, J.M., Berrigan, D., Vignieri, S.N., Hill, C.E., Hoang, A., Gibert, P., and Beerli, P. (2001) The strength of phenotypic selection in natural populations. *American Naturalist*, **157**, 245–261.

Kirkpatrick, M. (1987) Sexual selection by female choice in polygynous animals. *Annual Review of Ecology and Systematics*, **18**, 43–70.

Kotiaho, J.S. (2001) Costs of sexual traits: a mismatch between theoretical considerations and empirical evidence. *Biological Reviews*, **76**, 365–376.

Kwiatkowski, M.A. (2003) Variation in conspicuousness among populations of an iguanid lizard, *Sauromalus obesus* (=ater). *Copeia*, **2003**, 481–492.

Laidre, M.E. (2007) Vulnerability and reliable signaling in conflicts between hermit crabs. *Behavioral Ecology*, **18**, 736–741.

Laidre, M.E. and Vehrencamp, S.L. (2008) Is bird song a reliable signal of aggressive intent? *Behavioral Ecology and Sociobiology*, **62**, 1207–1211.

Lailvaux, S.P. and Irschick, D.J. (2006) A functional perspective on sexual selection: insights and future prospects. *Animal Behaviour*, **72**, 263–273.

Lailvaux, S.P. and Irschick, D.J. (2007) The evolution of performance-based male fighting ability in Caribbean *Anolis* lizards. *American Naturalist*, **170**, 573–586.

Lailvaux, S.P. and Kasumovic, M.M. (2011) Defining individual quality over lifetimes and selective contexts. *Proceedings of the Royal Society of London, Series B: Biological Sciences*, **278**, 321–328.

Lailvaux, S.P., Herrel, A., Vanhooydonck, B., Meyers, J.J., and Irschick, D.J. (2004) Performance capacity, fighting tactics and the evolution of life-stage male morphs in the green anole lizard (*Anolis carolinensis*). *Proceedings of the Royal Society of London, Series B: Biological Sciences*, **271**, 2501–2508.

Lailvaux, S.P., Hathway, J., Pomfret, J., and Knell, R.J. (2005) Horn size predicts physical performance in the beetle *Euoniticellus intermedius* (Coleoptera: Scarabaeidae). *Functional Ecology*, **19**, 632–639.

Lailvaux, S.P., Reaney, L.T., and Backwell, P.R.Y. (2009) Dishonest signalling of fighting ability and multiple performance traits in the fiddler crab *Uca mjoebergi*. *Functional Ecology*, **23**, 359–366.

Lailvaux, S.P., Hall, M.D., and Brooks, R.C. (2010) Whole-organism performance is no proxy for genetic quality: tradeoffs between locomotor performance, sexual attractiveness and life-history traits in the field cricket *Teleogryllus commodus*. *Ecology*, **91**, 1530–1537.

Lande, R. and Arnold, S.J. (1983) The measurement of selection on correlated characters. *Evolution*, **37**, 1210–1226.

Langerhans, R.B., Layman, C.A., and DeWitt, T.J. (2005) Male genital size reflects a tradeoff between attracting mates and avoiding predators in two live-bearing fish species. *Proceedings of the National Academy of Sciences of the United States of America*, **102**, 7618–7623.

Lappin, A.K. and Husak, J.F. (2005) Weapon performance, not size, determines mating success and potential reproductive output in the collared lizard (*Crotaphytus collaris*). *American Naturalist*, **166**, 426–436.

Lappin, A.K., Brandt, Y., Husak, J.F., Macedonia, J.M., and Kemp, D.J. (2006) Gaping displays reveal and amplify a mechanically-based index of weapon performance. *American Naturalist*, **168**, 100–113.

Leal, M. and Fleishman, L.J. (2004) Differences in visual signal design and detectability between allopatric populations of *Anolis* lizards. *American Naturalist*, **163**, 26–39.

Levinton, J.S. and Allen, B.J. (2005) The paradox of the weakening combatant: trade-off between closing force and gripping speed in a sexually selected combat structure. *Functional Ecology*, **19**, 159–165.

Marcus, J.M. and McCune, A.R. (1999) Ontogeny and phylogeny in the northern swordtail clade of *Xiphophorus*. *Systematic Biology*, **48**, 491–522.

Maynard Smith, J. and Harper, D.G.C. (1995) Animal signals, models and terminology. *Journal of Theoretical Biology*, **177**, 305–311.

Maynard Smith, J. and Harper, D. (2003) *Animal Signals*, Oxford University Press, Oxford.

Maynard Smith, J. and Parker, G.A. (1976) The logic of asymmetric contests. *Animal Behaviour*, **24**, 159–175.

McCullough, E.L., Weingarden, P.R., and Emlen, D.J. (2012) Costs of elaborate weapons in a rhinoceros beetle: how difficult is it to fly with a big horn? *Behavioral Ecology*, **23**, 1042–1048.

McCullough, E.L. and Tobalske, B.W. (2013) Elaborate horns in a giant rhinoceros beetle incur negligible aerodynamic costs. *Proceedings of the Royal Society of London, Series B: Biological Sciences*, **280**, 20130197.

Moczek, A.P. and Nijhout, H.F. (2004) Trade-offs during the development of primary and secondary sexual traits in a horned beetle. *American Naturalist*, **163**, 184–191.

Møller, A.P. (1987) Social control of deception among status signalling house sparrows *Passer domesticus*. *Behavioral Ecology and Sociobiology*, **20**, 307–311.

Møller, A.P. (1996) The cost of secondary sexual characters and the evolution of cost-reducing traits. *Ibis*, **138**, 112–119.

Møller, A.P., de Lope, F., and Saino, N. (1995) Sexual selection in the barn swallow *Hirundo rustica*. VI. Aerodynamic adaptations. *Journal of Evolutionary Biology*, **8**, 671–687.

Møller, A.P., Barbosa, A., Cuervo, J.J., de Lope, F., Merino, S., and Saino, N. (1998) Sexual selection and tail streamers in the barn swallow. *Proceedings of the Royal Society of London, Series B: Biological Sciences*, **265**, 409–414.

Morrell, L.J., Backwell, P.R.Y., and Metcalfe, N.B. (2005) Fighting in fiddler crabs *Uca mjoebergi*: what determines duration? *Animal Behaviour*, **10**, 653–662.

Murai, M., Backwell, P.R.Y., and Jennions, M.D. (2009) The cost of reliable signaling: experimental evidence for predictable variation among males in a cost-benefit trade-off between sexually selected traits. *Evolution*, **63**, 2363–2371.

Nicoletto, P.F. (1991) The relationship between male ornamentation and swimming performance in the guppy, *Poecilia reticulata*. *Behavioral Ecology and Sociobiology*, **28**, 365–370.

Nijhout, H.F. and Emlen, D.J. (1998) Competition among body parts in the development and evolution of insect morphology. *Proceedings of the National Academy of Sciences of the United States of America*, **95**, 3685–3689.

Nowicki, S., Peters, S., and Podos, J. (1998) Song learning, early nutrition and sexual selection in song birds. *American Zoologist*, **38**, 179–190.

Nowicki, S., Searcy, W.A., and Peters, S. (2002) Brain development, song learning and mate choice in birds: a review and experimental test of the "nutritional stress hypothesis.". *Journal of Comparative Physiology A*, **188**, 1003–1014.

Oufiero, C.E. and Garland, T. Jr. (2007) Evaluating performance costs of sexually selected traits. *Functional Ecology*, **21**, 676–689.

Oufiero, C.E., Meredith, R., Jugo, K., Tran, P., Chappell, M.A., Springer, M., Reznick, D.N. and Garland, T., Jr. (2014) The evolution of the sexually selected sword in *Xiphophorus* does not compromise aerobic locomotor performance. *Evolution*, **68**, 1806–1823.

Panhuis, T. and Wilkinson, G. (1999) Exaggerated male eye span influences contest outcome in stalk-eyed flies (Diopsidae). *Behavioral Ecology and Sociobiology*, **46**, 221–227.

Park, K.J., Rosén, M., and Hedenström, A. (2001) Flight kinematics of the barn swallow (*Hirundo rustica*) over a wide range of speeds in a wind tunnel. *Journal of Experimental Biology*, **204**, 2741–2750.

Parker, G.A. (1974) Assessment strategy and the evolution of fighting behaviour. *Journal of Theoretical Biology*, **47**, 223–243.

Perry, G., Levering, K., Girard, I., and Garland, T. Jr. (2004) Locomotor performance and dominance in male *Anolis cristatellus*. *Animal Behaviour*, **67**, 37–47.

Petrie, M. (1994) Improved growth and survival of offspring of peacocks with more elaborate trains. *Nature*, **371**, 598–599.

Podos, J. (2001) Correlated evolution of morphology and vocal signal structure in Darwin's finches. *Nature*, **409**, 185–188.

Pough, F.H. (1989) Organismal performance and Darwinian fitness: approaches and interpretations. *Physiological Zoology*, **62**, 199–236.

Reaney, L.T., Milner, R.N.C., Detto, T., and Backwell, P.R.Y. (2008) The effects of claw regeneration on territory ownership and mating success in the fiddler crab, *Uca mjoebergi*. *Animal Behaviour*, **75**, 1473–1478.

Ribak, G. and Swallow, J.G. (2007) Free flight maneuvers of stalkeyed flies: do eye-stalks affect aerial turning behavior? *Journal of Comparative Physiology A*, **193**, 1065–1079.

Ribak, G., Pitts, M.L., Wilkinson, G.S., and Swallow, J.G. (2009) Wing shape, wing size, and sexual dimorphism in eye-span in stalk-eyed flies (Diopsidae). *Biological Journal of the Linnean Society*, **98**, 860–871.

Robson, M.A. and Miles, D.B. (2000) Locomotor performance and dominance in male Tree Lizards, *Urosaurus ornatus*. *Functional Ecology*, **14**, 338–344.

Roff, D.A. (1997) *Evolutionary Quantitative Genetics*, Chapman and Hall, New York.

Roff, D.A. (2001) *Life History Evolution*, Sinauer Associates, Sunderland, MA.

Rohwer, S. (1977) Status signaling in Harris' sparrows: some experiments in deception. *Behaviour*, **61**, 107–129.

Royle, N.J., Metcalfe, N.B., and Lindstrom, J. (2006) Sexual selection, growth compensation and fast-start swimming performance in Green Swordtails, *Xiphophorus helleri*. *Functional Ecology*, **20**, 662–669.

Ryan, M.J. (1988) Phenotype, genotype, swimming endurance and sexual selection in a Swordtail (*Xiphophorus nigrensis*). *Copeia*, **1988**, 484–487.

Ryan, M.J., Tuttle, M.D., and Rand, A.S. (1982) Sexual advertisement and bat predation in a Neotropical frog. *American Naturalist*, **119**, 136–139.

Searcy, W.A. and Nowicki, S. (2005) *The Evolution of Animal Communication: Reliability and Deception in Signaling Systems*, Princeton University Press, Princeton.

Searcy, W.A., Anderson, R.C., and Nowicki, S. (2008) Is bird song a reliable signal of aggressive intent? A reply. *Behavioral Ecology and Sociobiology*, **62**, 1213–1216.

Sneddon, L.U., Huntingford, F.A., Taylor, A.C., and Orr, J.F. (2000) Weapon strength and competitive success in the fights of shore crabs. *Journal of Zoology (London)*, **250**, 397–403.

Stearns, S.C. (1992) *The Evolution of Life Histories*, Oxford University Press, Oxford, UK.

Swallow, J.G., Wilkinson, G.S., and Marden, J.H. (2000) Aerial performance of stalk-eyed flies that differ in eye span. *Journal of Comparative Physiology B*, **170**, 481–487.

Swallow, J.G., Hayes, J.P., Koteja, P., and Garland, T. Jr. (2009) Selection experiments and experimental evolution of performance and physiology, in *Experimental Evolution: Concepts, Methods, and Applications of Selection Experiments* (eds T. Garland Jr. and M.R. Rose), University of California Press, Berkeley, CA, pp. 301–351.

Taigen, T.L. and Wells, K.D. (1985) Energetics of vocalization by an anuran amphibian, *Hyla versicolor*. *Journal of Comparative Physiology B*, **155**, 163–170.

Taylor, P.W. and Elwood, R.W. (2003) The mismeasure of animal contests. *Animal Behaviour*, **65**, 1195–1202.

Taylor, P.W., Hasson, O., and Clark, D.L. (2000) Body posture and patterns as amplifiers of physical condition. *Proceedings of the Royal Society of London, Series B: Biological Sciences*, **267**, 917–922.

Thomas, A.L.R. (1993) On the aerodynamics of bird tails. *Philosophical Transactions of the Royal Society of London B*, **340**, 361–380.

Tomkins, J.L., Kotiaho, J.S., and LeBas, N.R. (2005) Phenotypic plasticity in the developmental integration of morphological trade-offs and secondary sexual trait compensation. *Proceedings of the Royal Society of London, Series B: Biological Sciences*, **272**, 543–551.

Trivers, R.L. (1972) Parental investment and sexual selection, in *Sexual Selection and the Descent of Man 1871–1971* (ed B. Campbell), Aldine, Chicago, pp. 136–179.

Van Damme, R., Wilson, R., Vanhooydonck, B., and Aerts, P. (2002) Performance constraints in decathletes. *Nature*, **415**, 755–756.

Vanhooydonck, B., Van Damme, R., and Aerts, P. (2001) Speed and stamina trade-off in lacertid lizards. *Evolution*, **55**, 1040–1048.

Vanhooydonck, B., Herrel, A., Meyers, J.J., Van Damme, R., and Irschick, D.J. (2005a) The relationship between dewlap size and performance changes with age and sex in a Green Anole (*Anolis carolinensis*) lizard population. *Behavioral Ecology and Sociobiology*, **59**, 157–165.

Vanhooydonck, B., Herrel, A., and Irschick, D.J. (2005b) Does dewlap size predict male bite force in Jamaican *Anolis* lizards? *Functional Ecology*, **19**, 38–42.

Vanhooydonck, B., Van Damme, R., Herrel, A., and Irschick, D.J. (2007) A performance based approach to distinguish indices from handicaps in sexual selection studies. *Functional Ecology*, **21**, 645–652.

Vanhooydonck, B., Herrel, A., Meyers, J.J., and Irschick, D.J. (2009) What determines dewlap diversity in *Anolis* lizards? An among-island comparison. *Journal of Evolutionary Biology*, **22**, 293–305.

Vehrencamp, S.L. (2000) Handicap, index, and conventional signal elements of bird song, in *Animal Signals: Signalling and Signal Design in Animal Communication* (eds Y. Espmark, T. Amundsen, and G. Rosenqvist), Tapir, Trondheim, Norway, pp. 277–300.

Whiting, M.J., Nagy, K.A., and Bateman, P.W. (2003) Evolution and maintenance of status-signaling badges: experimental manipulation in lizards, in *Lizard Social Behavior* (eds S.F. Fox, J.K. McCoy, and T.A. Baird), Johns Hopkins University Press, Baltimore, MD, pp. 47–82.

Wilkinson, G.S. and Reillo, P.R. (1994) Female choice response to artificial selection on an exaggerated male trait in a stalk-eyed fly. *Proceedings of the Royal Society of London, Series B: Biological Sciences*, **255**, 1–6.

Wilkinson, G.S., Kahler, H., and Baker, R.W. (1998) Evolution of female mate preferences in stalk-eyed flies. *Behavioral Ecology*, **9**, 525–533.

Wilson, R.S., Angilletta, M.J. Jr., James, R.S., Navas, C., and Seebacher, F. (2007) Dishonest signals of strength in male slender crayfish (*Cherax dispar*) during agonistic encounters. *American Naturalist*, **170**, 284–291.

Wilson, R.S., James, R.S., Bywater, C., and Seebacher, F. (2009) Costs and benefits of increased weapon size differ between sexes of the slender crayfish, *Cherax dispar*. *Journal of Experimental Biology*, **212**, 853–858.

Worthington, A.M. and Swallow, J.G. (2010) Gender differences in survival and anti-predatory behavior in stalk-eyed flies. *Behavioral Ecology*, **21**, 759–766.

Zahavi, A. (1975) Mate selection – a selection for a handicap. *Journal of Theoretical Biology*, **53**, 205–214.

Zahavi, A. and Zahavi, A. (1997) *The Handicap Principle: a Missing Piece of Darwin's Puzzle*, Oxford University Press, Oxford.

Zuk, M. and Kolluru, G.R. (1998) Exploitation of sexual signals by predators and parasitoids. *Quarterly Review of Biology*, **73**, 415–438.

4

COGNITIVELY DRIVEN CO-OPTION AND THE EVOLUTION OF COMPLEX SEXUAL DISPLAYS IN BOWERBIRDS

Gerald Borgia and Jason Keagy

*Department of Biology and Behavior, Ecology, Evolution, and Systematics Program,
University of Maryland, College Park, MD, USA*

INTRODUCTION

Two of Darwin's greatest contributions were his work on sexual selection (Darwin, 1871) and on the evolution of mental processes (Darwin, 1872). Currently, sexual selection is among the most studied topics in evolutionary and behavioral biology (Andersson and Simmons, 2006; Cronin, 1991), but there has been little consideration of how mental ability affects sexual selection (Miller, 2001; Ryan *et al.*, 2009; Boogert *et al.*, 2011). This issue was addressed in the debate between Darwin and Wallace over the adequacy of females to carry out choice needed to produce sexually dimorphic traits in male birds. And while Darwin prevailed in his view that female choice was critical in the evolution of male sexual signals, there remained a widely held view that most species, with the exception of primates, had very limited cognitive ability (Herrick, 1924). Then, starting around 30 years ago, a number of studies showed unexpectedly high levels of mental development in other animals, particularly in birds. Generally, cognitive ability as we use the term here, refers to the ability of individuals to use mental processes to solve problems that under natural conditions leads to increased fitness. Various studies have shown a role for cognition in spatial memory (Kamil and Balda, 1985; Bennett, 1993; Healy and Hurly, 1995; Pravosudov and Clayton, 2002), food theft (Bugnyar and Kotrschal, 2002), vocal ability (Pepperberg, 1999), tool-making (Jones and Kamil,

Animal Signaling and Function: An Integrative Approach, First Edition.
Edited by Duncan J. Irschick, Mark Briffa, and Jeffrey Podos.
© 2015 John Wiley & Sons, Inc. Published 2015 by John Wiley & Sons, Inc.

1973; Taylor *et al.*, 2007), string-pulling (Werdenich and Huber, 2006), and many other contexts. Recent neuroanatomical studies have shown that birds have relatively large brains for their body size (Jarvis *et al.*, 2005). Emery and Clayton (2004) suggested that the particularly large-brained New Caledonian crow may be similar to chimpanzees in cognitive ability.

Despite these studies, there has been an almost inexplicable lag in developing the seemingly obvious connection between cognitive ability and sexual selection. It would seem that the ability to perform relatively sophisticated cognitive processing would be beneficial for the very important fitness-affecting role of enhancing reproduction through sexual selection. This may be especially so in lekking species (those with nonresource-based (NRB) mating systems) that commonly show large skews in male reproduction (see Höglund and Alatalo, 1995; Widemo and Owens, 1995; Johnstone and Earn, 1999; Wiley, 1991). With these large skews in mating success, even small differences in male performance can be amplified into very large fitness effects.

Evidence of significant cognitive capabilities in various species suggests that they might often be applied in mate choice and display where they can have a large effect on fitness. In this chapter, we identify instances where females and males appear to use cognitive abilities in mate choice and display (see also Boogert *et al.*, 2011). This includes complex patterns of mate searching and assessment by females, complex courtship communication between the sexes that allows males to adjust courtship to the needs of individual females, and for males, the acquisition of learned complex displays from other males, innovation of new types of display behavior, and age-related improvement in the quality of display.

Most current sexual selection models assume little or no complex cognitive ability affecting male display or female choice (Lande, 1981; Zahavi, 1975, 1977; Hamilton and Zuk, 1982; Ryan and Rand, 1993; Kokko *et al.*, 2002; Borgia, 1979). This may be attractive to evolutionary theoreticians because it allows their models to be applied across a wide variety of species, including those with very limited cognitive abilities. This approach replicates some of the issues raised in the Darwin–Wallace controversy in the sense that females are often perceived to have limited ability to make adaptive adjustments when choosing mates, and male use of cognitive ability to enhance attractiveness to females is not considered. Limitations on female ability to choose sires that provide high fitness for their offspring has led to the conclusion that mate choice does not result in adaptive outcomes (e.g., Kirkpatrick, 1987), which remains a commonly held view (e.g., Dawkins and Guilford, 1996; Prum, 2010). Several widely cited models are dependent on the constraint of low female mental ability and how it restricts them in profitably discriminating among males. For example, in their chase-away model, Holland and Rice (1998) argue that males evolve highly elaborate displays as part of selection to overcome repeated episodes of initial female attraction and then resistance to their display. Males are suggested to repeatedly overcome the adaptive choice behavior of females due to the inability of females to reject increasingly large and otherwise maladaptive displays. Cognitively capable females should have more options for overcoming male manipulations that might otherwise repeatedly override their interests, even when engaged with males who are also cognitively capable. The ability to make cognitively based mating decisions suggests that females have multiple options for mate choice and can express context-dependent preferences and decisions, rather than being locked into particular nonadaptive preferences, such as those suggested to evolve as side effects of other evolutionary processes (e.g., Ryan and Rand, 1990; Rodd et al., 2002).

There is now increasing evidence for behavioral flexibility in male display (King et al., 1996; Patricelli et al., 2002) and female choice (e.g., Morris et al., 2003; Kodric-Brown and Nicoletto, 2001; Coleman et al., 2004; Hebets, 2003) across a variety of species. This flexibility is consistent with adaptive mate choice influenced by cognitive processes, and does not support some popular models that require genetic correlations between a particular male trait and its corresponding female preference (e.g., runaway, Lande, 1981; some versions of good genes, Pomiankowski, 1988, or those that claim to integrate these two kinds of models, Kokko et al., 2002). In the past 30 years, there has been relatively little progress on the critical question of the relative importance of competing sexual selection models (Andersson, 1994; Cronin, 1991; Borgia, 2006). Establishing the important role for cognition in sexual selection offers the possibility of breaking this impasse by identifying models that are and are not compatible with cognitively dependent mate choice and display. This should lead to consideration of models tuned to the cognitive capacity of the set of species under study. We need to begin evaluating the connection between cognition and sexual selection, exploring how different elements of male display might be used to indicate general or possibly more specific male cognitive capabilities, and how cognition influences the evolution of sexual displays.

Females may choose males with greater cognitive ability for a variety of reasons. For example, (i) Cognitive ability may be a heritable good genes trait that allows choosing females to produce cognitively superior offspring. Offspring with higher cognitive ability may live longer by avoiding predation, finding more food, gaining better territories, or avoiding parasites. In addition, their sons may have better sexual displays (Airey et al., 2000a) and their daughters may exhibit more effective mate choice (Leitner and Catchpole, 2002) by better discriminating between males and spending less time mate-searching for high-quality males. (ii) In species with male parental care, males with better cognitive performance may be better at provisioning offspring (Isler and van Schaik, 2006, 2008). (iii) Males may be able to use their cognitive abilities to attract or coerce females into copulating with them in ways that may not be beneficial for females, but increase their own reproductive success. (iv) If females vary in their requirements for sexual display depending on their age (Coleman et al., 2004) or level of experience (Hebets, 2003), then cognitively capable males may respond better to their diverse needs. (v) Cognitively superior males may be better at acquiring and manipulating display sites that enhance mating success, such as courts with better lighting (Gomez and Théry, 2004), that are safer from predators, or which are located near higher concentrations of receptive females (see Bradbury, 1981). There are many other reasons why individuals with superior cognitive abilities might perform better in sexual selection.

A variety of studies have sought to demonstrate a relationship between cognition and sexual display. Comparative studies investigating the relationship between brain size and display trait elaboration show inconsistent results, with some finding a positive relationship between total brain size and trait size (e.g., Madden, 2001; but see Day *et al.*, 2005), a negative relationship (Pitnick *et al.*, 2006), or no relationship (Garamszegi *et al.*, 2005a, 2005b; Spencer *et al.*, 2005). Iwaniuk and Hurd (2005) and Healy and Rowe (2007) are critical of this approach, pointing out that numerous factors that affect brain size are not controlled in these studies. Additionally, brain size is a coarse measure of cognitive ability, which can be affected by morphological traits other than size. Highly detailed information on comparative brain anatomy and its relation to brain function, which is not yet available for most species, is necessary if neuroanatomical comparisons are to meaningfully address the relationship between cognitive ability and sexual selection.

Recent studies have shown that within-species differences in song complexity are correlated with the size of the key song control nucleus (HVC) (Spencer *et al.*, 2005; also see Nottebohm *et al.*, 1981; Canady *et al.*, 1984; Airey *et al.*, 2000b; Nowicki *et al.*, 2000, 2002). Additional evidence that the quality of male display is related to brain function comes from tests of the nutritional stress (Nowicki *et al.*, 2002) and developmental stress hypotheses (Buchanan *et al.*, 1999, 2003). In these studies, young birds either were fed highly restricted diets or were exposed to parasites and, compared to controls, had smaller song control nuclei and sang lower quality songs. These studies show a clear relationship between characteristics of particular brain structures and male ability to produce vocal displays. A more direct approach for assessing cognitive ability has been to measure it using tests of problem-solving ability (PSA) (Roth and Dicke, 2005). Boogert *et al.* (2008) assessed male performance on problem-solving tests involving foraging skills and found that scores from these tests were positively related to male vocal ability. In the only test relating cognitive ability directly to male reproductive success, we found (Keagy *et al.*, 2009, 2011) in two different PSA tests that in male satin bowerbirds (*Ptilonorhynchus violaceus*) cognitive ability was associated with male mating success.

COGNITION, CO-OPTION, AND COMPLEX DISPLAY

Biologists have long been interested in the causes of sexual display trait complexity and elaboration (Darwin, 1872; Fisher, 1915, 1930; Andersson, 1994; and many others). Here we argue that two elements, co-option and cognition, can greatly affect the evolution of these complex displays. As modern evolutionary biology has shown, co-option is an important ingredient for the evolution of complex adaptations. Co-option involves expressing already-evolved traits under novel conditions that may allow for new functions. This rapid evolution can shortcut the gradual process of building complex adaptations that might otherwise require many small changes over time. Co-option can have particularly dramatic effects when cognition directs this process, and it allows individuals to rapidly bring together different, already-evolved component traits to form novel complex adaptations.

Cognition can play a major role in co-option events because it can speed up the acquisition and assembly of traits, bypassing the slower process of genetic evolution. This can occur because cognitive processes like learning and the ability to anticipate future needs (i) do not require genetic change and allow acquisition of new traits within one generation, (ii) allow individuals to preferentially acquire traits that they recognize as functioning well in other contexts or individuals, and (iii) can integrate different acquired component traits to produce a high level of function.

Cognitive acquisition of traits through co-option has been used to explain rapid and dramatic human cultural evolution (Alexander, 1981) and to a more limited extent considered for primates (van Schaik et al., 1999; Tomasello, 1999). It has received less attention in the discussion of the behavior of other animals (but see Bugnyar and Kotrschal, 2002; Sasvari and Hegyi, 1998), and in particular, in the literature on sexual selection (Koksal and Domjan, 1998). As noted above, there has been relatively little progress in understanding the evolution of complex elaborate displays. We propose that in relatively large-brained animals such as passerine birds (Emery and Clayton, 2004), cognition may play an important role in the development of sexual display, by driving the co-option of existing behavioral elements and the assembly of these elements into novel complex displays. There is likely high overlap in the perceptual abilities and preferences of males and females, and males appear to assess elements of their own display (see below). We suggest that a high level of male cognitive ability may allow males to anticipate how females will react to display innovations and to use this information in crafting more effective displays.

The behavioral displays of bowerbirds (Family Ptilonorhynchidae) are among the most elaborate sexual displays of any species, and these displays provide an excellent model for considering the role of how cognition and co-option might shape these displays. In bowerbirds, there are many instances where cognition and co-option working separately and, at other times, together appear to have shaped aspects of male display behavior and related female responses, and we review some of these cases. Satin bowerbirds (P. violaceus) have received particular attention in studies of male display and female choice. Courtship and mating occur at their display sites called bowers. The bower structure is made of sticks with an adjacent decorated display court. Because courtship and mating occur at the bower, we have been able to position automated video cameras to monitor the bower, providing a complete record of all behavior of marked individuals at bowers through the mating season (Borgia, 1995a). Analysis of these detailed records has shown a consistent pattern in which males and females appear to use their cognitive abilities in ways that influence male mating success.

Male satin bowerbirds locate display sites on an exploded lek where individual males build and maintain decorated stick bowers separated by more than 100 m, and court and mate with visiting females (Borgia, 1985a; Uy et al., 2001a; Hansell,

2009). The display court on the north end of the bower is predominantly decorated with blue objects placed on a carpet of yellow straw and leaves. Decorations are collected from the surrounding habitat or stolen from other males (Borgia and Gore, 1986; Hunter and Dwyer, 1997; Wojcieszek *et al.*, 2007). Male mating success is affected by multiple display elements including bower decorations (Borgia, 1985a, 1993; Coleman *et al.*, 2004), bower quality (Borgia, 1985a), vocal display (Loffredo and Borgia, 1986; Coleman *et al.*, 2007), and dancing display (Patricelli *et al.*, 2002, 2003, 2004, 2006). Several other aspects of display may also be important, including the holding of specific types of bower decorations in males' beaks during courtship (Borgia, unpublished data) and the painting of bower walls with masticated plant fiber by males (Bravery *et al.*, 2006; Cendes, 2009; Hicks *et al.*, 2013).

Other bowerbird species show many similarities with satin bowerbirds, but there are also significant differences. Males of most polygynous bowerbird species build bowers; exceptions are toothbilled (*Scenopoeetes dentirostris*) and Archbold's bowerbirds (*Amblyornis* (*Archboldia*) *papuensis*) that display on cleared courts with no bower, and the monogamous catbirds that do not have either courts or bowers (see Borgia, 1995b). Satin bowerbirds belong to one clade of bowerbirds that erect two-walled bowers made of sticks with a display court at one or both ends (Kusmierski *et al.*, 1997). Some of these bowers are raised up on tall stick platforms (fawn breasted (*Chlamydera cerviniventris*) and yellow-breasted (*Chlamydera lauterbachi*) bowerbirds). The other clade includes maypole builders that build a stick maypole around a sapling surrounded by a display court. In some species (streaked (*Amblyornis subalaris*) and Vogelkop (*Amblyornis inornatus*)) a hut-like roof is built over part of the court (see Gillard, 1969; Cooper and Forshaw, 1979; Frith and Frith, 2004). Males of all court clearing species display to visiting females with dancing displays. These species display decorations collected from the habitat on their courts and sometimes on the bower. Species differ in color, number, and types of objects used as decorations.

Bowerbirds have all of the attributes that Emery (2006) suggests are associated with high intelligence species. We believe that evidence from highly detailed studies of bowerbird courtship offers a unique set of information supporting a large role for cognition affecting mate choice in bowerbirds. When we started this research, we expected to test sexual selection models that relied on genetic correlations such as runaway and some good genes models. It became obvious that bowerbirds have many traits indicating a high level of cognitive involvement, for example, learning in display trait acquisition (Collis and Borgia, 1993) and complex decision-making (Uy *et al.*, 2001a, 2001b; Patricelli *et al.*, 2002), suggesting that the co-evolutionary genetic models that dominate much of sexual selection theory do not adequately explain sexual behavior in satin and other bowerbird species. Instead, models dependent on a high level of cognitive ability and co-option seem more appropriate. Here we review observations of bowerbird display that support this view.

Cognition can affect the acquisition and utilization of bowerbird display elements in several ways. Some of these mechanisms are (i) learning, (ii) innovation, (iii) flexibility, and (iv) anticipation. Each of these requires, at least for traits expressed in bowerbirds, a reasonably sophisticated mental framework for their successful operation. In the following sections, we discuss how these mechanisms may be involved in male display and female mate choice. Most of these observations have been obtained from our video camera system that has provided highly detailed records of male and female behavior at bowers. We also include some anecdotal results that reflect extraordinary cases that are not suited to statistical analysis, but give important insights into the cognitive capabilities of some exceptional individuals (see Lefebvre *et al.*, 2004).

DELAYED MALE MATURITY, MALE–MALE COURTSHIP, AND DISPLAY TRAIT ACQUISITION

Male sexual maturity is delayed in bowerbirds, and this period appears to provide an opportunity for learning complex elements of sexual displays. A progressive pattern of age-related plumage changes in male satin bowerbirds appears to affect interactions among males at display sites. At 5 years old, their neck and upper chest plumage changes from mottled yellow to dark green and their beak changes from black to yellow (Vellenga, 1980). At age 7, males produce a blue-black plumage and around this age attempt to acquire a permanent display site where they build a bower. Male satin bowerbirds produce sperm when they are 5 years old. Young males build temporary, poorly constructed practice bowers where they court other young males (Collis and Borgia, 1993). Before the start of the mating season, these males often visit adult, bower-holding males who court them. Adult males who had decorations experimentally added to their bowers were preferred for visits by young males, suggesting that these young males may use decorations, as do females, for identifying adult males with high-quality displays (Coleman, 2005). Such males might serve as more effective tutors for young males.

Young males implanted with testosterone up to 4 years before their natural transition to adult plumage changed to adult plumage at their next molt (Collis and Borgia, 1993). These implanted males constructed low quality bowers, suggesting that their normal delayed maturation is necessary to allow them to learn bower building and other aspects of display. The specialized plumage of older juvenile males may allow them to attract less threat from adult males than if they had the adult male plumage. This may improve their opportunity to learn displays when they are courted at adult males' bowers.

Young males visiting adult male bowers and receiving courtship appear to learn the content of display (Collis and Borgia, 1993). For example, as part of their vocal displays, male satin bowerbirds mimic up to five other species, and males with higher-quality mimicry (Loffredo and Borgia, 1986; Coleman *et al.*, 2007) and more species mimicked (Coleman *et al.*, 2007), obtain more matings. The complexity of these displays and their mimetic nature leave little doubt that they are learned as might be the initial portion of the courtship vocalization that involves buzzes similar to those of cicadas, a food source for young birds (Donaghey, 1996). The delayed maturation of young males that visit the bowers of adult males, and courtship involving learned components are consistent with the hypothesis of cultural transmission of male display traits (Collis and Borgia, 1993; Loffredo and Borgia, 1986; Coleman et al., 2007; see also Madden, 2008) and a critical role for cognition in the acquisition and effective display of these traits.

Adult male satin bowerbirds often display to empty bowers and, before the mating season, to young males. Since successful males courting females must be able to adjust their display in response to signals from female courtship partners (see next section), males receiving courtship may be useful surrogates providing feedback to courting males on how to modulate their display in response to the receiver's reactions. These courtships appear to be important to courting males because of the high frequency of these male–male courtships, and the use of calls, including begging calls, to attract other males to receive same-sex displays. Once the mating season starts and there is the potential for interference with male–female courtships by visiting males, adult male bower owner behavior changes and visiting males are no longer tolerated at bowers. This reversal in the behavior by bower-holding males appears to be a beneficial cognitive adaptation for improving display quality while reducing costly side effects during the mating season.

FEMALE SIGNALING TO AFFECT MALE DISPLAY INTENSITY: AN INNOVATION THAT IMPROVES COURTSHIP SUCCESS

Female bowerbirds are attracted to intense male courtship displays that contain elements similar to male aggressive displays that can be threatening to females (Borgia and Presgraves, 1998; Patricelli et al., 2002, 2004). In satin bowerbirds, females signal their level of comfort with display to the courting male. Females may make multiple visits to a male's bower, and females comfortable with the male's display commonly return for additional courtship and move into a deeper crouch. As females assess potential mates, female tolerance for intense display increases during successive courtships with preferred males. Females then signal their readiness for copulation (Patricelli et al., 2002, 2004). Other females, less comfortable with a male's display, lower into a crouch more slowly or not at all, or if already in a crouch, may rise up rapidly showing a "startle" response. Female crouching is inversely related to the frequency of female startling during courtship and thus signals the degree of display intensity that females will tolerate from a male without startling (Patricelli et al., 2002). Female age and condition affect the level of female comfort (Coleman et al., 2004). To test the hypothesis that successful males reduce the intensity of their courtship displays after females indicate their discomfort, we used robotic female bowerbirds that mimicked female behavior in the bower. Supporting this hypothesis, males displayed with significantly lower intensity toward robots when they crouched at a lower rate (Patricelli et al., 2002) and when the female robot startled (Patricelli et al., 2006) in experimental treatments, compared to treatments with faster crouching and no startling. Thus, female signals help direct males to produce displays of appropriate intensity. Males able to respond by adjusting display intensity relative to the level of female comfort with the male's display are attractive to a wider range of females and thus gain a mating advantage from their ability to respond to female signals (Patricelli *et al.*, 2002). Females benefit from this communication by receiving displays at an intensity they are willing to tolerate, thus reducing startles and the possible disruption of courtship, and allowing them to more effectively choose their mate (Patricelli *et al.*, 2004).

An additional cognitive component in bowerbird intersexual communication signals by females is suggested by age-related changes in female preferences for components of male courtship display in mate choice. All age classes of female satin bowerbirds use bower decorations in early assessment of males in initial courtship visits to bowers, but in later courtship visits, older females select males based on male dancing and vocal elements of courtship display. Younger females appear to be more threatened by the high intensity male courtships and base mate choice more on bower decorations than courtship displays (Coleman *et al.*, 2004). When bower decorations were manipulated by placing an increased number of blue plastic squares on bowers, this had a greater effect on the mating decisions by young females than their older counterparts (Coleman *et al.*, 2004). Cognitive ability appears important because young females must learn not to be threatened by intense male courtship displays, and how to use these displays in choosing mates.

Cognition affects male ability to react to female signals of comfort and this has an important effect on male mating success (Patricelli *et al.*, 2002, 2004, 2006). The intensive practice by males both before sexual maturity and as adults in displays with other males may allow fine tuning of male ability to read the reactions of courtship partners. The large amount of time and effort devoted to attracting same-sex courtship partners and displaying to them suggests that the refinement of courtship abilities is important in male success in courtships with females. The significant age-related change in female response to male display suggests that learning and cognitive ability affect female ability to respond to male display.

MATE SEARCHING AND FLEXIBILITY IN ADAPTIVE DECISION-MAKING

Mate searching is a critical process for females attempting to maximize the fitness benefits of mate choice. Females can apply their cognitive abilities to mate searching to enhance its benefits by allowing them to locate high-quality mates while limiting the costs of mate searching. Using time and date information on video tapes of female visitation at bowers, we (Uy *et al.*, 2000, 2001a, 2001b) reconstructed female satin bowerbird mate-searching patterns. We found that females commonly repeatedly visited bowers and sampled courtships of multiple males before mating. Females tended to have two distinct sampling bouts, the first involving a larger set of males with some rejected before nest building, followed by a second bout during which a mate was chosen from a more limited set of males. Females varied in the number of males sampled, with some visiting the bower of only one male. Some of this variation can be explained by differences among females in past experience with males. Females who chose males attractive to a large number of other females (top males) typically mated with that same male in the following year, sampling relatively few other males. Females who had mated with less attractive males typically rejected these males as mates the next year. These females searched among a relatively large set of males, usually finding a mate of higher quality than their previous mate (Uy *et al.*, 2000). Females mated to top males who died sampled more males the next year. These results indicate that the loss of attractive mates forces females to increase their search and provides evidence that past experience in sampling males shapes subsequent mate-searching behavior (Uy *et al.*, 2001b). This suggests that the cognitive ability of females can provide adaptive flexibility in mate-searching tactics that allows females to obtain higher-quality mates while limiting their search costs. Previous mate-searching models had not considered the effect of long-term memory on female flexibility in mate-searching tactics that is clearly dependent on having the cognitive capability to make complex fitness-enhancing decisions.

FEMALE UNCERTAINTY AND FLEXIBILITY IN ACTIVE MATE ASSESSMENT

Female assessment of male display behavior is often difficult to quantify because females often passively view male display. However, if females actively sample aspects of male display, this can provide valuable information about how females assess male display components. In satin bowerbirds, active and quantifiable female sampling behavior is associated with the assessment of paint males apply to their bower walls. Males paint their bowers by applying masticated vegetation to the sticks on the inside of their bower walls, and females sample paint by nipping at the painted bower wall. The intensity of female sampling of paint can be related to the stage of courtship females are involved in and other characteristics of females that relate to their mate choice decision process. Females sample more in early courtship visits to males (Bravery *et al.*, 2006; Cendes, 2009). This suggests that paint sampling is more common when females are less certain about which male they will choose. Cendes (2009) tested this hypothesis by using three different measures of female uncertainty and then relating them to the degree of paint sampling. First, she found that the proportion of visits with paint sampling by females is positively correlated with the number of males visited and number of visits to males. Second, females mating with multiple males had visits with a high level of paint sampling. Third, paint sampling was positively related to the proportion of switches between males visited. Thus, paint sampling is higher among females who need additional information to resolve mate choice decisions. These cognitively based adaptive adjustments in paint sampling should allow females to improve their assessment of males in mate choice. Also, it may allow females to make up for deficiencies in male display, such as when bower decorations are scarce (see Bravery *et al.*, 2006).

LONG-TERM AGE-RELATED IMPROVEMENT IN DECORATION DISPLAY: SYMMETRICAL DECORATION DISPLAYS ON OLDER MALES' BOWERS

One of the most attractive features of bowerbird display to human observers involves elaborate patterns of bower decoration. Some bowers have spectacular decoration displays, like a spotted bowerbird (*Chlamydera maculata*) bower we found near the home of a stained glass artisan. The male bower owner collected fragments of stained glass and laid them out in a beautiful symmetrical design on the display court with white sheep vertebrae (bones) near the bower encircled by neatly laid out arcs of orange and blue glass that covered an area 4 m wide. This marked male disappeared between mating seasons and was replaced by a different and probably younger, less experienced male the next year. The colored glass decorations the former owner used were still present, but the new bower owner created a far less elaborate display with the colored decorations mixed in a pile with white vertebrae near the bower. The new male's failure to artfully display those colorful decorations may be due to it having less experience in bower decoration display. This hypothesis is supported by observations of increased complexity of bower decoration displays in cases where we have monitored individual males at bowers for 10 or more years. These males have a more colorful set of decorations, often with a high degree of symmetry in the placement of these decorations. These examples suggest that it may take years for males to develop the skills and obtain the materials to produce these displays. Often males "try out" locations for new decorations, changing their position over days and sometimes weeks. This suggests that developing complex displays is a cognitive process in which birds are sensitive to small changes in the placement of decorations on their display courts. With decoration displays composed of thousands of decorations, a male bower owner must make numerous decisions about which decorations to include in his display and where on the court to place them. Successful spotted and great bowerbirds commonly sort decorations into piles of like-colored objects, with certain colors placed at particular locations around the bower (Borgia and Presgraves, 1998), for example, flat green glass near the bower entrance, small pieces of cubic auto window glass or sometimes small green fruits in the bowl located in the middle of the bower avenue, and red plastic next to the outside of the bower walls near the bower entrance. For these decoration types, males commonly add similar kinds of decorations to already existing piles. The symmetrical arrangement of different colors far out from the bower are far less common, and appears to require the benefit of prolonged decorating experience. We do not know if highly ornate symmetrical decoration displays result in increased male attractiveness to females. We hypothesize that it takes much of a male bird's lifetime to develop the ability to produce these elaborate displays. Interestingly, the decoration display itself can serve as a repository of the products of past decisions that the bird can manipulate and build upon with slight improvements that eventually lead to ornate displays. The ability to build symmetrical displays may result from the same sense of symmetry applied to bower building. Successful male bowerbirds tend to build more symmetrical bowers (Borgia, 1985a) and this may reflect a general preference for symmetry that is extended to these ornate decoration displays. The placement of decorations in patterns could be functional in aiding female mate searching. Patterned displays make bowers more recognizable from a distance as a designed display rather than an accidental grouping of bright objects. These patterns could also provide females information about the bower owner such as his age, based on the complexity of the display, and his identity, by indicating through a unique display that the same male is present at the bower over successive years. Highly complex symmetrical displays might also indicate a male's cognitive ability.

ANTICIPATION OF MALE ROUTES DURING COURTSHIP: PATHS ON DISPLAY COURTS OF SPOTTED BOWERBIRDS

In most bowerbird species, males have specific paths around the bower that they travel on during courtship. Large decorations can potentially obstruct male movements along these paths as they court females. In spotted bowerbirds, males using numerous large sheep vertebrae, snail shells, or stones as decorations leave paths clear of large decorations near the bower where they travel during courtship. Bower owners may carpet these paths with flat stones, a stick mat, or leave them bare. Beside the path, vertebrae may be stacked into piles three high, providing a white background (see Figure 4.1) for courting males to display their lilac crests. The narrow, neatly sculpted pathway between the bower wall and piles of vertebrae leaves few other options for movement during courtship near the bower. As with other aspects of decoration display, there are clear cognitive aspects to the patterning of these display elements. When we placed vertebrae or other large decorations on the path, they were quickly cleared away (Borgia, personal observation). These behaviors suggest that males anticipate the movements they will make during courtship and clear objects that might impede their display.

One method males use to advantageously place decorations in positions that will not interfere with courtship is through practice courtship displays. Practice displays to empty bowers or to visiting males are common among a variety of

Figure 4.1.   A wall of sheep vertebrae is positioned as a backdrop for males displaying in front of the bower. (*See insert for color representation of this figure.*)

bowerbird species. In satin bowerbirds, males often begin practice displays and then suddenly stop, and then rearrange decorations on the bower. Thus, by rehearsing their display movements, males can identify problems with the position of display items and reposition them for a more effective display. Movement of decorations on display courts also occurs independently of practice courtships (see Borgia *et al.*, 1987; Borgia and Keagy, 2006). Male bowerbirds are able to use a great diversity of objects, including many novel man-made objects, and choose those of appropriate size, shape, and color for different functions on the bower, placing them in appropriate positions, often in piles of similar color or type of decorations (Borgia, 1995a; Uy and Borgia, 2000; Doerr, 2010). This suggests attention to fine detail in developing their decoration display and the ability to classify objects by color and other features.

A different explanation for the positioning of decorations on display courts has been offered by Endler *et al.* (2010) and Kelley and Endler (2012). They suggest that male great bowerbirds use the placement of decorations of different size to create a forced perspective optical illusion that may keep female attention longer during courtship, for example by causing the displaying male to look larger to her. Anderson (2012) points out that Kelley and Endler (2012) fail to consider the effects of other display elements that could account for a positive relationship between decoration gradients and mating success. Our own analysis of their data fails to support their conclusion that the proposed illusion was related to male mating success (Borgia *et al.*, 2012) and in the populations we observe, we see little evidence of the linear gradations of decoration size they describe, necessary to create the proposed optical illusion (see Figure 4.2). An alternative hypothesis that explains why decoration displays often show a stepwise pattern of size change is that smaller decorations on the display court near the bower allow the male to display near the bower entrance and larger decorations placed farther away are less likely to impede male movement during courtship.

SOME OTHER POSSIBLE COGNITIVE DISPLAY-RELATED BEHAVIORS OF BOWERBIRDS

A variety of other studies have suggested a possible cognitive function affecting male display in bowerbirds. Mikami *et al.* (2010) found that great bowerbird bowers tend not to be affected by fire and they argue that males have learned to build them away from grassy spots that are likely to burn. However, there are a variety of reasons why males may not build bowers in grassy areas, including avoiding areas where predators may not be easily detected, or areas where it is difficult for females to see the bower. Each of these alternatives is also consistent with a role for cognition in bower site selection.

Madden (2002) suggests that less dominant, spotted bowerbird males remove decorations added to their bower in order to avoid destructions by more dominant males. Borgia (1993) tested for this effect in satin bowerbirds that have closer neighboring bowers and higher destruction rates and found no tendency for any males to remove preferred types of decorations from their bowers. Doerr (2010)

Figure 4.2. The bower of a great bowerbird near Mareena, Queensland. This bower is unusual because of its large size, the way the male positioned it within a set of thick branches, and the uneven slope on which it is built. The pattern of decorations seen here and in most other great bowerbird bowers we have seen does not show a size gradient of decorations necessary to create the forced perspective illusion. In this case, decorations arrayed at the front of the bower are approximately the same size. More-over, because these decorations are on a downward slope, they cannot be seen by a female being courted in the bower avenue to produce a forced perspective illusion even if there were a size gradient. Thus, the decorations appear to function in attract-ing females to the bower before they arrive in the bower avenue. (*See insert for color representation of this figure.*)

found similar results studying great bowerbirds. Bravery *et al.* (2006) suggest that decoration removal leads to higher levels of painting to compensate for the loss of this component of display. However, males with fewer bower decorations to arrange may simply be using time that might have been used for bower decoration instead for bower building.

CONSTRUCTION OF SUCCESSIVE SCENES FOR FEMALES VISITING THE BOWER

The presence of two-walled bowers, like those of satin, great, and spotted bowerbirds, contributes greatly to the evolution of display complexity. Females commonly fly into the vicinity of a bower and may perch in a tree looking at the bower site from above, and if attracted by the decoration display, move toward the bower and then enter into the bower for courtship. The bower creates a series of special display environments (scenes) that add to the complexity and design of bower displays. The bower owner can create scenes appropriate for engaging females at each particular stage of courtship and mating. Decorations of different colors, compositions, and sizes are placed in the most advantageous positions in each scene to appeal to the visiting female's preferences. Males orient bowers and decorations to enhance illumination of the display site at critical times when females are likely to visit. The narrow avenue between the bower walls where the female stands during courtship dictates her orientation and allows the male to affect what she sees as he courts her. Males of several species decorate courts with colors that contrast with the male's plumage that may enhance the visibility of males to females.

In spotted bowerbirds, decorations up to 2 m in front of the bower are used for long distance attraction of females to the bower (Borgia, 1995a). Red, pink, orange, and black (less commonly yellow and light blue) plastic and piles of white bones are placed beyond a bush that might hide the bower from above. As females move close to the bower entrance, a different set of decorations become visible inside the bower, in the central bowl, and lining the inner walls. These were hidden by the bower walls when she was outside the bower. The decorations in the bowl are typically smaller than those outside the bower and slightly smaller than those lining the bower walls. These often include small cubic pieces of fractured quartz or auto window glass (Figure 4.3) that glow like gems in sunlight (Borgia, 1995b), small green fruit (Madden, 2002, 2003), and the shiniest of the male's small metal

Figure 4.3. Shiny quartz rocks are positioned in the central avenue of a spotted bowerbird bower. (*See insert for color representation of this figure.*)

objects including coins and jewelry, with less shiny versions placed outside the bower. When we offered males silver foil balls of different sizes, males placed the small balls inside the bower but kept the larger balls on the bower platform, supporting the suggestion that small items are preferred for display in the central avenue. In addition, the bower walls often have small pieces of metal, pull tops, red thin wire and stems, and silver seed pods arrayed on them.

The number of bower decorations in the bower's bowl is a strong predictor of male courtship success (Borgia and Mueller, 1992; Borgia, 1995a; Madden, 2002, 2003). This is not surprising as the female in the bower looking out the entrance can see few other colorful decorations. The green and red decorations outside of the bower are nearly all out of sight, hidden by the bower walls and appear to function mostly in attracting the female to the bower.

Just as male bowerbirds may acquire vocal displays by receiving them from other males as if they are females, males building bowers often traverse the route traveled by females into the bower. Males share many of the same preferences for decorations and bower architecture with females (Borgia and Keagy, 2006; Borgia, 1985a). This allows them to view scenes in the same way that females do and to build and adjust them to match female preferences at each step. The sole exception to this process is that males cannot see their own dancing display, and for this they may need to rely on the reaction of other males and females to their courtship displays. Males may begin to learn courtship behavior, bower decoration, and building from other males on visits to bowers when they are young. An important cognitive component of these displays involves remembering the elements of display, including vocal mimicry, and learning how those receiving courtship react to particular elements of display behavior, for example, does the female move away from, or is she attracted, a particular display element?

COGNITIVE ASPECTS OF BOWER BUILDING: AGE-RELATED IMPROVEMENT IN CONSTRUCTION AND NOVEL TECHNIQUES FOR MAINTAINING SYMMETRY

Bowers have a variety of design features that suggest that male bowerbirds must learn to build. Typically in satin bowerbirds, the bowers of young males are poorly built, but as males mature, the bower becomes more refined. An important difference results from the sticks used in bower construction; mature male satin bowerbirds use sticks that are straighter, thinner, and less often forked compared to sticks used by young males (Borgia, 1985a). Mature males pack sticks together tightly, placing them more vertically and using bends in the sticks to curve the bower walls. Bowers of attractive males have a more sculptured appearance, with even curvature of bower walls with a slight widening toward the center, a slight narrowing on the inside near the bower entrance, and the top of the walls curving over the bower avenue. Stick lengths are more even, often placed so that the wall has a more even height both within the wall and between the two walls. The overall appearance of the mature male bower has a finer and tighter overall structure, with few if any sloppily placed sticks.

In satin bowerbirds, bower symmetry scores are consistently correlated with male mating success (Borgia, 1985a). These birds have a unique mechanism for rebuilding bowers that are asymmetrically damaged. When one bower wall is destroyed, males use a technique we call templating in rebuilding the destroyed wall. After choosing a stick they will use in the destroyed bower wall, they first place it in or against the standing wall while standing on the midline of the bower avenue parallel with the wall. Then, they remove the stick and using a mirrored pattern of movements they used to initially place that stick in the standing wall, they place it in the destroyed wall they are rebuilding. Males appear to be using the standing wall as a template to measure the position of the stick they will place in the destroyed wall, allowing them to build a new wall that is symmetrical with the remaining standing wall. Males who show more templating behavior have more symmetrical bowers (Keagy *et al.*, 2012).

The age-related pattern of differences we see in satin bowerbird bowers suggests that similar variation in bower construction seen in other species is also age-related. In spotted and great bowerbirds, bowers of individual males we have followed for 10 or more years are neater and more precisely built. In addition, these males place decorations in a small pile in a small bowl built of sticks woven into the avenue floor (Figure 4.4). In the fawn breasted bowerbird (*C. cerviniventris*), bowers vary greatly. Mature males build more precisely constructed bowers, using fine sticks closely packed. The bower itself is built on a platform of sticks reaching 20 cm tall with the bower and a short decorated display court built on top. More sloppily built bowers may have no or only a small stick platform under the bower. Bowers of mature male yellow-breasted bowerbirds (*C. lauterbachi*)

Figure 4.4. A spotted bowerbird bower with a sunken central stick bowl where preferred decorations are placed. (*See insert for color representation of this figure.*)

are also raised on stick platforms with two stick ramps running from the middle of the bower up toward the tiny court on each end of the bower. The end of the court is raised up to form a set of end walls unique to this species, creating four walls rather than the two common in all other species of this clade. These unique features are absent from the bowers of young males so that there is little or no platform that raises the bower, and there is no ramp or end walls, and these bowers show less detail, lacking fine sticks or sculpturing. An exception to the trend for bowers of young males to be built on lower platforms occurs in western bowerbirds (*C. guttata*) at Exmouth, Australia, where the neater bowers of older males are typically built close to the ground while those of young males are often built on large rounded piles of sticks. Other differences are that young males have raised bower avenues that abruptly drop off at the ends of the bower avenue. Also, unlike mature males, they fail to line the stick walls of their bowers with yellow straw, although when we placed straw we collected near their bowers, they placed it in the bower walls.

The recurrence of age-related differences in the quality of bower construction across bowerbird species suggests that there may be significant learning necessary for effective bower construction. This hypothesis is supported by observations of young male visitors placing sticks in bowers owned by older males, and young males building generally poor quality bowers with misaligned and forked sticks years before they are likely to attract and copulate with females.

Alternatively, young males may be capable of building high-quality bowers but choose not to because lower-quality bowers are suitable for their needs in practicing display with other males, so it may not be worth their extra effort. Borgia (1985b) also hypothesized that young males may not build high-quality bowers in order to reduce threat of bower destruction by more aggressive dominant males. Evidence against these hypotheses comes from an experiment in which we implanted young males with testosterone, causing them to moult into adult plumage. These males who held permanent bower sites much like untreated adults, built poor quality bowers, as if they were still in juvenile plumage, suggesting that testosterone implantation did not provide the skills needed to build high-quality bowers, (Collis and Borgia, 1993).

There is also significant variation in the quality of adult bowers, and while the very youngest adult males may have poorer quality bowers, some differences persist over years and seem not to be age-dependent. In addition, males develop different distinctive styles of bower building affecting bower length, verticality of sticks, degree of sculpturing and symmetry of walls, avenue width, avenue height, and other features that make individual bowers recognizable. Often males build bowers on sloping terrain that they are able to level with a stick platform or by incorporating limbs from shrubs into the bower wall, showing the adaptability of males in building bowers. In golden bowerbirds (*Prionodura newtoniana*), males typically find a pair of saplings spaced approximately 1 m apart connected with a cross branch. The male then builds a stick spire around each sapling and surrounds the ends of the cross branch with tightly woven sticks. In part, due to the irregular separation, size, and branching of these saplings, these bowers are quite variable in shape and illustrate the ability of these males to adjust bower building to meet the conditions imposed by the available buttressing saplings in the area where they will locate their bower. We suspect that as learning is important in affecting male building abilities, differences in male cognitive ability may be indicated by the variation in quality and style of bower building. It is also possible that these individual stylistic differences represent adjustments by males to enhance their unique patterns of courtship and may be used by females to recognize males.

COGNITIVE FLEXIBILITY AND INNOVATION IN DISPLAY

Male spotted bowerbirds have variable courtship displays at different locations in New South Wales and Queensland, Australia. Males of most populations (e.g., Nyngan, NSW and Hughenden, Queensland) we have surveyed, court with wide ranging displays but as they approach the bower they position themselves near the end of one bower wall (see also Madden, 2006). However, in one population we studied north of Thallon, in southern Queensland, bowers were rotated 90° from the N-S direction seen in other locations, with the avenue built extra wide between thin straw walls. Males gave especially high intensity displays and courted females through the unique see-through side walls of the bower. Courted females in the bower positioned themselves perpendicular to the main axis of the bower avenue, facing one bower wall. These males produced loud and rapid "Skraa" calls, throwing decorations as they courted females (Borgia and Presgraves, 1998; Borgia and Mueller, 1992). Birds in other spotted bowerbird populations have more subdued displays, with females standing parallel to the bower walls inside the bower avenue, facing the front entrance of the bower, and males standing at the end of one bower wall as he courts her. By building the bower rotated by 90°, with courtship behind see-through straw walls, Thallon males create a protective barrier for females that physically separates them as he courts her with highly aggressive displays. This novel bower design allows males to give high intensity displays that are preferred by females while reducing the threatening aspects of the display to females (Borgia and Presgraves, 1998; Figure 4.5). To test the hypothesis that the bower wall functions to reduce threat to the courted female, we predicted that if one bower wall were destroyed, Thallon males would position themselves behind the remaining standing wall during courtship. When either the N or S wall was left standing, males consistently courted females with the standing wall between them (Borgia and Presgraves, 1998). In intact bowers, males tended to more often court from the N side, so to remove any bias males had for courting from one side of the bower, we rotated the bower 180°, thereby changing the position of the standing wall. We found that males adjusted to the new position of the wall, using it to separate them from females. In one case, the female stood outside the bower and the male courted her from within the bower, keeping the standing wall between them. Females also consistently moved behind the lone standing wall for courtship, suggesting that they could use the unusually positioned wall to mitigate threat from males. Males infrequently courted the female from the destroyed wall side of the bower, with no

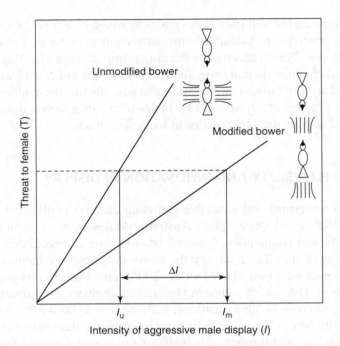

Figure 4.5. At Thallon in SE Queensland, male spotted bowerbirds have developed a special orientation and design to their bower. Unlike other spotted bowerbird populations and avenue builders, males court through a thin see-through wall that serves as a screen that appears to reduce threat to females from the males' especially intense display. This allows males to display at a much more intense level than males with unmodified bowers. (Figure taken from Borgia and Presgraves (1998).)

wall separating them from the female. In these cases, males significantly reduced display intensity (Borgia and Presgraves, 1998), showing that they were responsive to these unusual conditions. The flexible and highly appropriate reactions of males and females in this experiment suggest that they are cognizant of the threatening effect of male display or at least female reactions to it, and how the bower wall can be used to mitigate this threat. Males used this understanding to take up novel and effective courtship positions when bower walls were manipulated and then appropriately adjusted their display intensity when they did not have a protective wall present between them and the females they were courting. We did not anticipate that both sexes were capable of this high level of cognitive flexibility before conducting these experiments.

DECORATION STEALING: AN INNOVATION FOR DISPLAY TRAIT ACQUISITION

Male satin bowerbirds commonly use blue parrot feathers to decorate their bowers and the number of these feathers is positively related to male mating success (Borgia, 1985a; Patricelli *et al.*, 2004). Theft of feathers is critical for affecting the number of feathers on a male's bower. The most active feather thieves were from areas where bowers were close together and they were involved in reciprocal stealing with males at adjacent bowers (Borgia and Gore, 1986). The occurrence of reciprocal stealing suggests that males may be able to identify males who steal from them and then go to their bowers to retrieve decorations. Studies of female movements in mate searching show that females search in a limited area for mates, thereby making nearby males competitors for females who search in that area (Uy *et al.*, 2000). The focus on nearby bowers for stealing is beneficial to the stealing male because it requires less time for them to be away from their bower and it permits them to take decorations from males who are their most likely direct sexual competitors. Stealing typically occurs when owners are not at bowers, so, for males to steal effectively, they need to be able to monitor their neighbors' behavior around bowers, and in the case of reciprocal stealing, past experience may cause males to direct extra effort toward males who are likely stealing threats. Male ability to successfully steal appears to require the ability to track their neighbors' behavior, determine who has stolen from them, and limit time away from their own bower so that they can protect their own decorations. Male cognitive ability may be important in affecting these tasks. Given what we know about how age affects male display behavior and bower building ability, it would be interesting to explore if male skills in stealing are affected by experience.

COOPERATING WITH RELATIVES FOR DISPLAY: AN INNOVATION TO REDUCE SEXUAL COMPETITION

Bower destruction is a common feature of male sexual competition in satin bowerbirds. Bowers are used by females in mate assessment (Borgia, 1985a) and are commonly destroyed by males' two nearest neighbors (Borgia, 1985b). As noted for feather stealing, females search in a limited locality for mates, so near neighbors are typically competing for the same females as mates. Unlike feather stealing, males do not directly enhance their own display when they destroy a neighbor's bower, although they do decrease the quality of display of their competitor/neighbor (Borgia, 1985b). Male dominance in aggression at feeding sites strongly predicted success in bower destructions and unlike feather stealing, there was little reciprocal bower destruction (Borgia, 1985b). Males in lekking species sometimes aggregate with their kin, suggesting that kin relationships can affect sexual display. One common suggestion is that kin may aid relatives in display. When this was tested in the intricate cooperative displays of long-tailed manakins, this was not supported (McDonald and Potts, 1994; Loiselle *et al.*, 2006), but it has been shown in turkeys (Krakauer, 2005). Bowerbirds differ from these species in that males display individually on widely separated display sites, which rules out direct cooperation in display. We found that satin bowerbird kin associated disproportionately as first or second nearest neighbors. We then tested the hypothesis that the presence of related neighbors mitigated the negative consequences of bower destruction. We found that males direct fewer bower destructions toward kin than equidistant nonkin. Males with more relatives nearby also receive fewer bower destructions (Reynolds *et al.*, 2009). These results suggest that relatedness has a restraining effect on aggression and favors the close spatial association of related males. Male bower holders may repeatedly destroy the bowers of other less aggressive males and this behavior could affect the willingness of these victims to set up bowers in the neighborhood. The severe reduction in aggression in bower destruction seen by closely related males may facilitate close kin setting up bowers near their relatives. Male satin bowerbird cognitive ability may assist them in choosing which of their nearby neighbors' bowers to destroy and which others, owned by their relatives, not to destroy.

VOCAL MIMICRY: LEARNING AND INNOVATION IN USE OF CO-OPTED DISPLAYS

In satin bowerbirds, male courtship calls have two major components, first the aggressive "mechanical" portion in which males produce a whirring sound, followed by a trill accompanied by a rapid movement across the front of the bower while flashing their wings, and second, the "mimicry" portion of the call. Males mimic up to five species at our Wallaby Creek study site: Lewin's honeyeater (*Meliphaga lewinii*), laughing kookaburra (*Dacelo novaeguineae*), Australian raven (*Corvus coronoides*), sulphur-crested cockatoo (*Cacatua galerita*), and yellow-tailed black cockatoo (*Calyptorhynchus funereus*). Males show age-related improvement in the quality and completeness of mimetic calls, and males practice these calls as juveniles and adults, suggesting that learning is required to acquire these calls. Male age is related to the amount of practice, with younger males showing more and longer practice courtship bouts than older males. We found that both the mimetic and mechanical portions of the males' calls contributed to male courtship success (Loffredo and Borgia, 1986; Coleman *et al.*, 2007). Patricelli *et al.* (2004) found that, in satin bowerbirds, the mechanical portion of the display caused a higher level of discomfort in females, often causing females to startle, than the mimetic portions of male display. Male ability to modulate mechanical parts of the display in response to female discomfort was critical for their mating success. Coleman (2005) found that younger males induced a greater number of startles in male–male practice courtship displays. These results suggest that learning is important for the successful presentation of the mechanical calls at an intensity level that females or young males are comfortable viewing.

We hypothesize that male satin bowerbirds originally acquired mimetic calls by listening to model species. Now, however, because the same set of calls is used by males at Wallaby Creek and is also used in conspecific same-sex courtships of juvenile males by older males, it seems most likely that the calls are learned in these same-sex courtships. The alternative possibility that mimetic calls are genetically programmed is unlikely for several reasons. First, as passerine birds, bowerbirds are well-equipped to learn songs (e.g., Marler and Peters, 1981). Second, evolving the genetic programming to produce exact replicas of calls, some of which are complex, from five different species, is unlikely. Third, the observation that mimetic calls improve with age and males spend a great deal of time practicing mimetic displays is more consistent with a learning hypothesis (extensive observation of other males might not be necessary if the calls were inherited primarily by genetic programming). Fourth, several other bowerbird species mimic unique situation-specific sounds that are clearly learned from the habitat, suggesting that vocal learning is a commonly used mechanism across bowerbirds.

Two steps are involved in the co-option model we apply to the evolution of mimetic displays in satin bowerbird mate choice. First, the male trait evolves because it provides an initial benefit to the male. We propose that males added mimicry to their courtship display because it calmed females after threatening mechanical calls, which are similar to aggressive calls. Studies of satin (e.g., Patricelli et al., 2002; Coleman et al., 2004) and spotted bowerbirds (Borgia and Presgraves, 1998) indicate that successful courtships involve a balance between intense display and threat-reducing elements. The alternation of intense mechanical elements and less intense mimicry appears to contribute to this balance. By interspersing melodic mimetic laughing kookaburra and Lewin's honeyeater calls between episodes of harsh mechanical calls, males may calm females and improve the likelihood that females will stay for additional courtship and copulation. Patricelli et al.'s (2004) observation that the broad band mechanical portion of the courtship song is more threatening and accounts for a large proportion of female startling behavior suggests that the mimicry portion is less threatening and could enhance female comfort with male display.

Once mimicry became incorporated as a regular aspect of male display, females appear to have secondarily co-opted these calls for use in mate assessment. Because the quality of male mimetic calls varies, females could use them to assess differences in male vocal ability and possibly differences in male genetic quality. Highly developed mimetic ability could indicate well-developed neural circuitry (Leitner and Catchpole, 2002; Nowicki et al., 2000, 2002; Spencer et al., 2005) that may indicate the overall genetic quality of males.

Mimicry is widespread in the bowerbird family and is commonly used by tooth-billed, spotted, and streaked bowerbirds, but they do not use it during courtship at bowers like satin bowerbirds. This suggests that mimicry was a preexisting capability in bowerbirds that was available for use in courtship by satin bowerbirds. Male spotted bowerbirds use relatively simple "Skraa" calls co-opted from their aggressive calls (Borgia and Coleman, 2000) as the predominant vocal display element during courtship. Male vocal mimicry of other species is more idiosyncratic and shows a great deal of individual variation in the models mimicked. For example, when we approach spotted bowerbird bowers, some males mimic cat meows, portions of kookaburra songs, and calls of goshawks and other species that seem to be mixed signals of threat and discomfort (see also Kelley and Healy, 2011). Tooth-billed bowerbird males produce long streams of calls from numerous other bird species and the sounds of car engines, creaking trees, cicadas, frogs, and so on (Chaffer, 1984; Iredale, 1950; Marshall, 1954; Frith and McGuire, 1996). Male streaked bowerbirds mimic complex acoustical scenes. In one case, a male started with the mimicked sounds of a group of people talking as they moved through the forest with their machetes cutting bushes and dogs barking, and continued with the sound of machetes being used to fell a tree, complete with the rattle of shaking leaves after each blow, and eventually the sound of the tree falling and hitting the ground with a crash. Male Vogelkop bowerbirds (*A. inornatus*) in the Arfak Mountains are most like satin bowerbirds in regularly using mimicry during close courtship. Males display from the covered dark interior of their hut-like bower and vocally mimic the sounds of wing beating during courtship while their wings are held still. Spotted bowerbirds at Thallon, Queensland (see Borgia, 1995a) make a gurgling call during courtship that sounds like water in a stream running over stones that may or may not be mimicry. Male satin bowerbirds mimic ravens perhaps as a way of indicating a greater threat from a larger bird when attacked by groups of the cooperative breeding Australian bell miners (*Manorina melanophrys*). Bowerbirds clearly have the ability to mimic a large variety of sounds. While satin bowerbirds represent the only species that uses mimicry extensively in courtship, the numerous different contexts in which male bowerbirds use mimicry and the great diversity of sounds mimicked is consistent with the hypothesis that bowerbirds use cognitive processes to both learn the mimicry and then apply it in ways useful to them.

CO-OPTION MECHANISM

Co-option of preexisting male traits used in sexual display is widespread (Borgia, 2006). Some well-known examples include nests of weaverbirds (Lawes *et al.*, 2002) and sticklebacks (Barber *et al.*, 2001) that originally functioned in offspring rearing and are now also used in mate choice. The co-option process may start with an already existing male trait, for example, a neatly woven nest, that performs some nondisplay function and that can be assessed by females. If the male trait, for example, a well built nest, happens to be an effective indicator of male quality, then a female preference for that trait may evolve, resulting in co-option of the male trait and its use in male display. Co-option is also evident in several key aspects of bowerbird display.

Co-option is a critical and commonly unrecognized element in the acquisition of male display in a wide variety of species (Borgia, 2006). Initially it was proposed that aggressive displays were often borrowed for use in sexual display (Fisher, 1930; Borgia, 1979; Berglund and Rosenqvist, 2001; Berglund *et al.*, 1996; Borgia and Coleman, 2000). It was suggested that male aggressive signals are effective indicators of genome-wide good genes for fighting ability, and that females can use these same signals in mate choice to identify high-quality sires. Borgia and Coleman (2000) showed that the "Skraa" call that is used widely in bowerbirds as an aggressive call is also used as a courtship call in the bowerbird genus *Chlamydera*. This more limited distribution and the similarity of the call in the two contexts indicate that this call was co-opted for use in courtship from its previous use as an aggressive display. More recently, it has been recognized that many other preexisting male traits not associated with aggression can signal important information about males during courtship (Borgia, 2006).

The co-option of preexisting male traits resolves several difficult problems for sexual selection theory. For example, it explains how costly display traits might evolve. It is often argued that display traits must be costly to effectively indicate male quality (e.g., Zahavi, 1977; Grafen, 1990), but if there is a high cost to the trait, that reduces the net benefit of the trait (Borgia, 1979; Maynard Smith, 1978). However, if an expensive physical trait is already present, having evolved because of some other function, and it is then co-opted for a new function in sexual display, several problems related to trait cost-based honesty can be avoided. First, costs associated with the initial function of the trait can be effective in guaranteeing honesty related to the new trait function as a male quality indicator. Low quality males cannot pay for expensive traits, whether they evolved for sexual signaling or for a different original function. However, if this cost is already paid for, then the trait can be co-opted to serve as a mate choice signal that honestly distinguishes low- and high-quality males free of new added cost. This makes a co-opted signal more likely to evolve than a co-evolved trait and preference because of the greater net benefit (lower cost) to males and females of the co-opted trait. Second, the evolution of cost-based signals is complicated by the problem of low honesty for small inexpensive incipient signals in the early stages of display trait evolution. These signals are not likely to be effective in separating low- and high-quality males based on cost and as such signal evolution should be impeded. Co-option can bypass this problem by starting with an already enlarged and costly trait that allows only high-quality males to develop the displays preferred by females. While the reliability of all display traits may not always be cost-dependent (Borgia, 1993, 2006), co-option of existing traits may provide the best explanation for how expensive traits that use cost to ensure honesty can evolve.

Co-option models of sexual selection solve some additional issues in other common sexual selection models. First, they do not require difficult-to-initiate-and-maintain genetic correlations (Bakker and Pomiankowski, 1995) between male traits and female preferences that are required in many commonly considered models (e.g., Lande, 1981; Pomiankowski, 1988; Kokko *et al.*, 2002). If there is low genetic variation in either the male trait or female preference or females cannot effectively choose males with the appropriate trait, existing genetic correlations can be readily lost and may not be easily regenerated (Butlin, 1993). Co-option bypasses this problem because the male trait is already present and the female preference can evolve if females gain a benefit from choosing the male trait. Also, if the display trait and/or the female preference is learned, it is not clear that genetic correlations of the type needed to maintain the dynamic balance specified in these models can occur under all conditions. Second, in initial stages of co-evolutionary good genes models, it is necessary that both female preferences and male traits become simultaneously functional and operate with sufficient efficiency that there is net benefit for the choosing female. This is a much more difficult requirement than for a co-option model in which the fully developed male trait is already established and the female preference has time to evolve in the presence of the male trait. This less-complicated set of conditions with co-opted male traits should allow mate choice to more readily evolve than by co-evolutionary processes.

The multi-step process for co-option of existing male traits by females for mate choice may have occurred several times in bowerbirds. In the first step, a trait becomes established in males because it benefits males, but is not a product of selection by a direct female preference. For example, we hypothesize that in the evolution of bower decoration, males initially collected colorful objects to make their bowers more recognizable to mate-searching females as described by Parker's (1983) passive attraction hypothesis. Males who chose bright, distinctive, and rare colors of decorations for these displays produced a signal that was easier for females to recognize as a male display site in the clutter of the forest floor. Next, as males vied for limited distinctive decorations, male ability to collect and display these decorations, including stealing and defending them from one another (Borgia and Gore, 1986), coincidentally provided an indicator of differences in male physical and cognitive ability. Then, females that happened to evolve an active preference for males with more decorations could obtain high-quality mates, causing the preference to spread. Similarly, bowers originally evolved to reduce threat of forced copulation to females (see Borgia, 1995b). Male bowers differed in the quality of construction, reflecting differences in male motor and neurological quality. Females who attended to these differences in mate assessment effectively co-opted the bower for a secondary function in mate choice. This co-option process explains the dual function of traits like bowers and decorations and provides a relatively simple stepwise process by which complex signal systems can evolve without the need for genetic correlations difficult to develop and maintain. Other traits for which there may have been co-option in bowerbirds are mimetic calls used initially as a calming element in male display and then, as females recognized differences in male call quality, as a basis for female choice of mates (see section "Vocal Mimicry: Learning and Innovation in Use of Co-opted Displays"). Thus, several major components of satin bowerbird display appear to be co-opted from other preexisting traits that initially did not evolve through mate choice.

COGNITION IN DISPLAY TRAIT ACQUISITION

The co-option of complex bowerbird displays often appears to be dependent on cognitive ability. For females to discriminate among males based on decoration numbers, they have to recognize at least qualitative and perhaps quantitative differences in decoration numbers (Borgia, 1985a; Coleman et al., 2004), remember those differences as they move to separated bowers (Uy et al., 2000, 2001a, 2001b) and then select the most decorated among those visited (Patricelli et al., 2003, 2004; Borgia and Mueller, 1992; Uy et al., 2000; Madden, 2002). For this to occur, that females must have had the necessary cognitive abilities before their complex pattern of decoration choice evolved. Male courtship adjustments in response to female discomfort in courtship (Patricelli et al., 2002) suggest that males are able to appropriately perceive and respond to female behavior in the bower. Males may relate female behaviors to their own experiences resulting from courtship directed at them by other males (Collis and Borgia, 1993) and use this experience to appropriately modify their own display behavior.

Male bowerbirds appear to anticipate female reactions to novel display traits they produce. In several situations described already, males in the absence of females adjust their displays to match those preferred by females. This ability may have been important in the past integration of novel elements into male display. For example, male satin bowerbirds may have tried out different mimicked calls and learned to use those that calmed females after inserting them into their courtship display. But also intriguing is the possibility that when male satin bowerbirds initially chose calls to mimic during courtship, they found certain calls of other species calming to them and they anticipated that they would have a similar effect on females, and thus incorporated them into their displays. Male ability to anticipate female reactions based on shared reactions and experiences could be a potentially important mechanism for streamlining the evolution of complex displays. This could make the co-option process more effective in producing advantageous new displays.

In bowerbirds, a group with the most intensively studied pattern of sexual display, cognition appears to play a key role in the operation, acquisition, and assessment of these displays. Given that cognition is rarely mentioned in conjunction with sexual selection in most species and in NRB species in particular, there is a large discrepancy between what we have found and some of the most commonly cited models for sexual selection. This work suggests that a new set of models based on cognitively directed co-option that more directly considers cognitive processes in display trait evolution need to be developed. We believe that these models will more accurately describe how new behavioral signals are acquired and preferences evolve. This view sees females as sophisticated decision makers able to make complex fitness-enhancing mating decisions, and males as using their cognitive processes to creatively construct highly effective new displays.

REFERENCES

Alexander, R.D. (1981) *Darwinism and Human Affairs*, University of Washington Press, Seattle.

Airey, D.C., Castillo-Juarez, H., Casella, G., Pollak, E.J., and DeVoogd, T.J. (2000a) Variation in the volume of zebra finch song control nuclei is heritable: developmental and evolutionary implications. *Proceedings of the Royal Society of London, Series B: Biological Sciences*, **267**, 2099–2104.

Airey, L.C., Buchanan, K.L., Szekely, T., Catchpole, C.K., and Devoogd, T.J. (2000b) Song, sexual selection, and a song control nucleus (HVc) in the brains of European sedge warblers. *Journal of Neurobiology*, **44**, 1–6.

Anderson, B.L. (2012) Bird-brained illusionists. *Science*, **335**, 292–293.

Andersson, M. (1994) *Sexual Selection*, Princeton University Press, Princeton, New Jersey.

Andersson, M. and Simmons, L.W. (2006) Sexual selection and mate choice. *Trends in Ecology and Evolution*, **21**, 296–302.

Barber, I., Nairn, D., and Huntingford, F. (2001) Nests are ornaments: revealing construction by male sticklebacks. *Behavioral Ecology*, **12**, 390–396.

Bakker, T.C.M. and Pomiankowski, A. (1995) The genetic basis of female mate preferences. *Journal of Evolutionary Biology*, **8**, 129–171.

Bennett, A.T.D. (1993) Spatial memory in a food storing corvid. I. Near tall landmarks are primarily used. *Journal of Comparative Physiology A*, **173**, 193–207.

Berglund, A. and Rosenqvist, G. (2001) Male pipefish prefer dominant over attractive females. *Behavioral Ecology*, **12**, 402–406.

Berglund, A., Bisazza, A., and Pilastro, A. (1996) Armaments and ornaments: an evolutionary explanation of traits of dual utility. *Biological Journal of the Linnean Society of London*, **58**, 385–399.

Boogert, N.J., Giraldeau, L.A., and Lefebvre, L. (2008) Song complexity correlates with learning ability in zebra finch males. *Animal Behaviour*, **76**, 1735–1741.

Boogert, N.J., Fawcett, T.W., and Lefebvre, L. (2011) Mate choice for cognitive traits: a review of the evidence in nonhuman vertebrates. *Behavioral Ecology*, **22**, 447–459.

Borgia, G. (1979) Sexual selection and the evolution of mating systems, in *Sexual Selection and Reproductive Competition* (eds M.S. Blum and N.A. Blum), Academic Press, New York, pp. 19–80.

Borgia, G. (1985a) Bower quality, number of decorations and mating success of male satin bowerbirds (*Ptilonorhynchus violaceus*): an experimental analysis. *Animal Behaviour*, **33**, 266–271.

Borgia, G. (1985b) Bower destruction and sexual competition in the satin bowerbird (*Ptilonorhynchus violaceus*). *Behavioral Ecology and Sociobiology*, **18**, 91–100.

Borgia, G. (1993) The cost of display in the non-resource-based mating system of the satin bowerbird. *American Naturalist*, **141**, 729–743.

Borgia, G. (1995a) Complex male display and female choice in the spotted bowerbird: specialized functions for different bower decorations. *Animal Behaviour*, **49**, 1291–1301.

Borgia, G. (1995b) Why do bowerbirds build bowers? *American Scientist*, **83**, 542–547.

Borgia, G. (2006) Preexisting male traits are important in the evolution of elaborated male display. *Advances in the Study of Behavior*, **36**, 249–302.

Borgia, G. and Coleman, S.W. (2000) Co-option of male courtship signals from aggressive display in bowerbirds. *Proceedings of the Royal Society of London, Series B: Biological Sciences*, **267**, 1735–1740.

Borgia, G. and Gore, M.A. (1986) Feather stealing in the satin bowerbird (*Ptilonorhynchus violaceus*): male competition and the quality of display. *Animal Behaviour*, **34**, 727–738.

Borgia, G. and Keagy, J. (2006) An inverse relationship between decoration and food colour preferences in satin bowerbirds does not support the sensory drive hypothesis. *Animal Behaviour*, **72**, 1125–1133.

Borgia, G., Coyle, B., and Keagy, J. (2012) Comment on "Illusions Promote Mating Success in Great Bowerbirds". *Science*, **337** (6092), 292.

Borgia, G. and Mueller, U. (1992) Bower destruction, decoration stealing, and female choice in the spotted bowerbird (*Chlamydera maculata*). *Emu*, **92**, 11–18.

Borgia, G. and Presgraves, D.C. (1998) Coevolution of elaborated male display traits in the spotted bowerbird: an experimental test of the threat reduction hypothesis. *Animal Behaviour*, **56**, 1121–1128.

Borgia, G., Kaatz, I., and Condit, R. (1987) Flower choice and the decoration of the bower of the satin bowerbird (*Ptilonorhynchus violaceus*): a test of hypotheses for the evolution of display. *Animal Behaviour*, **35**, 129–139.

Bradbury, J.W. (1981) The evolution of leks, in *Natural Selection and Social Behavior* (eds R.D. Alexander and D.W. Tinkle), Chiron Press, New York, pp. 138–169.

Bravery, B.D., Nicholls, J.A., and Goldizen, A.W. (2006) Patterns of painting in satin bowerbirds *Ptilonorhynchus violaceus* and males' responses to changes in their paint. *Journal of Avian Biology*, **37**, 77–83.

Buchanan, K.L., Catchpole, C.K., Lewis, J.W., and Lodge, A. (1999) Song as an indicator of parasitism in the sedge warbler. *Animal Behaviour*, **57**, 307–314.

Buchanan, K.L., Spencer, K.A., Goldsmith, A.R., and Catchpole, C.K. (2003) Song as an honest signal of past developmental stress in the European starling (*Sturnus vulgaris*). *Proceedings of the Royal Society of London, Series B: Biological Sciences*, **270**, 1149–1156.

Bugnyar, T. and Kotrschal, K. (2002) Observational learning and the raiding of food caches in ravens, *Corvus corax*: is it 'tactical' deception? *Animal Behaviour*, **64**, 185–195.

Butlin, R.K. (1993) A comment on the evidence for a genetic correlation between the sexes in *Drosophila melanogaster*. *Animal Behaviour*, **45**, 403–404.

Canady, R.A., Kroodsma, D.E., and Nottebohm, F. (1984) Population differences in complexity of a learned skill are correlated with the brain space involved. *Proceedings of the National Academy of Sciences of the United States of America*, **81**, 6232–6234.

Cendes, L. (2009) Active female sampling of male display predicts female uncertainty in mate choice. Master's thesis. Department of Biology, University of Maryland, College Park, MD USA.

Chaffer, N. (1984) *In Quest of Bowerbirds*, Rigby, Melbourne.

Coleman, S.W., Patricelli, G.L., and Borgia, G. (2004) Variable female preferences drive complex male displays. *Nature*, **428**, 742–745.

Coleman, S.W. (2005) Variable female preferences and the evolution of complex male displays in the satin bowerbird (*Ptilonorhynchus violaceus*). Doctor of Philosophy Dissertation. Behavior, Ecology, Evolution, and Systematics Program, University of Maryland, College Park, MD USA.

Coleman, S.W., Patricelli, G.L., Coyle, B., Siani, J., and Borgia, G. (2007) A new role for vocal mimicry: the most attractive males are the best mimics. *Biology Letters*, **3**, 463–466.

Collis, K. and Borgia, G. (1993) The costs of male display and delayed plumage maturation in the satin bowerbird (*Ptilonorhynchus violaceus*). *Ethology*, **94**, 59–71.

Cronin, H. (1991) *The Ant and the Peacock: Altruism and Sexual Selection from Darwin to Today*, Cambridge University Press, Cambridge.

Cooper, W.T. and Forshaw, J.M. (1979) *The Birds of Paradise and Bowerbirds*, David R. Godine, Boston.

Darwin, C. (1871) *The Descent of Man and Selection in Relation to Sex*, D. Appleton Company, New York.

Darwin, C. (1872) *The Expression of the Emotions in Man and Animals*, D. Appleton Company, New York.

Dawkins, M.S. and Guilford, T. (1996) Sensory bias and the adaptiveness of female choice. *American Naturalist*, **148**, 937–942.

Day, L.B., Westcott, D.A., and Olster, D.H. (2005) Evolution of bower complexity and cerebellum size in bowerbirds. *Brain, Behavior and Evolution*, **66**, 62–72.

Doerr, N.R. (2010) Decoration supplementation and male–male competition in the great bowerbird (*Ptilonorhynchus nuchalis*): a test of the social control hypothesis. *Behavioral Ecology and Sociobiology*, **64**, 1887–1896.

Donaghey, R. (1996) Satin bowerbird, Regent Bowerbird, in *Finches, Bowerbirds and Other Passerines of Australia* (ed R. Strahan), Harper Collins Publishers, Sydney.

Emery, N.J. (2006) Cognitive ornithology: the evolution of avian intelligence. *Philosophical Transactions of the Royal Society of London, Series B*, **361**, 23–43.

Emery, N.J. and Clayton, N.S. (2004) The mentality of crows: convergent evolution of intelligence in corvids and apes. *Science*, **306**, 1903–1907.

Endler, J.A., Endler, L.C., and Doerr, N.R. (2010) Great Bowerbirds create theaters with forced perspective when seen by their audience. *Current Biology*, **20**, 1679–1684.

Fisher, R.A. (1915) The evolution of sexual preference. *Eugenics Review*, **7**, 184–192.

Fisher, R.A. (1930) *The Genetical Theory of Natural Selection*, Clarendon Press, Oxford.

Frith, C.B. and Mcguire, M. (1996) Visual evidence of vocal avian mimicry by male Tooth-Billed Bowerbirds *Scenopoeetes dentirostris* (Ptilonorhynchidae). *Emu*, **1**, 12–16.

Frith, C.B. and Frith, D.W. (2004) *The Bowerbirds: Ptilonorhynchidae*, Oxford University Press, Oxford.

Garamszegi, L.Z., Eens, M., Erritzoe, J., and Møller, A.P. (2005a) Sexually size dimorphic brains and song complexity in passerine birds. *Behavioral Ecology*, **16**, 335–345.

Garamszegi, L.Z., Eens, M., Erritzoe, J., and Møller, A.P. (2005b) Sperm competition and sexually size dimorphic brains in birds. *Proceedings of the Royal Society of London, Series B: Biological Sciences*, **272**, 159–166.

Gillard, E.T. (1969) The Birds of Paradise and Bowerbirds, Weidenfield and Nicholson, London.

Gomez, D. and Théry, M. (2004) Influence of ambient light on the evolution of colour signals: comparative analysis of a Neotropical rainforest bird community. *Ecology Letters*, **7**, 279–284.

Grafen, A. (1990) Biological signals as handicaps. *Journal of Theoretical Biology*, **144**, 517–546.

Hamilton, W.D. and Zuk, M. (1982) Heritable true fitness and bright birds: a role for parasites. *Science*, **218**, 384–387.

Hansell, M. (2009) *Built by Animals: The Natural History of Animal Architecture*, Oxford University Press, Oxford.

Healy, S.D. and Rowe, C. (2007) A critique of comparative studies of brain size. *Proceedings of the Royal Society of London, Series B: Biological Sciences*, **274**, 453–464.

Healy, S.D. and Hurly, T.A. (1995) Spatial memory in rufous hummingbirds (*Selasphorus rufus*): a field test. *Animal Learning and Behavior*, **23**, 63–68.

Hebets, E.A. (2003) Subadult experience influences adult mate choice in an arthropod: exposed female wolf spiders prefer males of a familiar phenotype. *Proceedings of the National Academy of Sciences of the United States of America*, **100**, 13390–13395.

Herrick, C.J. (1924) *Neurological Foundations of Animal Behaviour*, Hafner, New York.

Hicks, R.E., Larned, A., and Borgia, G. (2013) Bower paint removal leads to reduced female visits, suggesting bower paint functions as a chemical signal. *Animal Behaviour*, **85**, 1209–1215.

Höglund, J. and Alatalo, R.V. (1995) *Leks*, Princeton University Press, Princeton, New Jersey.

Holland, B. and Rice, W.R. (1998) Chase-away sexual selection: antagonistic seduction versus resistance. *Evolution*, **52**, 1–7.

Hunter, C.P. and Dwyer, P.D. (1997) The value of objects to satin bowerbirds *Ptilonorhynchus violaceus*. *Emu*, **97**, 200–206.

Iredale, T. (1950) *Birds of Paradise and Bowerbirds*, Georgian House, Melbourne.

Isler, K. and van Schaik, C. (2006) Costs of encephalization: the energy trade-off hypothesis tested on birds. *Journal of Human Evolution*, **51**, 228–243.

Isler, K. and van Schaik, C.P. (2008) Why are there so few smart mammals (but so many smart birds)? *Biology Letters*, **5**, 125–129.

Iwaniuk, A.N. and Hurd, P.L. (2005) The evolution of cerebrotypes in birds. *Brain, Behavior and Evolution*, **65**, 215–230.

Jarvis, E.D., Güntürkün, O., Bruce, L., Csillag, A., Karten, H., Kuenzel, W., Medina, L., Paxinos, G., Perkel, D.J., Shimizu, T., Striedter, G., Wild, J.M., Ball, G.F., Dugas-Ford, J., Durand, S.E., Hough, G.E., Husband, S., Kubikova, L., Lee, D.W., Mello, C.V., Powers, A., Siang, C., Smulders, T.V., Wada, K., White, S.A., Yamamoto, K., Yu, J., Reiner, A., and Butler, A.B. (2005) Opinion: Avian brains and a new understanding of vertebrate brain evolution. *Nature Reviews Neuroscience*, **6**, 151–159.

Johnstone, R. and Earn, D.J. (1999) Imperfect female choice and male mating skew on leks of different sizes. *Behavioral Ecology and Sociobiology*, **45**, 277–281.

Jones, T.B. and Kamil, A.C. (1973) Tool-making and tool-using in the northern blue jay. *Science*, **180**, 1076–1078.

Kamil, A. and Balda, R. (1985) Cache recovery and spatial memory in Clark's nutcrackers (*Nucifraga columbiana*). *Journal of Experimental Psychology and Animal Behavioral Processes*, **11**, 95–111.

Keagy, J., Savard, J.-F., and Borgia, G. (2009) Male satin bowerbird problem-solving ability predicts mating success. *Animal Behaviour*, **78**, 809–817.

Keagy, J., Savard, J.-F., and Borgia, G. (2011) Complex relationship between multiple measures of cognitive ability and male mating success in satin bowerbirds, *Ptilonorhynchus violaceus*. *Animal Behaviour*, **81**, 1063–1070.

Keagy, J., Savard, J.-F., and Borgia, G. (2012) Cognitive ability and the evolution of multiple behavioral display traits. *Behavioral Ecology*, **23**, 448–456.

Kelley, L.A. and Healy, S.D. (2011) The mimetic repertoire of the spotted bowerbird *Ptilonorhynchus maculates*. *Naturwissenschaften*, **98**, 501–507.

Kelley, L. and Endler, J. (2012) Illusions promote mating success in great bowerbirds. *Science*, **335**, 335–338.

King, A.P., Freeberg, T.M., and West, M.J. (1996) Social experience affects the process and outcome of vocal ontogeny in two populations of cowbirds (*Molothrus ater*). *Journal of Comparative Psychology*, **110**, 276–285.

Kirkpatrick, M. (1987) Sexual selection by female choice in polygynous animals. *Annual Review of Ecology and Systematics*, **18**, 43–70.

Kodric-Brown, A. and Nicoletto, P.F. (2001) Age and experience affect female choice in the guppy (*Poecilia reticulata*). *American Naturalist*, **157**, 316–323.

Kokko, H., Brooks, R., McNamara, J.M., and Houston, A.I. (2002) The sexual selection continuum. *Proceedings of the Royal Society of London, Series B: Biological Sciences*, **269**, 1331–1340.

Koksal, F. and Domjan, M. (1998) Observational conditioning of sexual behavior in the domesticated quail. *Animal Learning and Behavior*, **26**, 427–432.

Krakauer, A.H. (2005) Kin selection and cooperative courtship in wild turkeys. *Nature*, **434**, 69–72.

Kusmierski, R., Borgia, G., Uy, A., and Crozier, R. (1997) Molecular information on bowerbird phylogeny and the evolution of exaggerated male characters. *Proceedings of the Royal Society of London, Series B: Biological Sciences*, **264**, 307–313.

Lande, R. (1981) Models of speciation by sexual selection on polygenic traits. *Proceedings of the National Academy of Sciences of the United States of America*, **78**, 3721–3725.

Lawes, M., Slotow, R., and Andersson, S. (2002) Male nest building but not display behaviour directly influences mating success in the polygynous Red Bishop, *Euplectes orix*. *Ostrich*, **73**, 87–91.

Leitner, S. and Catchpole, C.K. (2002) Female canaries that respond and discriminate more between male songs of different quality have a larger song control nucleus (HVC) in the brain. *Journal of Neurobiology*, **52**, 294–301.

Lefebvre, L., Reader, S.M., and Sol, D. (2004) Brains, innovations and evolution in birds and primates. *Brain, Behavior and Evolution*, **63**, 233–246.

Loffredo, C.A. and Borgia, G. (1986) Male courtship vocalizations as cues for mate choice in the satin bowerbird (*Ptilonorhynchus violaceus*). *Auk*, **103**, 189–195.

Loiselle, B.A., Ryder, T.B., Duraes, R., Tori, W., Blake, J.G., and Parker, P.G. (2006) Kin selection does not explain male aggregation at leks of 4 manakin species. *Behavioral Ecology*, **18**, 287–291.

Madden, J.R. (2001) Sex, bowers and brains. *Proceedings of the Royal Society of London, Series B: Biological Sciences*, **268**, 833–838.

Madden, J. (2002) Bower decorations attract females but provoke other male spotted bowerbirds: bower owners resolve this trade-off. *Proceedings of the Royal Society of London, Series B: Biological Sciences*, **269**, 1347–1351.

Madden, J.R. (2003) Bower decorations are good predictors of mating success in the spotted bowerbird. *Behavioral Ecology and Sociobiology*, **53**, 269–277.

Madden, J.R. (2006) Interpopulation differences exhibited by Spotted Bowerbirds *Chlamydera maculata* across a suite of male traits and female preferences. *Ibis*, **148**, 425–435.

Madden, J.R. (2008) Do bowerbirds exhibit culture? *Animal Cognition*, **11**, 1–12.

Marler, P. and Peters, S. (1981) Sparrows learn adult song and more from memory. *Science*, **213**, 780–782.

Maynard Smith, J. (1978) *The Evolution of Sex*, Cambridge University Press, Cambridge.

McDonald, D.B. and Potts, W.K. (1994) Cooperative display and relatedness among males in a lek-mating bird. *Science*, **266**, 1030–1032.

Marshall, A.J. (1954) *Bower-Birds: Their Displays and Breeding Cycles*, Clarendon Press, Oxford.

Mikami, O.K., Katsuno, Y., Yamashita, D.M., Noske, R., and Eguch, K. (2010) Bowers of the Great Bowerbird (*Chlamydera nuchalis*) remained unburned after fire: is this an adaptation to fire? *Journal of Ethology*, **28**, 15–20.

Miller, G. (2001) *The Mating Mind: How Sexual Choice Shaped the Evolution of Human Nature*, Anchor Books, New York.

Morris, M.R., Nicoletto, P.F., and Hesselman, E. (2003) A polymorphism in female preference for a polymorphic male trait in the swordtail fish *Xiphophorus cortezi*. *Animal Behaviour*, **65**, 45–52.

Nottebohm, F., Kasparian, S., and Pandazis, C. (1981) Brain space for a learned task. *Brain Research*, **213**, 99–109.

Nowicki, S., Hasselquist, D., Bensch, S., and Peters, S. (2000) Nestling growth and song repertoire sire in great reed warblers: evidence for song learning as an indicator mechanism in mate choice. *Proceedings of the Royal Society of London, Series B: Biological Sciences*, **267**, 2419–2424.

Nowicki, S., Searcy, W.A., and Peters, S. (2002) Brain development, song learning and mate choice in birds: a review and experimental test of the "nutritional stress hypothesis". *Journal of Comparative Physiology A*, **188**, 1003–1014.

Parker, G.A. (1983) Mate quality and mating decisions, in *Mate Choice* (ed P. Bateson), Cambridge University Press, Cambridge, pp. 141–166.

Patricelli, G.L., Uy, J.A.C., Walsh, G., and Borgia, G. (2002) Male displays adjusted to female's response. *Nature*, **415**, 279–280.

Patricelli, G.L., Uy, J.A.C., and Borgia, G. (2003) Multiple male traits interact: attractive bower decorations facilitate attractive behavioural displays in satin bowerbirds. *Proceedings of the Royal Society of London, Series B: Biological Sciences*, **270**, 2389–2395.

Patricelli, G.L., Uy, J.A.C., and Borgia, G. (2004) Female signals enhance the efficiency of mate assessment in satin bower-birds (*Ptilonorhynchus violaceus*). *Behavioral Ecology*, **15**, 297–304.

Patricelli, G.L., Coleman, S.W., and Borgia, G. (2006) Male satin bowerbirds, *Ptilonorhynchus violaceus*, adjust their display intensity in response to female startling: an experiment with robotic females. *Animal Behaviour*, **71**, 49–59.

Pepperberg, I.M. (1999) *The Alex Studies: Cognitive and Communicative Abilities of Grey Parrots*, Harvard University Press, Cambridge, MA.

Pitnick, S., Jones, K.E., and Wilkinson, G.S. (2006) Mating system and brain size in bats. *Proceedings of the Royal Society of London, Series B: Biological Sciences*, **273**, 719–724.

Pomiankowski, A.N. (1988) The evolution of mate preferences for male genetic quality. *Oxford Surveys in Evolutionary Biology*, **5**, 136–184.

Pravosudov, V.V. and Clayton, N.S. (2002) A test of the adaptive specialization hypothesis: population differences in caching, memory, and the hippocampus in black-capped chickadees (*Poecile atricapilla*). *Behavioral Neuroscience*, **116**, 515–522.

Prum, R.O. (2010) The Lande–Kirpatrick mechanism is the null model of evolution by intersexual selection: implications for meaning, honesty, and design in intersexual signals. *Evolution*, **64**, 3085–3100. doi: 10.1111/j.1558-5646

Reynolds, S.M., Christman, M.C., Uy, J.A.C., Patricelli, G.L., Braun, M.J., and Borgia, G. (2009) Lekking satin bowerbird males aggregate with relatives to mitigate aggression. *Behavioral Ecology*, **20**, 410–415.

Rodd, F.H., Hughes, K.A., Grether, G., and Baril, C. (2002) A possible non-sexual origin of a mate preference: are male guppies mimicking fruit? *Proceedings of the Royal Society of London, Series B: Biological Sciences*, **269**, 475–481.

Roth, G. and Dicke, U. (2005) Evolution of the brain and intelligence. *Trends in Cognitive Sciences*, **9**, 250–257.

Ryan, M.J. and Rand, A.S. (1990) The sensory basis of sexual selection for complex calls in the tungara frog, *Physalaemus pustulosus* (sexual selection for sensory exploitation). *Evolution*, **44**, 305–314.

Ryan, M.J. and Rand, A.S. (1993) Sexual selection and signal evolution: the ghost of biases past. *Philosophical Transactions of the Royal Society, Series B*, **340**, 187–195.

Ryan, M.J., Akre, K.L., and Kirkpatrick, M. (2009) Cognitive mate choice, in *Cognitive Ecology II* (eds R. Dukas and J.M. Ratcliffe), University of Chicago Press, Chicago, pp. 137–155.

Sasvari, L. and Hegyi, Z. (1998) How mixed-species foraging flocks develop in response to benefits from observational learning. *Animal Behaviour*, **55**, 1461–1469.

Spencer, K.A., Buchanan, K.L., Leitner, S., Goldsmith, A.R., and Catchpole, C.K. (2005) Parasites affect song complexity and neural development in a songbird. *Proceedings of the Royal Society of London, Series B: Biological Sciences*, **272**, 2037–2043.

Taylor, A.H., Hunt, G.R., Holzhaider, J.C., and Gray, R.D. (2007) Spontaneous metatool use by New Caledonian crows. *Current Biology*, **17**, 1504–1507.

Tomasello, M. (1999) Emulation learning and cultural learning. *Behavioural and Brain Sciences*, **21**, 703–704.

Uy, J.A.C. and Borgia, G. (2000) Sexual selection drives rapid divergence in bowerbird display traits. *Evolution*, **54**, 273–278.

Uy, J.A.C., Patricelli, G.L., and Borgia, G. (2000) Dynamic mate-searching tactic allows female satin bowerbirds *Ptilonorhynchus violaceus* to reduce searching. *Proceedings of the Royal Society of London, Series B: Biological Sciences*, **267**, 251–256.

Uy, J.A.C., Patricelli, G.L., and Borgia, G. (2001a) Complex mate searching in the satin bowerbird *Ptilonorhynchus violaceus*. *American Naturalist*, **158**, 530–542.

Uy, J.A.C., Patricelli, G.L., and Borgia, G. (2001b) Loss of preferred mates forces female satin bowerbirds (*Ptilonorhynchus violaceus*) to increase mate searching. *Proceedings of the Royal Society of London, Series B: Biological Sciences*, **268**, 633–638.

van Schaik, C.P., Deaner, R.O., and Merrill, M.Y. (1999) The conditions for tool use in primates: implications for the evolution of material culture. *Journal of Human Evolution*, **36**, 719–741.

Vellenga, R.E. (1980) Moults of the satin bowerbird *Ptilonorhynchus violaceus*. *Emu*, **80**, 49–54.

Werdenich, D. and Huber, L. (2006) A case of quick problem solving in birds: string pulling in keas, *Nestor notabilis*. *Animal Behaviour*, **71**, 855–863.

Widemo, F. and Owens, I.P.F. (1995) Lek size, male mating skew and the evolution of lekking. *Nature*, **373**, 148–151.

Wiley, R.H. (1991) Lekking in birds and mammals: behavioral and evolutionary issues. *Advances in the Study of Behavior*, **20**, 201–291.

Wojcieszek, J.M., Nicholls, J.A., and Goldizen, A.W. (2007) Stealing behavior and the maintenance of a visual display in the satin bowerbird. *Behavioral Ecology*, **18**, 689–695.

Zahavi, A. (1975) Mate selection – selection for a handicap. *Journal of Theoretical Biology*, **53**, 205–214.

Zahavi, A. (1977) Cost of honesty – (further remarks on handicap principle). *Journal of Theoretical Biology*, **67**, 603–605.

5

INTEGRATING FUNCTIONAL AND EVOLUTIONARY APPROACHES TO THE STUDY OF COLOR-BASED ANIMAL SIGNALS

Darrell J. Kemp[1] and Gregory F. Grether[2]

[1]*Department of Biological Sciences, Macquarie University, Sydney, NSW, Australia*
[2]*Department of Ecology and Evolutionary Biology, University of California, Los Angeles, CA, USA*

INTRODUCTION

Conspicuous displays of color – that is, displays that are very bright, highly chromatic, or otherwise represent high background contrasts – constitute a major and functionally diverse class of animal signals. These displays, including the brilliant kaleidoscopic patterning of animals such as tropical birds, butterflies, and reef fishes, have inspired great cultural adulation and offered many icons to mainstream human society (e.g., the use of butterflies as mascots for tropical tourism campaigns). The enormous diversity of animal color patterns has also fueled lively scientific research programs, particularly in the fields of behavioral ecology, genetics, and evolution. As convenient phenotypic markers, color-based traits were particularly utilized in early studies of genetics and in classic documentations of evolutionary responses to natural selection in the wild (Endler, 1983). Evolutionary theories of speciation, adaptation, and mimicry were honed through the study of color-based traits. In more recent decades, workers have turned to explaining the colors and the color patterns themselves, across a range of functional, mechanistic, and evolutionary perspectives. As brilliantly exemplified by the peacock's train – the hallmark symbol of Darwin's (1871) sexual selection theory – much of

Animal Signaling and Function: An Integrative Approach, First Edition.
Edited by Duncan J. Irschick, Mark Briffa, and Jeffrey Podos.
© 2015 John Wiley & Sons, Inc. Published 2015 by John Wiley & Sons, Inc.

this interest has focused upon the evolution of color-based traits as exaggerated sexual signals. Our goal in this chapter is to review the recent advances in the field of color signal evolution, drawing upon several key study systems, as well as to provide a historic perspective on the evolution of the field itself.

The Production of Color

Color is a perceptual phenomenon that arises due to the detection of different wavelengths of light by at least two (often more) different photoreceptors in the eye, coupled with the subsequent processing of this information in the brain. An object could be perceived as "colorful" or "chromatic" if it reflects or emits light unevenly across the visible wavelength spectrum. In the animal world, this typically results from the presence of pigments (which selectively absorb certain light wavelengths) and/or surface structures (which selectively reflect certain light wavelengths). Less commonly, color signals may be produced through the emission of select wavelengths via the biochemical processes of fluorescence or bioluminescence (Pearn et al., 2001; Rivers and Morin, 2008). It is important to note that colorfulness of animal surfaces and structures need not necessarily serve any adaptive visual signal role; vertebrate blood appears red due to the oxygen-carrying pigment hemoglobin, while many nocturnal scorpions fluoresce brightly under ultraviolet (UV) light due to compounds located in their cuticular tissues (e.g., Kloock, 2008). Neither of these biological colors is thought to function adaptively as any type of visual signal, nor are they thought to have evolved for this reason. However, in most cases where animals bear visually conspicuous markings on their exterior surfaces, specialized cell layers and organelles are involved in color production, and the ultimate cause relates to the signaling of information to various combinations of predators, prey, conspecifics, and various other potential viewers. Readers interested in the techniques, materials, and technicalities involving the biological study of coloration are directed to several excellent recent reviews in the area (e.g., Endler and Mielke, 2005; Stevens et al., 2007).

Color Traits as Signals

Despite the complexities of color signals and color production, the field of visual signal evolution has been largely preoccupied with understanding how a few select pigments, such as carotenoids, may contribute to the signaling of mate quality (Kemp et al., 2012). Because carotenoids cannot be synthesized de novo (Goodwin, 1984), and therefore must be directly acquired from the environment, the expression of carotenoid-based yellows, oranges, and reds in male sexual displays has been argued as an especially revealing indicator of phenotypic and/or genetic mate quality (see review by Olson and Owens, 1998). In some animals, these compounds are also required for salient metabolic processes, such as antioxidant and/or immune function, and allocational trade-offs are thought to further enhance the information content of carotenoid-based signals (Chapter 2). Any female that used such traits to select mates would therefore receive greater than average "benefits" (such as access to nutrients, high quality territories, "good" or "attractive" genes), and therefore natural selection would favor female choosiness, which would in turn favor ever-increasingly extreme versions of the color trait. As we discuss later, carotenoid-based signals have been thought to offer particularly "honest" or "cheat-proof" traits and therefore to serve as evolutionarily stable quality indicators. Although there have been detractors (e.g., Griffith *et al.*, 2006; see later), this simple yet attractive idea has underpinned much of the conceptual and empirical work in the field of color signal ecology and evolution.

Workers in the field have increasingly realized, however, that this restricted focus on carotenoid pigments (and false dichotomies relating to the predicted signal content of carotenoids vs some other pigment, such as melanin; Griffith *et al.*, 2006) is extremely limited (Kemp *et al.*, 2012). This viewpoint not only overlooks the role of surface structures and structural colors (Rutowski *et al.*, 2005; Shawkey and Hill, 2005), but also completely ignores the fact that color signals often result from the concerted effects of diverse suites of pigments (e.g., Watt, 1964), which themselves may be costly or indicative of quality for various reasons (Griffith *et al.*, 2006; Stoehr, 2006; Ducrest *et al.*, 2008). Both of these issues are amply illustrated by the conspicuous "orange" markings of male guppies (*Poecilia reticulata*; Figures 5.1

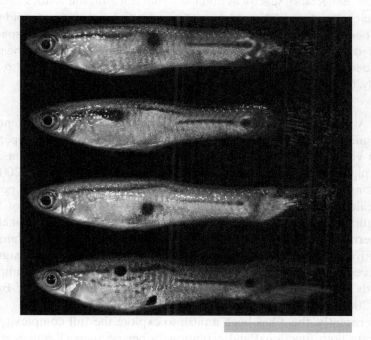

Figure 5.1. Four male guppies (*Poecilia reticulata*) from the Guanapo catchment of Trinidad, showing the complex mosaics of black, orange, blue, green, yellowish-bronze and silver markings that comprise the ornamental color pattern of this species. There is tremendous diversity among streams, populations, and individuals regarding the extent, placement, and spectral properties of each color element. The scale bar (lower right) indicates ~8 mm. (*See insert for color representation of this figure.*)

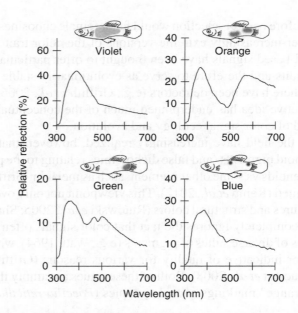

Figure 5.2. Reflectance spectra as obtained from scanning the various color spots of male guppies. The "orange" spots derive from the presence of carotenoid pigments, and have been treated as a classic pigment-based signal, but reflectance spectrometry has revealed a dramatic short-wave (UV) peak which likely results from the properties of the underlying iridophore layer. The violets, greens, and blues are also produced structurally. (Image reproduced with permission from Kemp *et al.* (2009).)

and 5.2), which, initially, served as the flagship example of carotenoid-based honest signaling (Endler, 1980). As we detail later, these spots appear orange to human viewers due not only to the presence of carotenoid pigments, but also pteridine pigments, or more specifically, drosopterins (Grether *et al.*, 2001, 2005). Furthermore, precise spectral measurements of these spots in several populations have revealed a reflectance peak in the UV (300–400 nm) region of the spectrum (Figure 5.2; Kemp *et al.*, 2008a; White *et al.*, 2003), which likely results from a dermal array of microstructures. Because guppies (unlike humans) are UV-sensitive, the presence of this short-wave peak would contribute significantly to the perception of the color of these markings, and may be just as important as carotenoids in contributing to their signal information content (Kodric-Brown and Johnson, 2002; although see: Smith *et al.*, 2002).

More recently, there has been a push to explore the full complexity of color signals from both functional and evolutionary perspectives (Kemp *et al.*, 2012). Along these lines, researchers have sought to define the full suites of pigments (e.g., Grether *et al.*, 2001; Hudon *et al.*, 2003) and precise structural color-producing devices (Prum *et al.*, 2006; Prum and Torres, 2003, 2004; Shawkey *et al.*, 2005; Vukusic and Sambles, 2003; Vukusic *et al.*, 2000, 2001) involved in the production of particular color traits. In recognition of the multicomponent nature of color signals (e.g., Grether *et al.*, 2004a), researchers have also sought to define the relative roles of pigments and structures in determining patterns of signal variation (Grether *et al.*, 2004a; Rutowski *et al.*, 2005; Shawkey and Hill, 2005). Behavioral and ecological studies have discovered hitherto unprecedented levels of variation, across a mechanistically and visually diverse breadth of color signals. This variation has been linked to salient selective contexts such as mate selection, male–male competition, and predator avoidance. Covariances between color parameters and various aspects of phenotypic quality have been sought (e.g., Kemp and Rutowski, 2007), and interpreted in light of mechanisms such as pleiotropy (e.g., Ducrest *et al.*, 2008). Evolutionary theoreticians have, in turn, drawn upon these knowledge advances to expand existing color signaling hypotheses and to generate new hypotheses for future evaluation (Fitzpatrick, 1998; Hausmann *et al.*, 2003).

In this chapter, we outline these advances, drawing upon exemplar cases across fishes, insects, and birds. Because this is an extremely large field, we adopt a necessarily non-comprehensive approach, focusing on the use of color in sexual communication. For more in-depth information and broader coverage of signaling contexts, interested readers are directed to excellent contemporary reviews of coloration in various animal groups (e.g., Birds and Mammals [Prum and Torres, 2003, 2004]; Insects [Thery and Gomez, 2010]; Butterflies [Prum *et al.*, 2006]; Beetles [Parker *et al.*, 1998; Luke *et al.*, 2010]; Spiders [Thery and Casas, 2009]). Further, due to the relative lack of emphasis historically given to structural colors, we place particular focus on these colors, their optical mechanisms of production, and their potential role in determining evolutionarily relevant properties of sexual displays. We begin with a consideration of the theoretical and mechanistic concepts of sexual signaling, prior to outlining three empirical study systems, which each demonstrate different aspects of the development of color signal-based research over the last several decades.

COLOR SIGNAL PRODUCTION IN MORE DETAIL

Most color in nature is generated by the presence of pigments or surface structures that, respectively, absorb or reflect select wavelengths of light. Perhaps as a matter of convenience, workers in the field routinely use the terms "pigment" and "structural" color, although these labels are rarely precisely defined. In reality, the production of pigment-based colors generally requires reflective structures because pigments primarily absorb light; something must be reflected from the surface for visible color. In the simplest such case, the underlying surface (which, paradoxically, may itself be constructed from pigment or materials containing pigment; Stavenga *et al.*, 2004; Morehouse *et al.*, 2007) may act as a broad-band reflector (much like a piece of white paper) and simply reflect white light which is subsequently "filtered" through one or several absorbing pigments. However, there are also cases (such as the "UV/oranges" of guppies and Coliadine butterflies – as described later) where dedicated color-producing structures either overlay or underlay layers of pigment, and the resultant color is a composite effect of one or several different pigments and/or structures (i.e., a truly multicomponent signal; Grether *et al.*, 2004a).

These issues have rarely (until recently) been explicitly considered by evolutionary and behavioral ecologists.

Functionally, structural layers produce color in a much less straightforward manner than pigments and the resultant visual signals are also generally more complex. Examples of color signals produced according to the presence of surface structures include the highly chromatic tail feathers of peacocks and the metallic wing colors of many tropical birds and butterflies. Such colors are often iridescent and polarized, and there is evidence that each of these properties can have unique signaling functions (see: Loyau et al., 2007; Mathger et al., 2009; Sweeney et al., 2003). Optically, the generation of structural colors ranges in complexity from the indiscriminate scattering of light from un-ordered surfaces (so-called "incoherent scattering") to the reflectance of highly specific wavelength bands due to processes such as constructive interference and diffraction (collectively, "coherent scattering"; Prum and Torres, 2003, 2004). In all cases, color is produced as a consequence of an optical interaction between incident light and surface architecture. The most elaborate structural arrays are found on butterfly wings, and recent research into this group has uncovered several surprising and hitherto undocumented optical mechanisms (Vukusic and Hooper, 2005; Vukusic et al., 2000, 2001). Although our understanding of the proximate optical workings of these structures has increased, our knowledge of their ultimate biological significance has lagged behind that of primarily pigment-based color signals, particularly those involving carotenoids.

SIGNALS, HONESTY, AND CONDITION-DEPENDENCE

The theory of honest signaling contends that in a signaler–receiver system only "cheat-proof" signals will be evolutionarily stable (Zahavi, 1975; Zahavi and Zahavi, 1997; Maynard Smith and Harper, 2003). Honesty is enforced through developmental, physiological, or social costs of generating or maintaining the signal (e.g., Rohwer, 1975), or if the signal trait is developmentally or ontogenetically difficult to produce (Maynard Smith and Harper, 2003). Key to this theory is the notion that not all signalers are able to display with the same intensity. Only individuals in good condition (Rowe and Houle, 1996) or of high genetic quality (Hunt et al., 2004) are able to achieve the most extreme expression of the signal trait(s). Receivers can therefore profitably use signal variation as a source of fitness-enhancing information, thus the signaling system will be favored by natural selection. The crucial contribution of honest signaling theory to our understanding of biological diversity is its ability to predict the nature of traits expected to evolve as components of signaler receiver systems.

Honest signaling theory has been called upon extensively in the study of mate choice and the evolution of male ornamentation. Honesty in this context means reliable information on aspects such as the quality of a potential mate's genes and/or the resources he (or she) could provide as a mating partner. The notion that signals should be costly to produce and/or maintain predicts that trait expression should be linked with individual condition, that is, that signal intensity or expression should be "condition-dependent" (Andersson, 1994). This notion has also forced an explicit consideration of how male ornamentation may accurately reflect inter-individual variation in genetic viability (i.e., "good genes"). Empirically, however, genetic quality-based hypotheses are often best addressed using some form of quantitative genetics. Because such designs are empirically demanding in many systems, most supporting evidence for signal information content therefore resides in a purely phenotypic domain, the exceptions being studies on flies (David *et al.*, 2000), guppies (Brooks, 2000; Brooks and Endler, 2001a), and butterflies (Kemp and Rutowski, 2007).

COLORATION AS AN HONEST ADVERTISEMENT

How might color-based displays act as honest signals of mate quality? One logically straightforward possibility, which is exemplified by the carotenoid argument, is if the biochemical building-blocks of the signal itself are difficult to obtain from the environment. This would see only high quality individuals have the ability to express a large, bright, and/or chromatic color signal. More broadly, however, any pigment-based color display could be costly (and therefore honest) if the pigment and/or its precursor compounds are difficult to obtain, retain, synthesize, or if they are also subject to demand for other metabolic processes (such as immune function; Lindsey and Altizer, 2009; Stoehr, 2006; Zuk, 1991). Honesty may be mediated at the genetic level via the pleiotropic effects of genes linking pigmentation with viability or quality traits (e.g., Ducrest *et al.*, 2008). In prey taxa, a conspicuous color trait will also increase the risk of predation (Endler, 1978), and may also subject an individual to increased social harassment (Martin and Forsman, 1999; Rohwer, 1975; Santos *et al.*, 2009), thereby increasing the costliness of signaling and enforcing signal honesty. As argued by Griffith *et al.* (2006), there are a multitude of potential physiological and behavioral routes by which particular color traits may exact viability costs on their bearer, and arguments of signal honesty need not hinge upon any existing well-known and/or logically simplistic paradigm (such as the carotenoid pigment argument).

In contrast to sexual displays based upon pigments, structural color-based displays have historically been treated as relatively invariant (Endler, 1980), and therefore, of little potential relevance as mate quality or viability indicators. Because these colors are generated as a consequence of a surface architecture, which may appear superficially as a structural re-arrangement of a pre-existing or non-colorful surface layer (Rutowski *et al.*, 2007; Shawkey *et al.*, 2006a, 2006b), it has also been tempting to conclude a priori that they should be easy and cheap to produce (and, hence, unlikely to honestly indicate mate quality). However, one consequence of the recent, more general approach to studying color traits has been the revelation that many structurally produced colors are highly phenotypically variable and condition-dependent (Kemp, 2006, 2008a; Kemp and Rutowski, 2007; Keyser and Hill, 1999; Lim and Li, 2007; Masello *et al.*, 2008; Siefferman and

Hill, 2005; Siitari *et al.*, 2007). Specifically, structurally cultured ornamentation reliably indicates age (Budden and Dickinson, 2009; Kemp, 2006; Lim and Li, 2007; Siitari *et al.*, 2007), resource acquisition (Kemp, 2008a; Kemp and Rutowski, 2007; Lim and Li, 2007), stability of the developmental environment (Kemp and Rutowski, 2007), parasite burden (Hill *et al.*, 2005; Mougeot *et al.*, 2005b), microbial load (Shawkey *et al.*, 2007), and testosterone levels (Siitari *et al.*, 2007), in many cases more reliably than coincident pigment-based colors. Variation in structural colors has also been strongly linked to fitness traits, such as mate attractiveness (Kemp, 2007, 2008b; Siefferman and Hill, 2005), parental effort (Siefferman and Hill, 2003), and to potential "environmental" costs such as predation risk (Kemp *et al.*, 2008a; Millar *et al.*, 2006; Lyytinen *et al.*, 2004) and social harassment (Santos *et al.*, 2009).

From a functional perspective, there is also a unique way in which structurally cultured ornaments, particularly those involving complex photonic devices, may indicate the condition and/or genetic viability of their bearer. This is based upon the fact that many visual properties of these signals, such as luminance, spectral purity (chromaticity), iridescence, angle-dependence, and polarization, will ultimately be determined by the individual's ability to construct a highly precise nano-structural array. Because the structures are extremely small (that is, with elements of a similar size to the wavelengths of light they are to interact with), an individual must construct an extremely large number of these in order to produce a macroscale color patch. If not all individuals can achieve the necessary precision during development, then variation in trait expression will be linked with underlying genetic viability (i.e., the ability of their genome to function for stable development). This concept is similar to that underlying fluctuating asymmetry (FA; defined as random deviations from perfect bilateral asymmetry) as an indicator of developmental stability. However, the use of structural reflectance has the major advantage that the necessary nano-scale variation expresses itself via a readily perceivable and measurable macroscale trait – the structural color-based display.

Whether color displays function in this way is, at this point, largely unknown. However, the fact that these (and similar) hypotheses are being injected into the literature stands as a testimony to the new, broader, non-typological perspective of color signaling. With this background, we now move to a more in-depth treatment of three systems (case studies) that have led the field in combining functional and evolutionary approaches to studying color signals.

TRINIDADIAN GUPPIES (*Poecilia reticulata*)

Guppies are small freshwater fish native to North-eastern South America and a few Caribbean islands, including Trinidad and Tobago. This species might never have attracted much attention if the males were not incredibly colorful (Figure 5.1). The skin of male guppies is packed with chromatophores that generate color through multiple mechanisms. Xanthophores and erythrophores contain carotenoids and pteridines, which preferentially absorb shortwave light.

Iridophores and leucophores contain sub-cellular structures that cause specular reflection, constructive interference, or scattering. Melanophores contain melanins, which absorb light across the human visible spectrum and into the UV. How these different cell types are layered largely determines the reflectance spectrum of the skin and hence the perceived color (reviewed in Grether *et al.*, 2004a). Besides displaying a wide range of both iridescent and non-iridescent colors, the color patterns of guppies are highly variable. Males collected from the same pool can have completely different arrangements of color spots, and while much of the variation is heritable, even full-sib brothers can usually be distinguished based on gross differences in spot number or location (Grether, 2000; Hampton *et al.*, 2009; Houde, 1997; Jirotkul, 1999a).

There are, however, strong geographic patterns to the color variation. One of the most striking patterns is that guppies are most colorful (i.e., have the largest and most numerous color spots) in small headwater streams, where they coexist with the weakly predatory fish Hart's rivulus (*Rivulus hartii*), and least colorful in lowland rivers, where they coexist with multiple species of piscine predators (e.g., the Pike cichlid *Crenicichla alta*; Endler, 1978; Haskins *et al.*, 1961). Endler (1978, 1980, 1983) demonstrated, through experimental introductions in the field and natural selection experiments in the laboratory, that changes in the predator community can cause the color patterns of guppies to evolve rapidly in the directions and to the extent of divergence observed in nature. For this to explain the observed geographic pattern, however, survival selection against colorful males must be counterbalanced by some other form of selection that favors colorful males. Thus, before any mate choice experiments were done, Endler (1980) deduced that females must prefer colorful males. To explain why such mate preferences might exist, he offered his now famous hypothesis that carotenoid coloration can serve as an honest indicator of mate quality. Specifically, he suggested that algae and other sources of carotenoids are scarce in Trinidadian streams and that the carotenoid spots of males are thus indicators of foraging ability.

Subsequent research confirmed key components of the carotenoid-indicator mechanism. Female guppies usually do express a preference for males with larger orange spots and higher orange spot carotenoid concentrations (Brooks and Endler, 2001a, 2001b; Endler and Houde, 1995; Houde, 1987; Houde and Torio, 1992; Jirotkul, 1999b; Karino and Shinjo, 2004; Kodric-Brown, 1985, 1989; Long and Houde, 1989; but see Hampton *et al.*, 2009; Schwartz and Hendry, 2007). Algal carotenoid availability in undisturbed Trinidadian streams limits carotenoid deposition in the orange spots (Grether *et al.*, 1999), the ability of guppies to find algae is heritable (Karino *et al.*, 2005), and orange spot chroma (i.e., color saturation) correlates positively with algal-foraging ability (Karino and Shinjo, 2007; Karino *et al.*, 2007). Besides being an indicator of algal-foraging ability, the chroma of the orange spots is reduced by infection with the ectoparasite *Gyrodactylus turnbulli* (Houde and Torio, 1992) and correlates positively with swimming performance (Nicoletto, 1991). Thus, females may benefit in multiple ways by preferentially mating with high-carotenoid males, including avoidance of sexually transmitted parasites. Similar results have been reported for several other species with carotenoid coloration (reviewed in Møller *et al.*, 2000; Olson and Owens, 1998).

But would such a signaling system be evolutionarily stable? Carotenoids were once viewed as a cheap way to produce color precisely because they are ubiquitous in animal diets (Goodwin, 1984). In recent years, however, it has become apparent that a trade-off often exists between displaying carotenoids in the integument versus using them for other biological functions (Lozano, 1994; Olson and Owens, 1998; Clotfelter *et al.*, 2007; McGraw and Ardia, 2003, 2004, 2007; McGraw and Klasing, 2004, 2006; Peters, 2007; Smith *et al.*, 2007; Grether *et al.*, 2004a). If carotenoids are scarce and have other important biological functions, why do animals rely on these pigments for attracting mates? The standard answer is that carotenoids attract mates precisely because they are scarce and biologically valuable (Olson and Owens, 1998). But this does not explain how cheating is prevented if alternatives to carotenoids exist. Carotenoid-limited populations would seem to be vulnerable to invasion by mutants displaying structural layers or synthesized pigments that mimic the spectral properties of carotenoids (Grether *et al.*, 2001).

Endler (1980) foreshadowed this problem by suggesting that a female preference for bright carotenoid spots might account for the evolution of bright structural colors through automimicry. That is, structural spots might be carotenoid spot mimics. To our knowledge, this hypothesis has not been tested explicitly, but some relevant data have been collected. First, while the preference for orange coloration is nearly universal, preferences for structural colors are highly variable among populations (Endler and Houde, 1995). Second, in replicate introduction experiments carried out in two different river drainages, the way in which guppies evolved under relaxed predation to become more colorful differed between replicates: orange coloration increased dramatically in one replicate but iridescent structural coloration increased dramatically in the other (Kemp *et al.*, 2009). Thus, it appears that female preferences for orange and structural coloration are at least somewhat independent and that mimicry of carotenoids by structural colors is not a plausible explanation for the existence of the latter.

Another place to look for automimicry is within the orange spots themselves. In fact, the orange spots contain red pteridines (drosopterins), in addition to carotenoids, and both types of pigments contribute positively to the chroma of the orange spots (Grether *et al.*, 2001). Unlike carotenoids, pteridines are synthesized *de novo* by animals, from carbohydrates and proteins (Hurst, 1980). This could pose a problem for the carotenoid-indicator hypothesis. If females prefer males with higher orange chroma, as several researchers have inferred (Grether, 2000; Houde and Torio, 1992; Karino and Shinjo, 2004; Kodric-Brown, 1989), then males could compensate for low-carotenoid intake by producing drosopterins, and the orange spots would no longer be reliable indicators of algal-foraging ability. If this were (still) happening, we would expect to find a negative correlation between the drosopterin and carotenoid content of the orange spots across sites differing in carotenoid availability.

Instead, precisely the opposite pattern was found. Across sites, the drosopterin and carotenoid contents of the orange spots are positively correlated, such that the ratio of the two types of pigments is roughly conserved (Grether *et al.*, 2001). Laboratory experiments have shown that most of the interpopulation variation in drosopterin production is genetic, drosopterin production is highly heritable in some populations, and neither food intake nor carotenoid intake has significant effects on drosopterin production within individual fish (Grether *et al.*, 2005a). Variation in the carotenoid:drosopterin ratio affects the shape of the reflectance spectrum and the perceived hue of the orange spots, because drosopterins and carotenoids have different absorption spectra. While orange spot chroma varies geographically, orange spot hue is relatively constant across sites differing in carotenoid availability (Grether *et al.*, 2005a). Thus, genetic divergence in drosopterin production counteracts the effects of the carotenoid availability gradient on the hue of the orange spots (Grether, 2005). Clearly, male guppies do not currently use drosopterins as carotenoid mimics, but why not, and what process could account for this countergradient geographic pattern?

One likely explanation for the countergradient pattern is that female guppies prefer a particular hue of orange that males can only achieve by combining drosopterins and carotenoids in a particular ratio. To test for a hue preference, Deere *et al.* (2012) crossed low- and high-drosopterin populations in the lab and mated full-sibs in the F_1 generation to generate an F_2 generation with high levels of genetic variation in drosopterin production. Mate choice tests with F_2 males raised on a low-carotenoid diet indicate that females prefer males with intermediate drosopterin levels, as predicted by the hue preference hypothesis and not by the prevailing idea that females always prefer higher orange chroma. A hue preference might merely be a property of a pre-existing "orange attraction" sensory bias that has no function in a mating context (Rodd *et al.*, 2002; but see Grether *et al.*, 2005b). Alternatively, a hue preference might be a female counteradaptation to males using drosopterins as carotenoid mimics in the past (Grether *et al.*, 2001). Carotenoids that appear yellow at low concentrations shift to orange at high concentrations (Hudon *et al.*, 2003), and thus a preference for orange, as opposed to yellow or red hues, might be optimal for identifying high-carotenoid males while avoiding being tricked by low-carotenoid/high-drosopterin males.

Regardless of why the female color preference evolved, we can ask why male guppies use the specific pigments that they use. Although animals cannot synthesize carotenoids from scratch, they can convert ingested carotenoids into pigments with other spectral properties. Guppies convert most of the carotenoids that they ingest into tunaxanthin, which is a yellow carotenoid with an even shorter wavelength of peak absorption (438 nm) than its dietary precursors (β carotene, 449 nm; zeaxanthin, 449 nm; lutein, 445 nm). Instead, why not convert ingested carotenoids into orange carotenoids (e.g., astaxanthin, 467 nm), as many animals do? No wild poeciliids have been reported to use orange 4-keto-carotenoids, but they have been found in laboratory strains (Rempeters *et al.*, 1981). Thus, the absence of 4-keto-carotenoids in guppies seems unlikely to reflect a deep phylogenetic constraint. The key to solving the puzzle of why male guppies use drosopterins and tunaxanthin may be to simultaneously consider all three axes of color perception that are likely to influence a male's attractiveness to females: hue, chroma, and brightness. First, because the shape of a pigment's absorbance spectrum changes with its concentration, a single pigment would not permit males to achieve a consistent hue across different levels of carotenoid availability. Second, based on computer simulations, while similar hues could be obtained by combining drosopterins with a 4-keto-carotenoid, higher chroma and brightness can be achieved by combining drosopterins with tunaxanthin (holding total pigment content constant; Hudon *et al.*, 2003). Thus, the drosopterin–tunaxanthin system may enable males to achieve the preferred hue while maximizing chroma and brightness.

Computer simulations based loosely on guppy coloration and vision have also been used to examine interactions between chromatophore layers in a signal evolution context (Grether *et al.*, 2004a). Briefly, these simulations have revealed, for example, that increases in the reflectivity of the iridophore layer, which lies immediately below the xanthophore layer, can have the spectral effect of amplifying perceived differences in the pigment content of the xanthophore layer. Likewise, increases in the absorbance of the melanophore layer, which lies below the iridophore layer, can amplify perceived differences in the ultrastructure of the iridophore layer. Because the wavelength composition and proportion of incident light reflected by an iridophore layer depend critically on the precise thickness and spacing of reflecting platelets at the nanometer scale (Denton and Land, 1971; Herring, 1994; Huxley, 1968), the ultrastructural perfection and uniformity of iridophores may be an indicator of developmental stability, as has been suggested for the structural color-producing mechanisms in bird feathers and butterfly wings (Fitzpatrick, 1998; Kemp and Rutowski, 2007; see Pierid butterflies (subfamily Coliadinae)). Thus far, to our knowledge, the evolution of interactions among chromatophore layers has not been studied empirically in any system, and stands as a key avenue for discovery in the era of multicomponent signal research.

PIERID BUTTERFLIES (SUBFAMILY COLIADINAE)

From the colorful guppy, which continues to rank as one of the single most informative species regarding color evolution, we now move to examine a group whose color patterns are equally striking, yet perhaps underutilized by behavioral and evolutionary ecologists: butterflies. These creatures are, in fact, one of nature's most colorful animal groups, and their wings display an unrivalled diversity of color-producing mechanisms. These displays have featured in the development of theories relating to mate selection (Jiggins *et al.*, 2001; Kronforst *et al.*, 2007), speciation (e.g., Chamberlain *et al.*, 2009), and aposematism and mimicry (e.g., Kapan, 2001; Kunte, 2009), but have not tended to feature in studies of mate quality indicators. This is surprising because males in this group routinely display brighter, more chromatic, and sometimes more iridescent markings than their female counterparts, a potentially strong signature of sexual selection (Andersson, 1994; Darwin, 1871).

Various model butterfly groups, such as *Heliconius* and *Pieris*, have been harnessed to examine color signals from perspectives of species/mate recognition, reinforcement, and speciation (e.g., Jiggins *et al*., 2001; Kronforst *et al*., 2007). By contrast, studies of the potential for exaggerated male ornamentation to indicate mate quality or viability have largely focused on several species within the subfamily Coliadinae; most notably *Colias eurytheme* (the "orange sulphur") and *Eurema hecabe* (the "large grass yellow"). These species are well suited to such work because their relatively simple color patterns exhibit three morphogenetically different colors (pteridine pigment-based yellow, melanin pigment-based black, and coherently scattered UV), which are sexually dimorphic, amenable to experimental manipulation, and suitable for mechanistic study (Ghiradella *et al*., 1972; Stavenga *et al*., 2004; Vukusic and Hooper, 2005; Vukusic *et al*., 2000). *Colias* in particular has served as a long-standing model for the study of visual signals, including the examination of UV sexual signals (Silberglied and Taylor, 1973, 1978) two decades before such phenomena became popularized across behavioral ecology via avian models (e.g., Bennett *et al*., 1996; Andersson and Amundsen, 1997; Andersson *et al*., 1998; Sheldon *et al*., 1999).

To a human viewer, the dorsal wing surface of males and females of these two coliadine butterfly species consists largely of orange and yellow framed at the wing margins by a narrow black band (Box 5.1). The yellow/orange is generated by a suite of pteridine pigments (Watt, 1964), and the black by melanin pigment. There are subtle differences between the sexes in the demarcation of these two color patches, and in the characteristics of the pteridine-generated yellow/orange color. But by far the greatest sexual dimorphism occurs in the UV (300–400 nm) bandwidth, where male wings reflect brightly across their dorsal surface, while female wings reflect no (*C. eurytheme*) or restricted (*E. hecabe*) UV. This color arises due to the presence of dedicated optical nanostructures situated above the pigmented wing scale layers, a phenomenon first described by Ghiradella *et al*. (1972) for the conspecific *Eurema lisa*. As would be expected if bright UV markings increase predation risk (e.g., Lyytinen *et al*., 2004), this trait does not appear on the ventral wing surfaces in either sex, which is the visible wing surface when the animal is at rest (the same is true across all coliadine species).

The male-limited UV coloration of *C. eurytheme* has been known since the work of Silberglied and Taylor (1973, 1978) to function for species and sex recognition. However, recent work has also revealed that females prefer males bearing the brightest dorsal UV wing markings (Papke *et al*., 2007). This has also been found true for *E. hecabe* (Kemp, 2008b), as well as *Hypolimnas bolina* (Kemp, 2007), a Nymphalid species in which male wings reflect UV due to a photonic mechanism identical to that of both Pierids (Kemp and Macedonia, 2006). These findings indicate that sexual selection, via female mate choice, is the selective agent responsible for the high visual exaggeration of these structurally generated signals. This also raises interesting questions about the links between function and evolution in color signals: Does this component of the male color pattern contain salient information regarding mate quality? Has such a signaling system evolved due to the

Box 5.1. The Multicomponent Nature of *Colias* Butterfly Coloration
(***See insert for color representation of this figure.***)

The male dorsal wing surface of the butterfly *C. eurytheme* (the "orange sulphur") perfectly illustrates the notion of a multicomponent color signal. These wings appear yellowish-orange to the human observer [Panel (a), below], but over a certain range of orientations they also reflect brightly in the UV bandwidth. Panels (b) and (c) show the wing photographed under UV light and showing the "UV-on" and "UV-off" positions, respectively (that is, with the wing situated in orientations in which the bright UV reflectance is either visible or not).

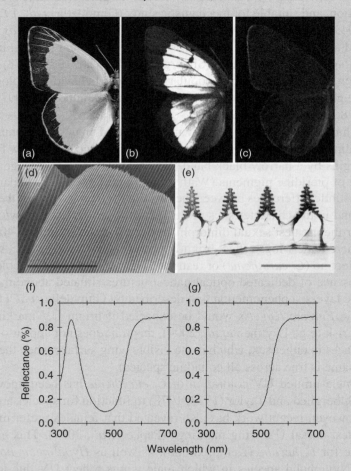

Functionally, the yellowish-orange coloration is generated by a suite of pteridine pigments, and the bright UV is generated by a nano-scale architecture on the surface of wing scales, depicted in Panels (d) and (e). As shown in (d), each scale is lined with longitudinal "ridges," which a cross-sectional view (e) shows are themselves adorned with a series of horizontal "lamellae." The whole structure is composed of semi-transparent insect cuticle (which has a refractive index of ~1.6; Ghiradella *et al.*, 1972). The alternate air-cuticle layering presented by these structures functions to coherently scatter light in UV wavelengths, thus producing very bright "mirror-like" reflectance over a limited range of orientations. Interestingly, as shown in (c), the wing is entirely dark in the "UV-off" position. This is because the pteridine pigments, which are situated in the body of each wing scale (below the UV-reflecting structures), are strongly UV-absorbing, and so "capture" any UV light that might otherwise have been diffusely scattered of the wing ultrastructure (Rutowski *et al.*, 2005).

Panels (f) and (g) show the normalized reflectance curves obtained for the wing situated in the "UV-on" and "UV-off" orientations, respectively (averaged across 60 lab-reared specimens). The chief difference between these curves is the reflectance peak situated in the UV range (centered upon ~348 nm), which is generated by the surface structures indicated in (e). The uniformly low reflectance below 500 nm seen in (g) is caused by the presence of a suite of pteridine pigments, which act essentially as high-pass filters.

Image reprinted in part from Kemp and Rutowski (2007) with permission (Rightslink License 3197970783626).

potentially unique revealing nature of optical surface nanostructures? What role – if any – might the coincident pigmentary component of coloration play in all of this?

Several of these questions have now been addressed. First, from the perspective of signal information content, the UV markings of male *C. eurytheme* have been shown to vary with adult age, larval nutrient acquisition, and the stability of the thermal environment experienced throughout pupation (Kemp, 2006; Kemp and Rutowski, 2007). Younger males have brighter UV markings, as do individuals that experienced a more nutritious larval diet and a more thermally stable period of pupal development. Hence, this trait encodes information on male phenotypic condition (and provides some insight into an individual's prior developmental "experiences"). Notably, the pigment-based coincident yellowish-orange also varies according to age and larval nutrition, but the magnitude of variation (and therefore signal information content) is much less than seen in the UV component of male dorsal coloration (Kemp, 2006; Kemp and Rutowski, 2007). Further, this phenotypic quality information is potentially salient because male butterflies donate both sperm and nutrients to females during mating, and the nutrients are known to contribute to female longevity and reproductive output (Boggs and Gilbert, 1979). Natural selection should favor – if at all possible – female choosiness for males capable of donating viable and nutritious ejaculates, which could potentially be achieved through choosing UV-bright mates (to the extent that such males experienced high quality juvenile environments). Still, whether this is the ultimate reason why females have evolved such preferences stands to be convincingly demonstrated in this group (see also Kemp *et al.*, 2008b).

Given that the UV component of these butterflies' dorsal coloration is a structural color of relatively optically complex provenance (see Box 5.1), this system also offers an opportunity to test the enticing idea that structural and/or iridescent coloration could signal viability through "visually magnifying" essential physiological and/or developmental processes (Fitzpatrick, 1998; Kemp and Rutowski, 2007). This hypothesis contends that males of higher genetic quality – that is, males in possession of less inbred, more adapted, or more internally harmonious genomes – should express brighter (or otherwise visibly different) color signals due to their greater capacity for acquiring limited resources from the environment and allocating those resources to the color structures. The latter point (allocation) includes being able to achieve nano-scale constructional precision with regularity in all structural arrays responsible for color generation. Environmental stress should decrease the level of color signal expression, but more viable individuals are predicted to be affected relatively less (i.e., exhibit heightened developmental stability) relative to their less-viable counterparts (Rowe and Houle, 1996).

Kemp and Rutowski (2007) used *C. eurytheme* to test two specific formulations of this hypothesis, each based around the fact that butterflies construct their entire adult phenotype (including wing structures) over a several day period during metamorphosis. The first formulation contended that more viable individuals should be able to achieve more highly precise structures (thus, bright UV) even when the available building-blocks to construct their adult phenotype are limited. The prediction here is that, among a sample of differentially viable genotypes, there should be variation in how well individuals are able to acquire resources, grow, and allocate to both soma and sexual trait expression (e.g., see David *et al.*, 2000). The second formulation contended that more viable individuals should be able to achieve greater architectural precision even when faced with physiological stress, such as a highly variable thermal environment. Kemp and Rutowski (2007) tested these predictions by looking for genotype-by-environment interactions (GEIs) in the expression of the male's dorsal UV markings. By rearing family groups in different nutritional environments, and subjecting them to heat and cold shocks during metamorphosis, they were able to use the GEIs as a formal test of whether some families (i.e., some genetic combinations) were able to withstand stress better than others, and whether this was signaled via their wing coloration. Importantly, each stressor was found to adversely affect the expression (brightness and angular visibility) of the male's iridescent coloration, while the coincident yellowish-orange was less strongly and less generally affected. All color elements were also strongly heritable, but the predicted GEIs were uniformly absent, which argues against the viability indicator hypothesis (at least as a signal of "good genes" for circumventing poor larval nutrition and/or thermally unstable pupal environments). This approach nevertheless stands as a key area of future investigation of signal content, particularly across the range of unstudied species and study systems best suited to quantitative genetic- and lab-based manipulation of condition.

Finally, researchers of coliadine butterflies have taken advantage of the fact that both structural and pigment-generated coloration exists on a single signaling surface (the dorsal wing surface) to contrast the variation, condition-dependence, and behavioral significance of these two different color types (Silberglied and Taylor, 1978; Kemp and Rutowski, 2007; Papke *et al.*, 2007; Kemp, 2008b). In reality, both of these mechanisms contribute to a single color patch, and would be perceived simultaneously (in the manner of a true multicomponent signal; Grether *et al.*, 2004a). Studies comparing the relative expression of each mechanism are therefore asking whether either one (structures or pigments) contributes more to the overall variation of the signal, and whether such variation is ecologically or evolutionarily salient. However, from a mechanistic perspective, recent work in *Colias* (Rutowski *et al.*, 2005) has also emphasized the full extent of how these two mechanisms may interact. Pteridine pigments absorb strongly in the UV range, but they are situated beneath the surface structures responsible for the bright, spectrally pure UV reflectance. The optical functionality and orientation of these structures across the wing surface is such that this coherently scattered UV is only visible from a limited range of viewing angles (given a fixed, point illumination source, such as the sun). At all other angles, the wing is UV-dark, but far more so than it otherwise would be, because the pteridine pigments extinguish any diffuse UV light that would normally be incoherently scattered off wing ultrastructures. The pigment therefore acts to amplify the difference between the wing when viewed in "UV-on" versus "UV-off" orientations, and would therefore accentuate the flash or beacon-like nature of the UV signal in butterflies flying in sunlight (Box 5.1; Rutowski *et al.*, 2005). This situation is directly analogous to that in the cultured skin of fishes and reptiles, in which lower chromatophore layers can magnify or amplify variation in upper chromatophore layers (Grether *et al.*, 2004a). The overall message is that different signal components are unlikely to be independent, or even have simple additive effects on signal perception; researchers should instead be aware of their full potential interactive complexity.

BIRDS

From the cultured wing scales of butterflies, we now move to the exquisitely cultured plumage of birds, a wildly popular group in behavioral and evolutionary ecology. This group has contributed greatly to our understanding of color signal evolution, and continues to provide highly original insights. Like butterflies and fishes, birds have harnessed an impressive spread of pigmentary and photonic mechanisms to generate color signals that span the entire visual spectrum. Regrettably, there has been comparatively little integration of concepts and approaches across major animal taxa, although commendable recent efforts (e.g., Ducrest *et al.*, 2008; Shawkey *et al.*, 2009) promise to inform a more general understanding of color signal mechanisms, function, and evolution.

The study of avian color signals experienced a period of rapid development in the 1990s. This decade saw an explosion of interest in two different types of signals: the first being colors whose variation is mostly due to underlying carotenoid pigment (so-called "carotenoid signals"), and the second being signals which reflect mainly or significantly in the UV wavelength range (so-called "UV signals"). The interest in carotenoids, which was sparked initially by Endler's (1980) thoughts on the potential honesty of such signals in guppies, saw the development of the house finch as a high-profile study system for color-based mate choice (Hill, 1990; Hill and Montgomerie, 1994; Hill *et al.*, 2002). It also spawned a volley of studies aimed at comparing carotenoid-generated colors with those generated by other pigments, such as melanin (Badyaev and Hill, 2000; McGraw and Hill, 2000; McGraw *et al.*, 2002), and more recently, structural colors (Hegyi *et al.*, 2007; Doucet and Hill, 2009). Such studies are widespread in birds (reviewed by Griffith *et al.*, 2006), partly due to the fact that this class of pigmentation is widely used in this taxon as a colorant of feathers, beaks, and skin.

The focus on carotenoid-generated coloration has now been seriously questioned (Kemp *et al.*, 2012), with at least one objective meta-review finding no evidence for heightened condition-dependence in these signals (Griffith *et al.*, 2006). Another motivator for this criticism is that the same acquisitional-based arguments claimed for carotenoid honesty should apply equally to other pigments, such as melanin, for which the precursors may be challenging to acquire or valuable in other metabolic realms (Stoehr, 2006). On the mechanistic level, work in vertebrates suggests that links between melanism and viability traits may be mediated at the quantitative genetic level due to the pleiotropic effects of genes affecting melanin regulation (Ducrest *et al.*, 2008). How and why such pleiotropy arises and is maintained – especially in consideration of selection for "cheating" by decoupling these covariances – offers an interesting direction forward in understanding the role of genetic architecture in signal design (see later).

Running parallel with the push to study pigment- (primarily carotenoid-) based signals, UV coloration was studied in a number of avian models, including finches (Bennett *et al.*, 1996; Andersson *et al.*, 1998), blue tits (Sheldon *et al.*, 1999), and blue-throats (Andersson and Amundsen, 1997). Significant effort was directed at answering the question of whether or not signals in this range of the color spectrum are in any way "special" (Bennett and Cuthill, 1994; Hunt *et al.*, 2001; Eaton and Lanyon, 2003; Hausmann *et al.*, 2003). Interestingly, similar questions were raised and examined over three decades ago in insects (Silberglied and Taylor, 1973, 1978; Silberglied, 1979), but dismissed when it became apparent that visual sensitivity to UV-A (315–400 nm) is extremely widespread in the animal kingdom. On the balance, while there are avian systems in which UV-based color signals are undoubtedly important, it stands to be demonstrated whether this wavelength channel carries any unique, over-arching significance for visual signal design and evolution (although signals in the UV-B range [280–315 nm] present a different, highly curious case; Li *et al.*, 2008). Such traits continue to be viewed in a slightly mysterious light, however, largely or perhaps only because they are invisible to the unaided human eye.

As suggested by these two historic foci, research on avian coloration has tended to emphasize pigments and surface structures as discrete, independent methods of color production. However, the truly concerted or interactive nature of these mechanisms is now being increasingly embraced (Shawkey and Hill, 2005), as it is by researchers of butterflies and other animals (e.g., Rutowski *et al.*, 2005; Steffen and McGraw, 2009). The field of avian coloration is also moving increasingly beyond the highly simplified, all-inclusive notion of "structural color" as a single class of signals. Valuable recent developments include efforts to better characterize the different photonic mechanisms of structural color production (e.g., Prum and Torres, 2003; Prum *et al.*, 1998; Shawkey *et al.*, 2003, 2005; Doucet *et al.*, 2006) and to understand how these mechanisms evolved (Shawkey *et al.*, 2006a, 2006b). In bluebirds, for instance, blue versus rusty coloration is apparently achieved via variation in the positioning of melanin pigment within the different layers of feather barbs. The barbs of both types of feathers have a "spongy layer," consisting of an irregular matrix of keratin, which Fourier analysis reveals is appropriate for coherently scattering blue light (Prum *et al.*, 1998). In blue barbs, this layer is underlain by a basal layer of melanin (which is similar to the arrangements seen in the structural blue wing scales of butterflies; e.g., Vukusic *et al.*, 1999, 2000). In rusty-cultured barbs, however, melanin instead occurs in granules spread throughout the spongy layer, and the spongy layer itself appears to have less demarcated air–keratin interfaces. These subtle differences suggest simple mechanisms underlying the production of blue versus rusty-cultured feathers in this group, thus hinting at precisely how such changes are effected through evolutionary time. A similarly simple re-arrangement of melanin granules is also implicated in evolutionary shifts between iridescent and matte black plumage in grackles and their allies (Shawkey *et al.*, 2006b). Studies of this nature promise excellent progress in our understanding of how selection acts upon available morphologies and materials to drive evolutionary innovations in color pattern (*sensu* Endler *et al.* 2005).

DISCUSSION/CONCLUSION/FUTURE WORK

Highly informative models for the study of animal signals are obviously not restricted to birds, butterflies, and fishes; we simply chose isolated examples from these three groups to illustrate the value of combining functional and evolutionary perspectives. The over-arching conceptual themes regarding coloration and visual signaling are common across all groups – color production mechanisms are often similar, there are striking commonalities in the physical nature of certain color signals (e.g., the UV-oranges of butterflies and guppies, compare Box 5.1 panel (g) with the "orange" guppy reflectance curve in Figure 5.1), and universal rules govern visual signal perception. However, as we have attempted to demonstrate, research into color and visual signaling has often proceeded independently in different groups, sometimes in a similar vein, sometimes in a very different vein, and sometimes leading to significant duplication of concepts (e.g., the UV literature in butterflies versus birds). We echo the recent calls for integration of research across taxa and believe that significant progress can be achieved by breaking down the traditional taxon-based boundaries that are an unfortunate hallmark of the animal coloration literature.

Despite considerable progress over the past several decades, it is still common for behavioral ecologists to treat color signals as simple, one- or two-dimensional traits. This, in some ways, may simply mirror the conceptual ethic of the field, which – arguably – has expressed a tendency to answer questions by focusing on single, well logically articulated hypotheses (such as the logic underlying carotenoids as an "honest" signal component). Arguably, some degree of over-simplification of empirical systems may be required, for example, to carry out comparative studies or to facilitate future meta-analyses. Moreover, it is difficult not to use some terminological and methodological shortcuts when dealing with traits as complex and multifaceted as color patterns. Nevertheless, as we have tried to show, some of the most interesting evolutionary questions can only be approached by measuring color variation along multiple dimensions, and by recognizing that individual color patches can comprise multiple, evolvable traits. It is no longer tenable to classify color signals into the simple categories of "structural" and "pigmentary," because most colors are produced by some combination of structural and pigmentary elements. It is also unwarranted to make inferences about the environmental sensitivity or condition-dependence of a color patch, based solely on knowledge of the underlying color-producing mechanisms (Griffith *et al.*, 2006).

While not all animal colors are necessarily signals, and not all signals are necessarily honest, honest signaling theory (Zahavi, 1975; Zahavi and Zahavi, 1997) nevertheless provides a useful organizing framework for studying animal coloration. This is particularly true of color traits that play a role in intrasexual competition and mate choice. While the best known version of the honest signaling hypothesis, at least as applied to coloration, is the carotenoid-indicator hypothesis, it should not be assumed that carotenoid coloration always serves as an honest signal or even that color patches that contain carotenoids are more likely to serve as honest signals than color patches devoid of these pigments. Animal coloration research has yet to arrive at any robust generalizations regarding the honesty or condition-dependence of different pigment systems. The hypothesis that structural coloration serves as an honest signal of mate quality because precisely spaced microstructures depend on a high degree of developmental stability remains plausible but should not be viewed as well supported. The most direct test of this hypothesis carried out to date, in butterflies, produced negative results (see Pierid butterflies (subfamily Coliadinae)).

We are only just beginning to understand how visual signals evolve at the level of color-producing pigments and structures. The wide diversity of color-producing mechanisms suggests that there are multiple alternative evolutionary routes to any given outcome. On the other hand, larger taxonomic patterns (e.g., a predominance of a particular pigment in the signals of a bird genus, or in an insect order, or a greater tendency for related animals for using certain iridescent signals as quality indicators; Kemp, 2008a, 2008b) may emerge due to phylogenetic inertia in the mechanisms underlying and linking broader (viability-based) metabolic processes with the use of certain colorants (e.g., Kemp and Rutowski, 2007; Ducrest *et al.*, 2008). Why particular routes are taken in particular groups and species remains poorly understood, but partial answers have been obtained in some systems (e.g., guppies and birds; see Pierid butterflies (subfamily Coliadinae; *Anolis* lizards; Macedonia *et al.*, 2000; Macedonia, 2001)).

Even though we have couched this chapter almost exclusively in terms of inter-sexual signaling, the concepts and principles apply to any signaling context in which individual quality is thought to be advertised or appraised. This includes agonistic, sexually selected contexts such as male–male competition and social status (e.g., Fitzstephens and Getty, 2000), and non-sexual behaviors such as begging (e.g., Cotton *et al.*, 1996). Questions such as those formulated above could equally be leveled at the visual signals used in contexts where individual quality is not being signaled or appraised – that is, where the costs of signaling are thought to be low – such as mimicry, brood deception, aposematism, and crypsis. Empirically, excellent opportunities exist for applying the latest "whole-color pattern" analytic approaches in these realms, and artificial selection offers an exciting tool for investigating the evolutionary costs and constraints of coloration and its individual elements.

ACKNOWLEDGMENTS

We thank M. Briffa, M. Herberstein, and one anonymous reviewer for helpful critiques of an earlier version of this manuscript. R.L. Rutowski originally provided the UV *Colias* image in Box 5.1, and D.N. Reznick provided the guppies photographed in Figure 5.1. We also thank the Royal Society of London for allowing the reproduction of Figure 5.2.

REFERENCES

Andersson, M.B. (1994) *Sexual Selection*, Princeton University Press, Princeton, New Jersey.

Andersson, S. and Amundsen, T. (1997) Ultraviolet colour vision and ornamentation in bluethroats. *Proceedings of the Royal Society of London, Series B: Biological Sciences*, **264**, 1587–1591.

Andersson, S., Ornborg, J., and Andersson, M. (1998) Ultraviolet sexual dimorphism and assortative mating in blue tits. *Proceedings of the Royal Society of London, Series B: Biological Sciences*, **265**, 445–450.

Badyaev, A.V. and Hill, G.E. (2000) Evolution of sexual dichromatism: contribution of carotenoid- versus melanin-based coloration. *Biological Journal of the Linnean Society*, **69**, 153–172.

Bennett, A.T.D. and Cuthill, I.C. (1994) Ultraviolet vision in birds – what is its function? *Vision Research*, **34**, 1471–1478.

Bennett, A.T.D., Cuthill, I.C., Partridge, J.C., and Maier, E.J. (1996) Ultraviolet vision and mate choice in zebra finches. *Nature*, **380**, 433–435.

Boggs, C.L. and Gilbert, L.E. (1979) Male contribution to egg production in butterflies – evidence for transfer of nutrients at mating. *Science*, **206**, 83–84.

Brooks, R. (2000) Negative genetic correlation between male sexual attractiveness and survival. *Nature*, **406**, 67–70.

Brooks, R. and Endler, J.A. (2001a) Direct and indirect sexual selection and quantitative genetics of male traits in guppies (*Poecilia reticulata*). *Evolution*, **55**, 1002–1015.

Brooks, R. and Endler, J.A. (2001b) Female guppies agree to differ: phenotypic and genetic variation in mate-choice behavior and the consequences for sexual selection. *Evolution*, **55**, 1644–1655.

Budden, A.E. and Dickinson, J.L. (2009) Signals of quality and age: the information content of multiple plumage ornaments in male western bluebirds *Sialia mexicana*. *Journal of Avian Biology*, **40**, 18–27.

Chamberlain, N.L., Hill, R.I., Kapan, D.D., Gilbert, L.E., and Kronforst, M.R. (2009) Polymorphic butterfly reveals the missing link in ecological speciation. *Science*, **326**, 847–850.

Clotfelter, E.D., Ardia, D.R., and McGraw, K.J. (2007) Red fish, blue fish: trade-offs between pigmentation and immunity in *Betta splendens*. *Behavioral Ecology*, **18**, 1139–1145.

Cotton, P.A., Kacelnik, A., and Wright, J. (1996) Chick begging as a signal: are nestlings honest? *Behavioral Ecology*, **7**, 178–182.

Darwin, C. (1871) *The Descent of Man and Selection in Relation to Sex*, John Murray, London.

David, P., Bjorksten, T., Fowler, K., and Pomiankowski, A. (2000) Condition-dependent signalling of genetic variation in stalk-eyed flies. *Nature*, **406**, 186–188.

Deere, K.A., Grether, G.F., Sun, A. and Sinsheimer, J.S. (2012) Female mate preference explains countergradient variation in the sexual coloration of guppies (Poecilia reticulata). Proceedings of the Royal Society B. **279**, 1684–1690 DOI: 10.1098/rspb.2011.2132.

Denton, E.J. and Land, M.F. (1971) Mechanism of reflexion in silvery layers of fish and cephalopods. *Proceedings of the Royal Society of London, Series A: Mathematical, Physical & Engineering Sciences*, **178**, 43–61.

Doucet, S.M., Shawkey, M.D., Hill, G.E., and Montgomerie, R. (2006) Iridescent plumage in satin bowerbirds: structure, mechanisms and nanostructural predictors of individual variation in colour. *Journal of Experimental Biology*, **209**, 380–390.

Doucet, S.M. and Hill, G.E. (2009) Do museum specimens accurately represent wild birds? A case study of carotenoid, melanin, and structural colours in long-tailed manakins *Chiroxiphia linearis*. *Journal of Avian Biology*, **40**, 146–156.

Ducrest, A.L., Keller, L., and Roulin, A. (2008) Pleiotropy in the melanocortin system, coloration and behavioural syndromes. *Trends in Ecology and Evolution*, **23**, 502–510. doi: 10.1016/j.tree.2008.06.001

Eaton, M.D. and Lanyon, S.M. (2003) The ubiquity of avian ultraviolet plumage reflectance. *Proceedings of the Royal Society of London, Series B: Biological Sciences*, **270**, 1721–1726.

Endler, J.A. (1978) A predator's view of animal color patterns. *Evolutionary Biology*, **11**, 319–364.

Endler, J.A. (1980) Natural selection on color patterns in *Poecilia reticulata*. *Evolution*, **34**, 76–91.

Endler, J.A. (1983) Natural and sexual selection on color patterns in poeciliid fishes. *Environmental Biology of Fishes*, **9**, 173–190.

Endler, J.A. and Houde, A.E. (1995) Geographic variation in female preferences for male traits in *Poecilia reticulata*. *Evolution*, **49**, 456–468.

Endler, J.A., Westcott, D.A., Madden, J.R. and Robson, T. (2005) Animal visual systems and the evolution of color patterns: sensory processing illuminates signal evolution. *Evolution*, **59**, 1795–818.

Fitzpatrick, S. (1998) Colour schemes for birds: structural coloration and signals of quality in feathers. *Annales Zoologici Fennici*, **35**, 67–77.

Fitzstephens, D.M. and Getty, T. (2000) Colour, fat and social status in male damselflies, *Calopteryx maculata*. Animal Behaviour **60**, 851–855. DOI: 10.1006/anbe.2000.1548.

Ghiradella, H., Aneshansley, D., Eisner, T., Silberglied, R.E., and Hinton, H.E. (1972) Ultraviolet reflection of a male butterfly: interference color caused by thin-layer elaboration of wing scales. *Science*, **178**, 1214–1217.

Goodwin, T.W. (1984) *The Biochemistry of Carotenoids*, Chapman & Hall, New York.

Grether, G.F. (2000) Carotenoid limitation and mate preference evolution: a test of the indicator hypothesis in guppies (*Poecilia reticulata*). *Evolution*, **54**, 1712–1724.

Grether, G.F. (2005) Environmental change, phenotypic plasticity and genetic compensation. *The American Naturalist*, **166**, E115–E123.

Grether, G.F., Hudon, J., and Millie, D.F. (1999) Carotenoid limitation of sexual coloration along an environmental gradient in guppies. *Proceedings of the Royal Society of London, Series B: Biological Sciences*, **266**, 1317–1322.

Grether, G.F., Hudon, J., and Endler, J.A. (2001) Carotenoid scarcity, synthetic pteridine pigments and the evolution of sexual coloration in guppies (*Poecilia reticulata*). *Proceedings of the Royal Society of London, Series B: Biological Sciences*, **268**, 1245–1253.

Grether, G.F., Kolluru, G.R., and Nersissian, K. (2004a) Individual colour patches as multicomponent signals. *Biological Reviews*, **79**, 583–610.

Grether, G.F., Cummings, M.E., and Hudon, J. (2005a) Countergradient variation in the sexual coloration of guppies (*Poecilia reticulata*): drosopterin synthesis balances carotenoid availability. *Evolution*, **59**, 175–188.

Grether, G.F., Kolluru, G.R., Rodd, F.H., de la Cerda, J., and Shimazaki, K. (2005b) Carotenoid availability affects the development of a colour-based mate preference and the sensory bias to which it is genetically linked. *Proceedings of the Royal Society of London, Series B: Biological Sciences*, **272**, 2181–2188.

Griffith, S.C., Parker, T.H., and Olson, V.A. (2006) Melanin-versus carotenoid-based sexual signals: is the difference really so black and red? *Animal Behaviour*, **71**, 749–763.

Hampton, K.J., Hughes, K.A., and Houde, A.E. (2009) The allure of the distinctive: reduced sexual responsiveness of female guppies to 'redundant' male colour patterns. *Ethology*, **115**, 475–481.

Haskins, C.P., Haskins, E.F., McLaughlin, J.J.A., and Hewitt, R.E. (1961) Polymorphism and population structure in *Lebistes reticulatus*, a population study, in *Vertebrate Speciation* (ed W.F. Blair), University of Texas Press, Austin, pp. 320–395.

Hausmann, F., Arnold, K.E., Mashall, N.J., and Owens, I.P.F. (2003) Ultraviolet signals in birds are special. *Proceedings of the Royal Society of London, Series B: Biological Sciences*, **270**, 61–67.

Hegyi, G., Szigeti, B., Torok, J., and Eens, M. (2007) Melanin, carotenoid and structural plumage ornaments: information content and role in great tits *Parus major*. *Journal of Avian Biology*, **38**, 698–708.

Herring, P.J. (1994) Reflective systems in aquatic animals. *Comparative Biochemistry and Physiology*, **109A**, 513–546.

Hill, G.E. (1990) Female house finches prefer colorful males: sexual selection for a condition-dependent trait. *Animal Behaviour*, **40**, 563–572.

Hill, G.E. and Montgomerie, R. (1994) Plumage color signals nutritional condition in the house finch. *Proceedings of the Royal Society of London, Series B: Biological Sciences*, **258**, 47–52.

Hill, G.E., Inouye, C.Y., and Montgomerie, R. (2002) Dietary carotenoids predict plumage coloration in wild house finches. *Proceedings of the Royal Society of London, Series B: Biological Sciences*, **269**, 1119–1124.

Hill, G.E., Doucet, S.M., and Buchholz, R. (2005) The effect of coccidial infection on iridescent plumage coloration in wild turkeys. *Animal Behaviour*, **69**, 387–394.

Houde, A.E. (1987) Mate choice based on naturally occurring color pattern variation in a guppy population. *Evolution*, **41**, 1–10.

Houde, A.E. (1997) *Sex, Color, and Mate Choice in Guppies*, Princeton University Press, Princeton.

Houde, A.E. and Torio, A.J. (1992) Effect of parasitic infection on male color pattern and female choice in guppies. *Behavioral Ecology*, **3**, 346–351.

Hudon, J., Grether, G.F., and Millie, D.F. (2003) Marginal differentiation between the sexual and general carotenoid pigmentation of guppies (*Poecilia reticulata*) and a possible visual explanation. *Physiological and Biochemical Zoology*, **76**, 776–790.

Hunt, S., Cuthill, I.C., Bennett, A.T.D., Church, S.C., and Partridge, J.C. (2001) Is the ultraviolet waveband a special communication channel in avian mate choice? *Journal of Experimental Biology*, **204**, 2499–2507.

Hunt, J., Bussiere, L.F., Jennions, M.D., and Brooks, R. (2004) What is genetic quality? *Trends in Ecology and Evolution*, **19**, 329–333.

Hurst, D.T. (1980) *An Introduction to the Chemistry and Biochemistry of Pyrimidines, Purines, and Pteridines*, John Wiley & Sons, New York.

Huxley, A.F. (1968) A theoretical treatment of the reflexion of light by multilayer structures. *Journal of Experimental Biology*, **48**, 227–245.

Jiggins, C.D., Naisbit, R.E., Coe, R.L. and Mallet, J. (2001) Reproductive isolation caused by colour pattern mimicry. *Nature*, **411**, 302–305. DOI: 10.1038/35077075.

Jirotkul, M. (1999a) Population density influences male–male competition in guppies. *Animal Behaviour*, **58**, 1169–1175.

Jirotkul, M. (1999b) Operational sex ratio influences female preference and male–male competition in guppies. *Animal Behaviour*, **58**, 287–294.

Kapan, D.D. (2001) Three-butterfly system provides a field test of Mullerian mimicry. *Nature*, **409**, 338–340. DOI: 10.1038/35053066.

Karino, K. and Shinjo, S. (2004) Female mate preference based on male orange spot patterns in the feral guppy *Poecilia reticulata* in Japan. *Ichthyological Research*, **51**, 316–320.

Karino, K. and Shinjo, S. (2007) Relationship between algal-foraging ability and expression of sexually selected traits in male guppies. *Zoological Science (Tokyo)*, **24**, 571–576.

Karino, K., Utagawa, T., and Shinjo, S. (2005) Heritability of the algal-foraging ability: an indirect benefit of female mate preference for males' carotenoid-based coloration in the guppy, *Poecilia reticulata*. *Behavioral Ecology and Sociobiology*, **59**, 1–5.

Karino, K., Shinjo, S., and Sato, A. (2007) Algal-searching ability in laboratory experiments reflects orange spot coloration of the male guppy in the wild. *Behaviour*, **144**, 101–113.

Kemp, D.J. (2006) Heightened phenotypic variation and age-based fading of ultraviolet butterfly wing coloration. *Evolutionary Ecology Research*, **8**, 515–527.

Kemp, D.J. (2007) Female butterflies prefer males bearing bright iridescent ornamentation. *Proceedings of the Royal Society of London, Series B: Biological Sciences*, **274**, 1043–1047.

Kemp, D.J. (2008a) Resource-mediated condition dependence in sexually dichromatic butterfly wing coloration. *Evolution*, **62**, 2346–2358.

Kemp, D.J. (2008b) Female mating biases for bright ultraviolet iridescence in the butterfly *Eurema hecabe* (Pieridae). *Behavioral Ecology*, **19**, 1–8.

Kemp, D.J. and Macedonia, J.M. (2006) Structural ultraviolet ornamentation in the butterfly *Hypolimnas bolina* L. (Nymphalidae): visual, morphological and ecological properties. *Australian Journal of Zoology*, **54**, 235–244.

Kemp, D.J. and Rutowski, R.L. (2007) Condition dependence, quantitative genetics, and the potential signal content of iridescent ultraviolet butterfly coloration. *Evolution*, **61**, 168–183.

Kemp, D.J., Reznick, D.N., and Grether, G.F. (2008a) Ornamental evolution in Trinidadian guppies (*Poecilia reticulata*): insights from sensory processing-based analyses of entire colour patterns. *Biological Journal of the Linnean Society*, **95**, 734–747.

Kemp, D.J., Macedonia, J.M., Ball, T.S., and Rutowski, R.L. (2008b) Potential direct fitness consequences of ornament-based mate choice in a butterfly. *Behavioral Ecology and Sociobiology*, **62**, 1017–1026.

Kemp, D.J., Reznick, D.N., Grether, G.F., and Endler, J.A. (2009) Predicting the direction of ornament evolution in Trinidadian guppies (*Poecilia reticulata*). *Proceedings of the Royal Society of London, Series B: Biological Sciences*, **276**, 4335–4343.

Kemp, D.J., Herberstein, M.E., and Grether, G.F. (2012) Unraveling the true complexity of costly color signaling. *Behavioral Ecology*, **23**, 233–236.

Keyser, A.J. and Hill, G.E. (1999) Condition-dependent variation in the blue-ultraviolet colouration of a structurally based plumage ornament. *Proceedings of the Royal Society of London, Series B: Biological Sciences*, **266**, 771–777.

Kloock, C.T. (2008) A comparison of fluorescence in two sympatric scorpion species. *Journal of Photochemistry and Photobiology B*, **91**, 132–136.

Kodric-Brown, A. (1985) Female preference and sexual selection for male coloration in the guppy (*Poecilia reticulata*). *Behavioral Ecology and Sociobiology*, **17**, 199–205.

Kodric-Brown, A. (1989) Dietary carotenoids and male mating success in the guppy: an environmental component to female choice. *Behavioral Ecology and Sociobiology*, **25**, 393–401.

Kodric-Brown, A. and Johnson, S.C. (2002) Ultraviolet reflectance patterns of male guppies enhance their attractiveness to females. *Animal Behaviour*, **63**, 391–396.

Kronforst, M.R., Young, L.G., and Gilbert, L.E. (2007) Reinforcement of mate preference among hybridizing *Heliconius* butterflies. *Journal of Evolutionary Biology*, **20**, 278–285.

Kunte, K. (2009) The diversity and evolution of batesian mimicry in *Papilio* swallowtail butterflies. *Evolution*, **63**, 2707–2716.

Li, J.J., Zhang, Z.T., Liu, F.X., Liu, Q.Q., Gan, W.J., Chen, J., Lim, M.L.M., and Li, D.Q. (2008) UVB-based mate-choice cues used by females of the jumping spider *Phintella vittata*. *Current Biology*, **18**, 699–703.

Lim, M.L.M. and Li, D.Q. (2007) Effects of age and feeding history on structure-based UV ornaments of a jumping spider (Araneae: Salticidae). *Proceedings of the Royal Society of London, Series B: Biological Sciences*, **274**, 569–575.

Lindsey, E. and Altizer, S. (2009) Sex differences in immune defenses and response to parasitism in monarch butterflies. *Evolutionary Ecology*, **23**, 607–620.

Long, K.D. and Houde, A.E. (1989) Orange spots as a visual cue for female mate choice in the guppy (*Poecilia reticulata*). *Ethology*, **82**, 316–324.

Loyau, A., Gomez, D., Moureau, B.T., Thery, M., Hart, N.S., Saint Jalme, M., Bennett, A.T.D., and Sorci, G. (2007) Iridescent structurally based coloration of eyespots correlates with mating success in the peacock. *Behavioral Ecology*, **18**, 1123–1131.

Lozano, G.A. (1994) Carotenoids, parasites, and sexual selection. *Oikos*, **70**, 309–311.

Luke, S.M., Hallam, B.T., and Vukusic, P. (2010) Structural optimization for broadband scattering in several ultra-thin white beetle scales. *Applied Optics*, **49**, 4246–4254.

Lyytinen, A., Lindstrom, L., and Mappes, J. (2004) Ultraviolet reflection and predation risk in diurnal and nocturnal Lepidoptera. *Behavioral Ecology*, **15**, 982–987.

Macedonia, J.M. (2001) Habitat light, colour variation, and ultraviolet reflectance in the Grand Cayman anole, *Anolis conspersus*. *Biological Journal of the Linnean Society*, **73**, 299–320.

Macedonia, J.M., James, S., Wittle, L.W., and Clark, D.L. (2000) Skin pigments and coloration in the Jamaican radiation of *Anolis* lizards. *Journal of Herpetology*, **34**, 99–109.

Martin, J. and Forsman, A. (1999) Social costs and development of nuptial coloration in male *Psammodromus algirus* lizards: an experiment. *Behavioral Ecology*, **10**, 396–400.

Masello, J.F., Lubjuhn, T., and Quillfeldt, P. (2008) Is the structural and psittacofulvin-based coloration of wild burrowing parrots *Cyanoliseus patagonus* condition dependent? *Journal of Avian Biology*, **39**, 653–662.

Mathger, L.M., Denton, E.J., Marshall, N.J., and Hanlon, R.T. (2009) Mechanisms and behavioural functions of structural coloration in cephalopods. *Journal of the Royal Society Interface*, **6**, S149–S163.

Maynard Smith, J. and Harper, D. (2003) *Animal Signals*, Oxford University Press, Oxford.

McGraw, K.J. and Ardia, D.R. (2003) Carotenoids, immunocompetence, and the information content of sexual colors: an experimental test. *The American Naturalist*, **162**, 704–712.

McGraw, K.J. and Ardia, D.R. (2004) Immunoregulatory activity of different dietary carotenoids in male zebra finches. *Chemoecology*, **14**, 25–29.

McGraw, K.J. and Ardia, D.R. (2007) Do carotenoids buffer testosterone-induced immunosuppression? An experimental test in a colourful songbird. *Biology Letters*, **3**, 375–378.

McGraw, K.J. and Hill, G.E. (2000) Differential effects of endoparasitism on the expression of carotenoid- and melanin-based ornamental coloration. *Proceedings of the Royal Society of London, Series B: Biological Sciences*, **267**, 1525–1531.

McGraw, K.J. and Klasing, K.C. (2004) Dietary carotenoids mediate a trade-off between egg quantity and quality in quail. *Integrative and Comparative Biology*, **44**, 600.

McGraw, K.J. and Klasing, K.C. (2006) Carotenoids, immunity, and integumentary coloration in red junglefowl (*Gallus gallus*). *Auk*, **123**, 1161–1171.

McGraw, K.J., MacKillop, E.A., Dale, J., and Hauber, M.E. (2002) Different colors reveal different information: how nutritional stress affects the expression of melanin- and structurally based ornamental plumage. *Journal of Experimental Biology*, **205**, 2343–2347.

Millar, N.P., Reznick, D.N., Kinnison, M.T., and Hendry, A.P. (2006) Disentangling the selective factors that act on male colour in wild guppies. *Oikos*, **113**, 1–12.

Møller, A.P., Biard, C., Blount, J.D., Houston, D.C., Ninni, P., Saino, N., and Surai, P.F. (2000) Carotenoid-dependent signals: indicators of foraging efficiency, immunocompetence or detoxification ability? *Avian and Poultry Biology Reviews*, **11**, 137–159.

Morehouse, N.I., Vukusic, P., and Rutowski, R. (2007) Pterin pigment granules are responsible for both broadband light scattering and wavelength selective absorption in the wing scales of pierid butterflies. *Proceedings of the Royal Society of London, Series B: Biological Sciences*, **274**, 359–366. doi: 10.1098/rspb.2006.3730

Mougeot, F., Redpath, S.M., and Leckie, F. (2005b) Ultra-violet reflectance of male and female red grouse, *Lagopus lagopus scoticus*: sexual ornaments reflect nematode parasite intensity. *Journal of Avian Biology*, **36**, 203–209.

Nicoletto, P.F. (1991) The relationship between male ornamentation and swimming performance in the guppy, *Poecilia reticulata*. *Behavioral Ecology and Sociobiology*, **28**, 365–370.

Olson, V.A. and Owens, I.P.F. (1998) Costly sexual signals: are carotenoids rare, risky or required? *Trends in Ecology and Evolution*, **13**, 510–514.

Papke, R.S., Kemp, D.J., and Rutowski, R.L. (2007) Multimodal signalling: structural ultraviolet reflectance predicts male mating success better than pheromones in the butterfly *Colias eurytheme* L. (Pieridae). *Animal Behaviour*, **73**, 47–54.

Parker, A.R., McKenzie, D.R., and Large, C.J. (1998) Multilayer reflectors in animals using green and gold beetles as contrasting examples. *Journal of Experimental Biology*, **201**, 1307–1313.

Pearn, S.M., Bennett, A.T.D., and Cuthill, I.C. (2001) Ultraviolet vision, fluorescence and mate choice in a parrot, the budgerigar *Melopsittacus undulatus*. *Proceedings of the Royal Society of London, Series B: Biological Sciences*, **268**, 2273–2279.

Peters, A. (2007) Testosterone and carotenoids: an integrated view of trade-offs between immunity and sexual signalling. *Bioessays*, **29**, 427–430.

Prum, R.O. and Torres, R. (2003) Structural colouration of avian skin: convergent evolution of coherently scattering dermal collagen arrays. *Journal of Experimental Biology*, **206**, 2409–2429.

Prum, R.O. and Torres, R.H. (2004) Structural colouration of mammalian skin: convergent evolution of coherently scattering dermal collagen arrays. *Journal of Experimental Biology*, **207**, 2157–2172.

Prum, R.O., Torres, R.H., Williamson, S., and Dyck, J. (1998) Coherent light scattering by blue feather barbs. *Nature*, **396**, 28–29.

Prum, R.O., Quinn, T., and Torres, R.H. (2006) Anatomically diverse butterfly scales all produce structural colours by coherent scattering. *Journal of Experimental Biology*, **209**, 748–765.

Rempeters, G., Henze, M., and Anders, F. (1981) Carotenoids and pteridines in the skin of interspecific hybrids of *Xiphophorus. Comparative Biochemistry and Physiology*, **69B**, 91–98.

Rivers, T.J. and Morin, J.G. (2008) Complex sexual courtship displays by luminescent male marine ostracods. *Journal of Experimental Biology*, **211**, 2252–2262.

Rodd, F.H., Hughes, K.A., Grether, G.F., and Baril, C.T. (2002) A possible non-sexual origin of a mate preference: are male guppies mimicking fruit? *Proceedings of the Royal Society of London, Series B: Biological Sciences*, **269**, 475–481.

Rohwer, S. (1975) Social significance of avian winter plumage variability. *Evolution*, **29**, 593–610.

Rowe, L. and Houle, D. (1996) The lek paradox and the capture of genetic variance by condition dependent traits. *Proceedings of the Royal Society of London, Series B: Biological Sciences*, **263**, 1415–1421.

Rutowski, R.L., Macedonia, J.M., Morehouse, N., and Taylor-Taft, L. (2005) Pterin pigments amplify iridescent ultraviolet signal in males of the orange sulphur butterfly, *Colias eurytheme. Proceedings of the Royal Society of London, Series B: Biological Sciences*, **272**, 2329–2335.

Rutowski, R.L., Macedonia, J.M., Kemp, D.J., and Taylor-Taft, L. (2007) Diversity in structural ultraviolet coloration among female sulphur butterflies (Coliadinae, Pieridae). *Arthropod Structure and Development*, **36**, 280–290.

Santos, E.S.A., Maia, R., and Macedo, R.H. (2009) Condition-dependent resource value affects male–male competition in the blue-black grassquit. *Behavioral Ecology*, **20**, 553–559.

Schwartz, A.K. and Hendry, A.P. (2007) A test for the parallel co-evolution of male colour and female preference in Trinidadian guppies (*Poecilia reticulata*). *Evolutionary Ecology Research*, **9**, 71–90.

Shawkey, M.D. and Hill, G.E. (2005) Carotenoids need structural colours to shine. *Biology Letters*, **1**, 121–124.

Shawkey, M.D., Estes, A.M., Siefferman, L.M., and Hill, G.E. (2003) Nanostructure predicts intraspecific variation in ultraviolet-blue plumage colour. *Proceedings of the Royal Society of London, Series B: Biological Sciences*, **270**, 1455–1460.

Shawkey, M.D., Estes, A.M., Siefferman, L., and Hill, G.E. (2005) The anatomical basis of sexual dichromatism in non-iridescent ultraviolet-blue structural coloration of feathers. *Biological Journal of the Linnaean Society*, **84**, 259–271.

Shawkey, M.D., Balenger, S.L., Hill, G.E., Johnson, L.S., Keyser, A.J., and Siefferman, L. (2006a) Mechanisms of evolutionary change in structural plumage coloration among bluebirds (*Sialia* spp.). *Journal of the Royal Society Interface*, **3**, 527–532.

Shawkey, M.D., Hauber, M.E., Estep, L.K., and Hill, G.E. (2006b) Evolutionary transitions and mechanisms of matte and iridescent plumage coloration in grackles and allies (Icteridae). *Journal of the Royal Society Interface*, **3**, 777–786.

Shawkey, M.D., Pillai, S.R., Hill, G.E., Siefferman, L.M., and Roberts, S.R. (2007) Bacteria as an agent for change in structural plumage color: correlational and experimental evidence. *The American Naturalist*, **169**, S112–S121.

Shawkey, M.D., Morehouse, N.I., and Vukusic, P. (2009) A protean palette: colour materials and mixing in birds and butterflies. *Journal of the Royal Society Interface*, **6**, S221–S231.

Sheldon, B.C., Andersson, S., Griffith, S.C., Ornborg, J., and Sendecka, J. (1999) Ultraviolet colour variation influences blue tit sex ratios. *Nature*, **402**, 874–877.

Siefferman, L. and Hill, G.E. (2003) Structural and melanin coloration indicate parental effort and reproductive success in male eastern bluebirds. *Behavioral Ecology*, **14**, 855–861.

Siefferman, L. and Hill, G.E. (2005) Evidence for sexual selection on structural plumage coloration in female eastern bluebirds (*Sialia sialis*). *Evolution*, **59**, 1819–1828.

Siitari, H., Alatalo, R.V., Halme, P., Buchanan, K.L., and Kilpimaa, J. (2007) Color signals in the black grouse (*Tetrao tetrix*): signal properties and their condition dependency. *The American Naturalist*, **169**, S81–S92.

Silberglied, R.E. (1979) Communication in the ultraviolet. *Annual Review of Ecology and Systematics*, **10**, 373–398.

Silberglied, R.E. and Taylor, O.R. (1973) Ultraviolet differences between sulfur butterflies, *Colias eurytheme* and *C. philodice*, and a possible isolating mechanism. *Nature*, **241**, 406–408.

Silberglied, R.E. and Taylor, O.R. (1978) Ultraviolet reflection and its behavioral role in the courtship of the sulphur butterflies *Colias eurytheme* and *C. philodice* (Lepidoptera, Pieridae). *Behavioral Ecology and Sociobiology*, **3**, 203–243.

Smith, E.J., Partridge, J.C., Parsons, K.N., White, E.M., Cuthill, I.C., Bennett, A.T.D., and Church, S.C. (2002) Ultraviolet vision and mate choice in the guppy (*Poecilia reticulata*). *Behavioral Ecology*, **13**, 11–19.

Smith, H.G., Raberg, L., Ohlsson, T., Granbom, M., and Hasselquist, D. (2007) Carotenoid and protein supplementation have differential effects on pheasant ornamentation and immunity. *Journal of Evolutionary Biology*, **20**, 310–319.

Stavenga, D.G., Stowe, S., Siebke, K., Zeil, J., and Arikawa, K. (2004) Butterfly wing colours: scale beads make white pierid wings brighter. *Proceedings of the Royal Society of London, Series B: Biological Sciences*, **271**, 1577–1584.

Steffen, J.E. and McGraw, K.J. (2009) How dewlap color reflects its carotenoid and pterin content in male and female brown anoles (*Norops sagrei*). *Comparative Biochemistry and Physiology. B*, **154**, 334–340.

Stevens, M., Parraga, C.A., Cuthill, I.C., Partridge, J.C. and Troscianko, T.S. (2007) Using digital photography to study animal coloration. Biological Journal of the Linnean Society **90**, 211–237. DOI: 10.1111/j.1095-8312.2007.00725.x

Stoehr, A.M. (2006) Costly melanin ornaments: the importance of taxon? *Functional Ecology*, **20**, 276–281.

Sweeney, A., Jiggins, C., and Johnsen, S. (2003) Polarized light as a butterfly mating signal. *Nature*, **423**, 31–32.

Thery, M. and Casas, J. (2009) The multiple disguises of spiders: web colour and decorations, body colour and movement. *Philosophical Transactions of the Royal Society of London B*, **364**, 471–480. doi: 10.1098/rstb.2008.0212

Thery, M. and Gomez, D. (2010) Insect colours and visual appearance in the eyes of their predators, in *Advances in Insect Physiology: Insect Integument and Colour*, Academic Press Ltd.–Elsevier Science Ltd., London, pp. 267–353.

Vukusic, P. and Hooper, I. (2005) Directionally controlled fluorescence emission in butterflies. *Science*, **310**, 1151.

Vukusic, P. and Sambles, J.R. (2003) Photonic structures in biology. *Nature*, **424**, 852–855.

Vukusic, P., Sambles, J.R., Lawrence, C.R., and Wootton, R.J. (1999) Quantified interference and diffraction in single *Morpho* butterfly scales. *Proceedings of the Royal Society of London, Series B: Biological Sciences*, **266**, 1403–1411.

Vukusic, P., Sambles, J.R., and Lawrence, C.R. (2000) Structural colour: colour mixing in wing scales of a butterfly. *Nature*, **404**, 457.

Vukusic, P., Sambles, J.R., Lawrence, C.R., and Wootton, R.J. (2001) Structural colour: now you see it, now you don't. *Nature*, **410**, 36.

Watt, W.B. (1964) Pteridine components of wing pigmentation in the butterfly *Colias eurytheme*. *Nature*, **201**, 1326–1327.

White, E.M., Partridge, J.C., and Church, S.C. (2003) Ultraviolet dermal reflexion and mate choice in the guppy, *Poecilia reticulata*. *Animal Behaviour*, **65**, 693–700.

Zahavi, A. (1975) Mate selection: a selection for a handicap. *Journal of Theoretical Biology*, **53**, 205–214.

Zahavi, A. and Zahavi, A. (1997) *The Handicap Principle: a Missing Piece of Darwin's Puzzle*, Oxford University Press, New York.

Zuk, M. (1991) Parasites and bright birds: new data and a new prediction, in *Bird–Parasite Interactions: Ecology, Evolution and Behaviour* (eds J.E. Loye and M. Zuk), Oxford University Press, Oxford, pp. 317–327.

6

AGONISTIC SIGNALS: INTEGRATING ANALYSIS OF FUNCTIONS AND MECHANISMS

Mark Briffa

Marine Biology and Ecology Research Centre, Plymouth University, Plymouth, UK

ANIMAL CONTESTS AND THE EVOLUTION OF AGONISTIC SIGNALS

Contest and Agonistic Signals

The study of aggressive behavior is an important component of a wide range of biological disciplines including areas such as conservation biology (e.g., Blanchet *et al.*, 2008), zoo biology (e.g., Davis *et al.*, 2009), animal welfare (e.g., Huntingford and Adams, 2005), and neuroscience (Huber *et al.*, 1997). Aggression has also been extensively studied from an evolutionary perspective by behavioral ecologists interested in analyzing the adaptive value (or "function") of agonistic behavior. Although such studies approach aggression from the perspective of the first of Tinbergen's "Four Questions" (Tinbergen, 1963), contest behavior is a prime example of how understanding animal behavior from both ultimate and proximate viewpoints is desirable. For behavioral ecologists, agonistic behavior occurs in the context of "contests" over the ownership of limited resources (Briffa and Sneddon, 2010). Contests often occur in pair-wise dyads but may also occur between rival groups, for example, between different colonies of social insects such as the wood ant *Formica xerophila* (Tanner, 2008). While competition may occur in a variety of ways, contest behavior is a special case of competition, where the rivals engage in direct interactions over the ownership of an indivisible resource. Contests are often described as "fights" but the range of agonistic behaviors that may occur during a contest include not only intentional attempts to inflict harm on the opponent, but also "trials of strength" such as wrestling or grappling, and the use of agonistic signals. Regardless of the types of agonistic behavior used, a key feature of a contest is that there is usually a clear outcome; at the end of the encounter it is usually possible to identify a winner, the contestant that has secured access to the resource, and a loser, the contestant that has been excluded from the resource (although see Jennings *et al.*, 2005 for a study on "drawn" contests in fallow deer *Dama dama*).

Contests resolved through the use of agonistic signals are considered to be more common than dangerous fighting where injuries might occur; certainly, agonistic signals have been a major focus of study, both from theoretical and empirical standpoints, compared to contests that involve injury or fatalities. The emphasis on non-injurious contest behavior is perhaps understandable from an animal welfare perspective but ethical consequences alone do not account for the disproportionate number of studies that have investigated non-injurious contests. Agonistic signals are extremely diverse in modality, ranging from visual displays such as claw waving in fiddler crabs, *Uca* spp. (Zeil *et al.*, 2006), dewlap extension in lizards (Vanhooydonck *et al.*, 2005), and oppercular flaring in Siamese fighting fish *Betta splendens* (Abrahams *et al.*, 2005); acoustic displays such as "roaring" in red deer, *Cervus elaphus* (Clutton-Brock and Albon, 1979), and "groaning" in fallow deer; tactile signals such as shell rapping in hermit crabs (Briffa *et al.*, 1998); urine-borne chemical signals as in crayfish *Astacus leptodactylus* (Breithaupt and Atema, 2000); ornaments such as horns in dung beetles *Euoniticellus intermedius* (Lailvaux *et al.*, 2005); and combinations of iridescent coloration and displays such as the aerial contests between rival pairs of male blue moon butterflies, *Hypolimnas bolina*, (Kemp, 2002). Thus, agonistic signals occur in many modalities, and involve the display of ornaments and weapons produced during development as well as active performances of dynamic signaling activities during the encounter; in consequence, agonistic displays are at least as diverse, and interesting to observe, as examples of communication used in other situations such as courtship. This diversity alone would justify the vast number of studies that have focused on agonistic signals but the majority of studies have been stimulated by one central question: In a contest situation, why use communication at all? Contests that involve communication can often be protracted and the production of ornaments and performance of agonistic signals will clearly involve some sort of cost. It could be argued that the stronger of the two opponents would maximize its fitness by simply injuring or killing its weaker rival, rather than wasting time and energy on producing signals. In other words, the key question that has motivated a large proportion of work on agonistic signals is this: What is the fitness benefit of using agonistic signals?

Why do Fighting Animals Use Displays?

On a superficial level, explanations for agonistic signals based on selection operating at the individual level might seem counter-intuitive; in the past, non-injurious agonistic behaviors have therefore attracted group-selective explanations (for example, see studies cited in Chapter 7 of Krebs and Davies, 1993). The argument behind such group-selective explanations is that groups that can avoid wasteful injuries and fatalities by resolving conflicts of interest through communication would be more efficient than groups that experienced high levels of intra-group aggression. Groups that avoided this source of mortality would be able to exploit resources more effectively than would groups that experienced high mortality rates due to fighting. Groups that could settle disputes among members though communication would therefore be favored over groups that settled disputes through injurious fighting. Thus, the frequency of groups using communication to settle disputes would increase and the adaptive value of agonistic signals would be to increase the efficiency of the group as a whole. Although selection at group level (rather than at the level of the individual) is not a theoretical impossibility (see Wilson, 1975), it is unlikely that group selection would not have a strong effect on the evolution of agonistic behavior. Briefly, the two determinants of the strength of selection, the degree of variation between selection units and the rate at which selection units suffer mortality, will both be greater for selection at the level of individuals in comparison to selection at the level of groups. Furthermore, a group composed of individuals that were acting in ways that increased the efficiency of the group as a whole would be very vulnerable to invasion by a mutant behavior of acting in a way that maximized individual benefits regardless of what this would mean for the group as a whole. Therefore it is unlikely that agonistic signals are maintained by group selection and Parker (1974) suggested that settling contests by "convention" must be the result of natural selection, occurring at the level of individuals.

The key insight concerning exactly how natural selection could result in agonistic signals was one adapted from a branch of economics called *game theory*. This approach of *classical game theory* was originally (and continues to be) used for modeling the decisions that humans should make in situations where there is a conflict of interests between interacting entities, such as trade or warfare. The key assumption is that the best strategy is not fixed but will vary according to the strategy used by the opponent, as well as with the costs (C) and potential benefits (the value of the contested resource, V) of engaging. In *evolutionary game theory*, the core assumption of frequency dependent benefits, and the question now becomes: "Which strategy should evolution 'choose' out of a range of alternatives?" In order to answer this question, the strategy that is immune to invasion by alternative strategies must be calculated; this is called the *evolutionarily stable strategy* or *ESS*. The evolutionary game theory approach has been applied to a wide range of questions in animal behavior (see Dugatin and Reeve, 1998 for a detailed overview and Kokko, 2013 for an overview of models used to explain contests) but the first area it was applied to was that of non-injurious fighting (Maynard Smith and Parker, 1976). This study contained two very influential models, the original War of Attrition or "waiting game" and the Hawk-Dove game. The War of Attrition is possibly more relevant to situations of scramble competition than to encounters that can easily be thought of as direct contests or fights, although more recent developments of the War of Attrition are very applicable to direct contests (see next section). The

Hawk-Dove game, on the other hand, provides the basic answer to our question of "Why use agonistic displays over injurious fighting?"

The model considers two alternative strategies and asks which should be favored by natural selection by considering the costs and benefits for each strategy when fighting against both strategies in pay-off matrix. The two alternative strategies are "Hawk," which is to try to injure the opponent and "Dove," which is to use agonistic signals. Hawks keep fighting until they get an injury, at which point they concede, whereas Doves always concede when faced with a Hawk. If a Hawk fights another Hawk then, if equal in all other respects, there is a 50% chance of winning but a 50% chance of picking up an injury. This means that the average pay-off to a Hawk fighting a Hawk is half the value of the resource, a positive pay-off, plus half the cost of an injury, a negative pay-off. A Hawk fighting against a Dove, on the other hand, will always win, so the pay-off to the Hawk is the full value of the resource (with no chance of injury). Therefore the opposite pay-off, to a Dove fighting a Hawk, is always zero. Finally, when a Dove fights against another Dove there is a 50% chance of winning but no chance of picking up an injury, so the average pay-off is half the value of the resource. By inserting values for C and V into the pay-off matrix, it is now possible to determine which of Hawk and Dove should be the ESS. The results of the model are that the ESS will vary with the difference between C and V. If $V > C$ in magnitude, then the average pay-off for Hawks will be greater than that for Doves. A population of Hawks would be immune to invasion from a Dove mutant but a mutation for Hawk behavior would quickly spread through a population of Doves until the Dove strategy was completely replaced. If $V < C$ in magnitude, however, neither strategy would be immune from invasion by the alternative and rather than a "pure ESS" of Hawk or Dove the model predicts evolution toward a "mixed ESS" of Hawks and Doves where the average pay-offs of each strategy are at equilibrium. The proportion of Hawks and Doves at equilibrium will be determined by the disparity between C and V. Although the model is very simple, its basic predictions seem to hold true. In cases where V is so high that it outbalances the fitness cost of an injury dangerous and indeed fatal fighting can occur. A sufficiently high value resource might be a one-off chance of mating and notable examples include fights between elephant seals, which are only likely to achieve "beach master" status for one reproductive season (Haley, 1994), and short lived and wingless male fig wasps which decapitate their rivals for the chance to mate with winged females before they disperse away from the fig (Hamilton, 1967). Resources other than a one-off chance to mate are seldom so valuable that the positive fitness units gained though ownership of V would outbalance the negative fitness costs arising from the potential of receiving an injury, C. Thus, in the vast majority of agonistic encounters, such as contests over access to mates when there is more than one chance of mating, food, territory, and shelters, the Hawk-Dove game shows how natural selection should favor the evolution of agonistic signals as well as the evolution of dangerous fighting. The mixed ESS could produce either a mixture of individuals that always play Hawk or Dove or individuals that play Hawk for a proportion of their time and Dove for the remainder. In both cases, the balance between Hawks and Doves at any one time will be determined by the ratio of C:V.

Unless the contest is over an extremely valuable resource, such as a genetic life or death situation when there is only one chance to secure a mating, outright injurious fighting is rare, and aggression involves the use of non-injurious agonistic behaviors. Rather than ending when one opponent is killed or badly injured, the encounter is resolved by a decision made by one of the opponents to withdraw or "give up." The opponent that gives up will lose the contest and will be excluded from the resource. The opponent that has not given up will have won the contest and secured access to the resource. It is an important feature of contests that the decision to quit is made by the loser such that losers, rather than winners, determine the duration of the contest.

Contest asymmetries refer to any differences between the two opponents that could have an influence on the outcome (which opponent decides to give up). A wide range of factors could influence the outcome, but broadly they can be grouped into two classes. First, although resources are of fixed value in absolute terms – for example, the number of calories in a food item, the size of a territory, or the fecundity of a female – in relative terms they may vary between opponents. A hungry individual, for example, might place greater value on a food item than would a satiated individual, or a male that has successfully mated might value a female less than a male that has yet to mate. Thus, although the contested resource will have a fixed value in absolute terms the subjective resource value (RV) for each opponent can be asymmetric. Second, although not considered in the simple Hawk-Dove game, the opponents may differ in fighting ability or *resource holding power* (sometimes called *resource holding potential*, both forms abbreviated to "RHP") (Parker, 1974). These asymmetries determine the outcome of agonistic encounters. In the absence of a difference in RHP, the winner should be the individual with greater subjective RV and in the absence of a difference in subjective RV, the winner should be the individual of greater RHP. Often RHP will simply vary with body size, the larger opponent usually being of greater RHP. However, RHP may also vary with a wide range of other traits such as weapon size, strength, physiological condition, and endurance. To a large extent understanding agonistic signal function requires elucidating the causal links between these factors and RHP.

The Information Contained in Agonistic Displays

Having demonstrated how agonistic signals could be an evolutionarily stable result of natural (rather than group) selection, the next goal of theory was to produce models that could explain the specific functions of agonistic signals; in other words to determine how agonistic signals would induce the opponent to give up. A signal has been defined as a behavior that has evolved as an information carrier, in the sense that it is used in order to reduce the receiver's uncertainty (McGregor, 2005) so the key question is: What information might these agonistic signals contain? Subjective RV should influence the "willingness" of an individual to enter into or persist with a fight (e.g., Hofmann and Schildberger, 2001), also referred to as *motivation* (e.g., Elwood *et al.*, 1998; Briffa and Elwood, 2001a). Knowing about the opponent's motivation would be very valuable to a fighting animal, but it is unlikely that this information would be revealed intentionally through the use of a signal. Turner and Huntingford (1986) pointed out that if a contestant reveals its motivation it is in effect revealing for how much longer it will persist in the contest, before making the decision to withdraw. In order to win the contest, a contestant would therefore only need to persist for the time its opponent is prepared to fight + 1 time unit. In the War of Attrition model, where encounters take the form of scramble competition, the ESS solution to this problem is to avoid revealing information about persistence by varying persistence time in an unpredictable way, a prediction that is supported by data from contests in dung flies, *Scatophaga stercoraria*, over the ownership of areas of cow pat used for egg-laying (Parker, 1970). In direct contests, the need to avoid revealing information on motivation state, suggests that agonistic signals advertising motivation should not be selected for. On the other hand, studies on fighting in mouth brooders (Turner and Huntingford, 1986) and studies using motivational probes in hermit crabs (Elwood *et al.*, 1998; Briffa and Elwood, 2001a) indicate that some information on motivation may be revealed "unintentionally" by agonistic signals. Nevertheless, evolutionary game theory assumes that at least the primary function of these signals should be to advertise the RHP of the sender.

Honesty and Deception

It makes logical sense that fighting animals should not reveal their motivation whereas the Hawk-Dove game shows that selection should favor the evolution of agonistic signals. The information carried by these should allow one of the opponents to end the contest by deciding to "give up." In the absence of information about the opponent's motivation, it seems reasonable that the decision might be based on the opponent's RHP; if the opponent is assessed as being stronger this should trigger a giving up decision. But this leads to another problem for theory; if fighting animals should avoid revealing their motivation, it could also be argued that the weaker individual should also avoid revealing its RHP. Indeed, it seems to make little sense to perform a signal that indicates that you are weaker than your rival and one option to overcome this would be to perform a signal that exaggerates RHP, leading to the opponent making an erroneously high assessment of the sender's fighting ability. Such exaggeration has also been called *bluffing*, *cheating*, or using a *dishonest signal*. The question of signal honesty applies to a much wider range of behavior than contests, and arises in any situation where the signal carries information on some aspect of the sender's quality, which the receiver uses as the basis a decision made in respect of the sender. In a fight this decision is whether to give up; in the context of courtship the decision could be whether to accept a copulation attempt.

Two types of theoretical explanation for the honesty of signals of individual quality have been devised. First, there is the possibility that signals of individual quality are "cost free" (Vanhooydonck *et al.*, 2007) in the sense that they do not significantly impact the ability to invest in other requirements. (Hasson, 1997; Taylor *et al.*, 2000). These signals should nevertheless be honest because there is a direct link between the size or appearance of the structure used in the signal and the trait that the receiver of the signal assesses. Such cost-free signals have been called *indices* and the direct link between the index and the trait being advertised means that, by definition, they cannot be bluffed. Second, the Handicap Principle, which is discussed in detail elsewhere in this volume, was described in verbal arguments by Zahavi (1975, 1977) and several different versions were then formally modeled by Grafen (1990). The core concept is that, even for signals that are not direct indices of the trait being advertised, high levels of bluffing should not be adaptive because signals of quality are extremely costly to produce, to the extent that they "handicap" the ability to allocate resources to other requirements. An example of this handicap signaling includes stotting behavior in Thompson's gazelle *Gazella thomsoni*. This form of prey to predator signaling, involves the gazelle jumping up and down at the same time as trying to flee from a coursing predator such as a wild dog *Lycaon pictus*. But stotting detracts from the ability to open up a safe distance between the gazelle and the pursuing dog, or in other words performing the signal handicaps the ability to run away. Remarkably, wild dogs choose not to pursue the gazelles that stot at the highest rates, instead choosing to pursue those that can only stot at a low rate (FitzGibbon and Fanshawe, 1988). Although high rates of stotting lead to less distance between the quarry and its predator, stotting seems to be an honest signal of running ability that is assessed by the wild dog. The costs of producing such a signal that indicates high quality can only be met by individuals that genuinely meet the quality levels indicated by the signal. For individuals of lower quality, the handicap costs would be so high that they outbalance the fitness benefits of bluffing. The handicap principle was originally formulated to explain elaborate sexually selected traits but further developments showed how it could apply to other situations where signals of individual quality are used such as predator–prey interactions and agonistic encounters. Relatively few studies have directly applied the idea of handicap signaling to agonistic displays but the idea of very costly signals seems to fit well with some of the key models of non-injurious fighting (see next section).

While indices are not susceptible to bluffing, signals where honesty is guaranteed only by the high costs of production (Zahavi, 1975, 1977; Grafen, 1990), rather than a direct link with the trait being advertised, are susceptible to low levels of cheating in populations where, on average, signals are honest (Szamado, 2000). In crustaceans such as the mantis shrimp, *Gonodactylus oerstedii* (Adams and Caldwell, 1990), fiddler crabs, *Uca* spp. (Backwell *et al.*, 2000; Lailvaux *et al.*, 2009) agonistic signals are produced by postural displays involving the chelipeds (claw bearing appendages), with the size of the chelae providing information either about the strength and therefore effectiveness of the claw itself as a weapon, or because it provides information on overall body size. It appears that low levels of cheating are possible because these weapons are subject to periods of reduced effectiveness, due to recent moult or regeneration, but do not change significantly in terms of appearance during these periods. A newly moulted claw, although soft and ineffective as a weapon, would not look different from an inter-moult claw of the same size that was fully hardened. Similarly, a regenerated claw that had been produced through rapid growth would be weaker than a non-regenerated claw but would look very similar.

Perhaps, such examples of signal dishonesty that depend on a temporary mismatch between the visual appearance of a weapon used in a display and its actual effectiveness as a weapon (and therefore the sender's RHP) should be regarded as exceptions where cheating is possible for a proportion of the population at any one time. On the other hand, low levels of cheating might be possible in the absence of such temporary mismatches in cases where there is sufficient variation within the population, in the relationship between a correlate of RHP such as overall body size and the visual appearance of a structure displayed as a signal of RHP such as claw size. In crustaceans, chelar size increases with body size but some individuals will have larger than average chelae for their body size and some individuals will have smaller than average chelae for their body size. Hughes (2000) described an analytical approach for determining whether low levels of cheating take place in such systems. This was based on looking at the residuals from a relationship between the magnitude of the agonistic signal to a known aspect of the sender's quality such as body size. If significant cheating occurs, individuals that use a bigger signal than expected for their body size (i.e., those with positive residuals) should produce more performances than individuals that do not exaggerate in this way (Figure 6.1). In snapping shrimp *Alpheus heterochaelis* (Hughes, 2000), and hermit crabs, *Pagurus bernhardus* (Arnott and Elwood, 2010) analyses of signal residuals indicate that there may be low but significant levels of cheating through the use of exaggerated agonistic displays involving the chelipeds, in that individuals with abnormally large chelipeds for their body size display them more frequently than individuals with average sized chelipeds.

These signals where weapons, although costly to produce during development and perhaps costly to maintain, are used in relatively static postural displays can be thought of as *conventional* signals (*sensu* Hurd, 1997). Conventional signals may be distinguished from *costly signals* (*sensu* Hurd, 1997), where some sort of dynamic activity is used. In this case, the cost is paid in terms of energy expenditure within the timescale of the agonistic encounter and the immediate post-fight recovery period (of course, in many cases agonistic signals might involve both types of display, such as the example of male fiddler crabs waving their enlarged major chelus). In an analysis of signal residuals of costly *shell rapping* signals, used during agonistic encounters in hermit crabs, there was no evidence that senders could exaggerate their RHP by signaling at an erroneously high level (Briffa, 2006). Indeed, although contests often involve static postures such as cheliped displays, dynamic and repetitive displays such as shell rapping appear to be a key feature of agonistic encounters. The validity of the distinction between indices, where the honesty of the signal is derived from a direct link to the trait being advertised, and other types of signal where honesty is derived from high production or maintenance costs that prohibit cheating, is the subject of ongoing debate (see Szmado 2003; Vanhooydonck *et al.*, 2007). Nevertheless the overall question of signal honesty seems just as relevant to agonistic signals, which may or may not be a result of sexual selection, as to examples of courtship displays. On the basis of currently

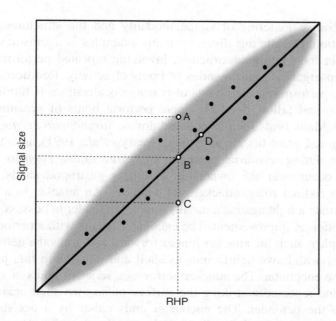

Figure 6.1. A hypothetical relationship between RHP and the magnitude (e.g., size, shape, color, rate) of an agonistic signal. (Adapted from Hughes 2000). If there is sufficient scatter in the relationship low levels of cheating could be maintained in an otherwise honest signaling system. Individuals A, B, and C all have the same RHP but individual A produces a signal that is larger than predicted by the relationship for its RHP. If fighting animals assess one another on the basis of this relationship individual A would appear be of greater RHP than individual B or C, and of individual D, whose true RHP is greater than that of individual A. Individuals such as A could cheat by using their signal often. If low levels of cheating were a significant feature of the signaling system, we would expect individuals with positive signal residuals (e.g., A) to use the signal more often than those with negative signal residuals (e.g., B).

available evidence, it appears that cheating is impossible when cost-free *indices* are used in agonistic displays and unlikely when *costly* (during the time-scale of the encounter) dynamic agonistic displays are used; however, there is the potential for cheating to be stable at low frequencies when *conventional* agonistic displays (where the cost of the signal is paid during development) are used. Thus far, the approach of residual analysis has been applied to fighting in crustaceans (Hughes, 2000; Briffa, 2006; Arnott and Elwood, 2009) but it also offers the potential to test for the presence of low frequencies of bluffing in populations of other animals that use conventional signals of RHP, for example, display of horns or mandibles in insects, of plumage in birds or dewlaps in lizards.

Why do Fighting Animals Repeat Displays?

If fighting animals should advertise their RHP there are a number of ways through which this could be achieved. As noted earlier, agonistic signals appear to be incredibly diverse in terms of signal modality and the structures involved in their production. Despite this diversity, many examples of agonistic signals show similarities in their temporal structure, involving repeated performances of the display, often organized into a series of bouts of activity. Red deer, *C. elaphus*, for example, perform repeated bouts of roaring vocalizations (Clutton-Brock and Albon, 1979) and fallow deer, *D. dama*, perform bouts of groaning (Jennings *et al.*, 2005). Many birds such as collared doves, *Streptopelia decaocto* (ten Cate *et al.*, 2002) and great tits *Parus major* (Weary *et al.*, 1991), also use repeated vocalizations during territorial behavior. In hermit crabs, *Pagurus bernhardus*, shell fights occur over the ownership of empty gastropod shells. Encounters involves two distinct roles, attacker and defender. An attacker in a sub-optimal shell will initiate a fight against a usually smaller defender in possession of a shell that would offer an improvement. The encounter begins with a period of postural cheliped displays then the attacker lunges forward and grabs the defender's shell. The defender withdraws tightly into its shell and remains in this position until the end of the encounter. The attacker performs a series of bouts of *shell rapping* by rapidly and repeatedly striking the surface of its own shell against the shell occupied by the defender. The encounter ends either by a decision from the defender to relinquish its shell, allowing the attacker to evict it by pulling it out through the aperture; or the attacker may give up, and release the defender, without effecting an eviction. In attempting to induce the defender to give up, shell rapping is the key activity that attackers must perform. Compared to attackers that give up without evicting the defender, attackers that effect an eviction perform more raps and bouts of rapping in total and rap more vigorously by performing more raps per bout, leaving shorter pauses between bouts (Briffa *et al.*, 1998), rapping more rapidly within a bout (i.e., shorter intra-bout intervals between individual raps) (Briffa and Elwood, 2000a) and supplying more power to individual raps (Briffa and Elwood, 2000b, 2002; Briffa *et al.*, 2003) (Figure 6.2a). Furthermore, successful attackers escalate the vigor of rapping from bout to bout toward the end of the fight while attackers that give up de-escalate during the final bouts of the fight (Figure 6.2b).

Since the mid 1980s the question addressed by models of agonistic behavior has concerned the specific function of such repeated performances of signals assumed to advertise RHP, in other words, how they advertise RHP. Again, it is worth noting that although signal repetition has primarily been studied in the context of agonistic signals, much of the theoretical framework may be applicable to other situations when repeated signals are used to advertise an aspect of individual quality. Indeed in a review article focusing on contests Payne and Pagel (1997) nevertheless posed the question in more general terms of "Why do animals repeat displays?" Payne and Pagel (1997) provide an overview of three different reasons for repeated signals, which is still a useful starting point to which subsequent developments in theory can be appended. They suggested three "reasons" why repeated signals could be adaptive: (i) to confirm the previous actions, (ii) to augment previous actions, and (iii) to replace the previous actions. Each of these reasons is therefore associated with a different "rule" on the part of the receiver for deciding whether to give up. These *assessment rules* (although this may be a slightly misleading term, see below, this section) also called *decision rules* are the core assumptions

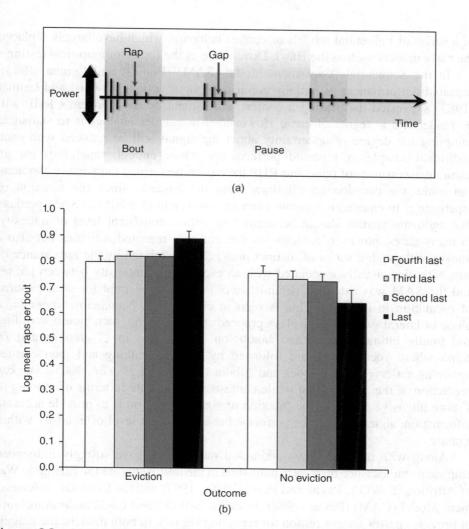

Figure 6.2. (a) Parameters of the pattern of shell rapping in hermit crab fights, an example of dynamic and "costly" repeated signal. Shell rapping is performed by the attacker during shell fights in hermit crabs. Successful attackers rap more intensely than those that give up, performing more raps per bout, leaving shorter pauses between bouts, rapping more quickly within a bout (such that intra-bout intervals or "gaps" are of shorter duration), and performing more powerful raps. For all attackers, the intensity de-escalates within bouts and between successive bouts at the start of the fight; but toward the end of the fight successful attackers re-escalate by performing more raps in each successive bout whereas attackers that give up perform progressively fewer raps in each bout. (b) Differences in the temporal pattern of change in shell rapping signals between attackers that evict the defender and attackers that do not effect an eviction. Over the last four bouts of rapping, successful attackers escalate by performing more raps in each successive bout whereas attackers that give up de-escalate by performing fewer raps in each bout. Such within-phase changes in intensity are not compatible with the sequential assessment model but are compatible with models based on "self assessment" such as the energetic War of Attrition or Cumulative Assessment Model. (Redrawn from Briffa *et al.* (1998).)

of a series of influential models of contest behavior, which have largely replaced the early models such as the Hawk-Dove game as the focus for empirical testing.

In the Sequential Assessment Model (SAM) (Enquist and Leimar, 1983), repeated performances essentially confirm previous actions. Enquist and Leimar (1983) suggested that during a contest, assessment of an opponent's RHP will be made with a degree of error. However, in a manner analogous to statistical sampling the degree of uncertainty about the signal will be reduced with each additional sample or repeated performance. Once enough repetitions for an accurate assessment of opponent RHP have been performed the weaker opponent can make the decision to withdraw from the contest. Since the function of repetition is to enable an accurate average assessment of RHP, the SAM predicts that agonistic signals should be performed with a consistent level of intensity. In many cases, however, contests involve not only repeated activities but also a transition through a series of distinct phases, characterized by the appearance of new activities, usually associated with an escalation in intensity between phases and the SAM was extended (Enquist *et al.*, 1990) to account for such patterns of escalation in intensity. This is seen in cichlid fish, *Nannacara anomala*, a phase of lateral swimming displays proceeds to tail beating, then mouth wrestling and finally biting (Enquist and Jakobsson, 1986); and in *C. elephas* and *D. dama* where vocalizations are followed by parallel walking and then clashes involving antlers (Clutton-Brock and Albon, 1979; Jennings *et al.*, 2005). The key prediction of the SAM is that while contests may escalate in terms of moving to a more intensive phase, as the function of signal repetition is to provide accurate information, signals should be performed at a consistent level of intensity within a phase.

Along with the SAM, two additional models that have strongly influenced empirical studies since they were published in the mid 1990s are the Energetic War of Attrition (E-WOA; Payne and Pagel, 1996a, 1997) and the Cumulative Assessment Model (CAM) (Payne, 1998). Both the E-WOA and CAM address the sum of previous performances reason for repeating signals. In both models, the magnitude of the signal of RHP is assumed to increase with each subsequent performance such that the signal accumulates as the contest progresses. Indeed, both the E-WOA and the CAM have been described as involving cumulative assessment rules. The E-WOA is the simpler of the two models and extends the simple WOA by replacing time as the main currency of cost, with energy. Thus the opponent that is able to allocate the greater amount of energy to the contest will win. In both models, the idea of assessment occurring is potentially misleading (Payne, personal communication) as there is no assumption that the loser bases its decision to give up on any direct information about the opponent (note however, that assuming equal RV the outcome of the contest will still be determined by the asymmetry in RHP between opponents, even though they have no means of gaining information on this during the encounter). Instead, the loser will give up when its own level of energetic costs, accrued through performing repeated agonistic displays, cross an individual-specific switching line of accumulated energy costs. Thus, if any assessment does occur, it is by each opponent of its own energetic state; these models have therefore been called *self-assessment* models (Taylor and Elwood, 2003) in order to distinguish them from *mutual assessment* models such as the SAM, where there is the assumption of information transfer between the two opponents. In theory, the different classes of models make different predictions about how contest duration should vary with relative and absolute differences between the opponent's RHP. It has become clear however, that it is difficult to distinguish between alternative assessment modes on the basis of these relationships alone (Taylor and Elwood, 2003; Gammell and Hardy, 2003) and it is also necessary to analyze the temporal structure of agonistic behavior as the contest progresses and the differences in costs accrued by winners and losers (Briffa and Elwood, 2009).

If giving up is based on self-assessment and there is no transfer of information on RHP between opponents, how is it that the agonistic behaviors used can be described as signals? As noted above, the widely accepted definition of a signal is an information carrier. For contests to be settled by self-assessment, the opponents must match one another's level of performance of agonistic behavior. In a contest without such matching, one individual could cheat by paying a lower cost while its opponent proceeds toward its giving up threshold. This matching implies that opponents monitor each other's behavior to some degree, so that the agonistic display signals that the sender is keeping up with its rival and is still in the contest. Such signals still do not reveal any direct information about the RHP of the sender that could be assessed independently of contest duration and matched rates of performance. In a self-assessment threshold model, the winner will be the individual that can keep going for longest and RHP should be determined by two factors: the position of this switching line and the rate at which costs accrue toward it for a given intensity of agonistic behavior. Therefore, it has been stated (Payne and Pagel, 1997) that when there is a cumulative assessment rule the RHP-linked trait advertised by agonistic displays is stamina.

The final possibility was that repeated performances replace previous performances. Here, senders are expected to begin with a low intensity signal, then perform a signal of greater magnitude if the low intensity performance is not sufficient to cause the opponent to give up. This process of replacing the previous performance with new performances of increasing magnitude has been called the *best so far* rule (Payne and Pagel, 1996b, 1997) and would continue until one of the opponents failed to increase the magnitude of its signal. In this respect, the best so far rule has some similarities with the E-WOA and CAM in that the giving up decision is based on an individual threshold. The predictions, however, are rather different as this model predicts only escalation in intensity whereas the cumulative models predict escalation, de-escalation, or no change depending on the rate at which costs (energy or injuries) accrue with time. Curiously, the signal replacement model has been less influential, in terms of the number of empirical studies that have tested its assumptions and predictions, than the models described earlier. A key prediction of this model, however, is that contests would be quite short, containing a low number of repetitions, or even a one–off display (Payne and Pagel, 1997) before the weaker opponent decides to give up. While some examples of agonistic encounters may fit this prediction (in their 1997 review, Payne and Pagel suggest male–male contests in anurans), the majority of study systems used in empirical work appear to involve protracted encounters. Thus, many empirical workers may have focused on alternative models as these seem more likely to fit their study system than does the best so far rule. Indeed, theoretical developments of the 1980s and 1990s continue to exert a major influence on how behavioral ecologists view contest behavior; they have led to a strong emphasis on investigations of protracted contests where, in addition to correlates of RHP such as body size and strength, the outcome of a contest is assumed to be driven by asymmetries in stamina (a slightly loose term, which can be defined more formally, see below, next section) between the opponents.

EMPIRICAL APPROACHES TO TESTING THEORY: "PHYSIOLOGICAL COSTS," "STAMINA," AND "PERFORMANCE"

Notwithstanding the possibility that cost-free indices of RHP may play a role in some contests, the assumption at the core of most models of contest behavior is that fighting should entail costs and that the opponent who can afford to pay the greater cost will win. In the Hawk-Dove game, these costs related to the negative fitness consequences of picking up an injury, and in the simple War of Attrition they relate to the negative fitness consequences of spending time on a contest. Later models added in the possibility that the energy expended during a contest could represent a significant cost. In the CAM two sources of cost, injuries (or some form of harm inflicted by the opponent) and energy expenditure, are assumed. Indeed it is possible to conceive of a suite of decision rules based on costs accruing through different combinations of time, injuries, and energy expenditure, that could be explored through future modeling work. There are also the possibilities that two opponents in the same fight may use different decision rules, if they adopt strongly asymmetric roles (Briffa and Elwood, 2005) and that decision rules might change between different phases of the encounter (Hsu et al., 2008). Irrespective of these possibilities, as noted above, the idea that the ability to pay energy costs should be a key correlate of RHP has captured the imagination of many researchers engaged in empirical work on agonistic behavior. In both theoretical and empirical studies, the capacity to allocate energy to a contest per unit of contest duration has been described as stamina. Stamina means endurance capacity, in other words the maximum duration for which an individual can continue to perform a given task at a given rate. Endurance is one component of what functional ecologists would term *whole body performance capacity*. Other examples of whole body performance capacity include various measures of strength (e.g., maximum bite or pulling force) and maximum speed. These might also have an influence on RHP and analysis of various aspects of whole body performance has also featured in a large body of work on agonistic behavior.

Of these performance capacities, current theory, as outlined already, has focused on the role of endurance. While endurance is a property of the whole organism, it will vary with a raft of underlying physiological factors such as energy stores, endocrine status, and aerobic capacity. These factors can vary temporally, over daily, seasonal and developmental timescales, and in the case of energy and hormones, during the time-scale of an agonistic encounter. Thus there are two potential approaches for testing hypotheses about the role of endurance during contests. One is to investigate the physiological underpinnings of endurance, by assaying for concentrations of metabolites or metabolic by-products associated with energy status. Such measures are typically made at the end of fights and focus on variables that could change as a result of engaging in agonistic behavior. Thus, the energy status approach (or "evolutionary physiology"; Irschick *et al.*, 2007) analyses dynamic variables that have the potential to change in the timescale of an agonistic encounter. This approach of linking agonistic signals to physiological status, has been applied to a wide range of sexually and naturally selected agonistic signals. It has also been applied to contests involving other agonistic assessment activities such as wrestling. A central aim of this work has been to discover the mechanisms underlying (and indeed the existence of) the behavioral switching lines assumed by theory to promote the giving up decision.

Such physiological variables, and the rates at which they change for a given rate of activity, may be linked to stamina, and between-individual variation in post-fight energetic status may reflect differences in stamina. But they do not represent stamina *per se*. A second approach, therefore, is to attempt to measure either stamina or other potential whole body correlates of RHP directly, through the route taken by functional ecologists of investigating whole body performance capacities. This approach was originally applied to the investigation of signals used in sexual displays (Irschick and Garland, 2001) but has been extended to analyze sexually selected traits that are also used for direct male–male competition (see Irschick *et al.*, 2007). Measures of the performance of a dynamic ecologically relevant task, such as locomotion, jumping, or biting are used to gauge some aspect of the individual's whole body performance capacity, such as stamina or strength. These measures may then be analyzed in relation to the contestant's performance during an agonistic encounter, or to an individual's expression of a trait (e.g., an ornament or weapon) that could either be useful in an agonistic encounter or is likely to be an actual adaptation for fighting. Studies adopting this approach are in general not primarily concerned with testing the predictions of the accumulated theoretical models of fighting, which are of so much interest to behavioral ecologists. Rather than focusing on the possible physiological basis of the decision rules suggested by theory, performance-based studies aim to determine the mechanisms underlying frequently observed correlations between factors such as the expression of a given trait and an individual's success in situations such as courtship and contests. In contrast to studies of energy status that focus on dynamic, repeated agonistic displays, the primary focus of performance studies has been on the evolution of the ornaments or weapons used, particularly if such traits are considered to be sexually selected. As these structures are paid for over developmental time (rather than during the timescale of an agonistic encounter), this approach brings an ontogenetic dimension to the study of signals.

Nevertheless, such an approach could also be useful for testing the key assumption of the theory that dynamic repeated agonistic signals should be related to stamina, and the prediction that winners should have greater stamina than losers.

Both of these approaches are concerned with linking the evolution of signals of individual quality with underlying proximate mechanisms and both have addressed the question of agonistic signals. Furthermore, both approaches have been used to investigate the possibility that stamina might be an important determinant of the magnitude of the agonistic signal that an individual can produce. In the following sections, I review the evidence linking agonistic success with stamina provided by both types of approach. I then highlight current links between the two approaches and discuss the potential for combining them. But first there is one further point to make: In addition to these two approaches for analyzing the energy constraints on agonistic behavior, it is important not to forget that new insights can be gained from direct ethological observation. At the very least, this is necessary in order to determine which opponent should be deemed the winner during subsequent data analysis; in some examples this will be very clear, but in others some sort of scoring system is required. Detailed behavioral observations are also useful for reasons other than applying scoring systems. In addition to making predictions about the costs of engaging in agonistic encounters, the various assessment models described above make predictions about how the intensity or vigor of the performance of agonistic behavior should vary during the fight (note that it is often useful to maintain a distinction between the performance of agonistic behavior during the fight, and although potentially linked to it, whole body performance capacities of ecologically relevant tasks outside of a fight). Briefly, the SAM predicts that agonistic signals should be performed at a constant rate within a phase of a contest and contests will escalate by progressing through a succession of phases containing increasingly costly behaviors; and threshold-based models predict that the intensity of agonistic signals will escalate if the rate of cost-increase is super-linear (i.e., more than a doubling in cost units for every doubling in time units), de-escalate if it is sub-linear, and stay the same if the rate of increase is linear. The reasons for these links between the time—cost relationship and temporal changes in agonistic behavior have been detailed elsewhere (see Payne and Pagel, 1997 or Briffa and Sneddon, 2010), but it appears that the best approach to understanding the functions of agonistic signals is to couple some measure relating to the proximate mechanisms involved in performing an agonistic signal (such as metabolite assays or whole body performance) with behavioral observations that are detailed enough to pick up changes in intensity.

ENERGY STATUS AND AGONISTIC SIGNALS

Theory assumes that non-injurious fighting should be demanding in the sense of requiring high rates of energy expenditure. This assumption is based on observations of fighting animals; studies on red deer *C. elaphus* reported apparently exhausted males collapsing after intense bouts of roaring (e.g., Clutton-Brock and

Albon, 1979) and it has been suggested that performing vocalizations may exhaust male birds (Weary *et al.*, 1991; ten Cate *et al.*, 2002) and anurans (Ryan, 1988) during agonistic encounters. Crustaceans such as portunid crabs, *Necora puber* and *Carcinus maenas*, show elevated ventilation (Smith and Taylor, 1993) and heartbeat (Rovero *et al.*, 2000) rates during fights, indicating that even the performance of postural agonistic displays can be demanding. In *C. maenas*, losers but not winners also showed increased ventilation rates during a post-contest recovery phase, indicating that losers were subject to greater energy demands than were winners. Joint respiration rates (by the winner and loser), investigated using respirometry in house crickets *Acheta domesticus* (Hack, 1997) and sierra dome spiders *Neriene litigiosa* (DeCarvalho *et al.*, 2004) show that metabolic rate increases as the contests escalate in intensity from displays to physical combat. This is predicted by the SAM, but in *N. litigiosa* there is also a progressive increase in respiration rates within phases. Several studies support the predictions of the SAM (e.g., Enquist *et al.*, 1990) but these within-phase changes in respiration in *N. litigiosa* make it unlikely that agonistic behaviors are repeated in order to enhance the accuracy of assessment in this case.

The alternative explanation for costly agonistic behavior is that the contest is settled on the basis of individual cost thresholds (E-WOA, CAM). While respirometry allows costs to be tracked through the encounter, it does not allow the costs that accrue to each opponent to be measured independently. An alternative approach is to analyze post-fight metabolites and by-products (although this does not allow the continuous tracking of energy that can be achieved with respirometry). Energy costs will be due primarily to the units of glucose expended during the contest. Under aerobic conditions, metabolism of one molecule of glucose will generate a theoretical maximum of 38 molecules of ATP (2 ATP from anaerobic glycolysis and 36 ATP from aerobic respiration). ATP is the molecule that interacts with muscle filaments, in the final steps of converting the chemical energy stored in food to the kinetic energy used in behavior. The glucose used for ATP production may originate from molecules freely circulating in the hemolymph but demanding activity may also involve the mobilization of muscular glycogen stores. Investigation of aerobic metabolism will not necessarily reveal all of the energy costs of fighting. If the agonistic behavior is performed with high intensity, the ventilation and circulatory systems may not be able to supply oxygen in sufficient quantities to allow the energy demands to be met through aerobic respiration. In this case, the shortfall in ATP generation can be achieved in the absence of oxygen but such anaerobic respiration is more costly than aerobic respiration. First, the process is far less efficient in terms of glucose depletion, yielding only two molecules of ATP from glycolysis. Second, pyruvate, the end product of glucolysis, which under aerobic conditions is converted to Acetyl CoA (thereby transferring the bulk of the energy stored in the original glucose molecule to the citric acid cycle to produce ATP aerobically) is instead fermented to produce large amounts of lactic acid, which accumulates in muscle tissue. This reaction is necessary to produce NAD^+, which is needed to maintain glycolysis under anaerobic conditions. Although the exact mechanism is still not fully understood accumulated lactate (the lactic acid will rapidly dissociate to lactate and H^+ ions) is known to be associated with impaired muscle function and a major cause of fatigue (depleted ATP levels as a cause of fatigue are unlikely; low ATP would permanently damage muscle tissue and levels are buffered by a pool of intermediary high energy phosphates such as arginine phosphate in most metazoans and creatine phosphate in vertebrates). The effects of prolonged anaerobic respiration appear to be particularly marked in crustaceans. The oxygen debt of accumulated lactate results in a post-activity recovery phase, which constrains activity rates until lactate levels have been reduced to resting levels in the presence of oxygen. Investigations of contest energetics have therefore focused on assays of lactate in addition to assays of metabolites including circulating glucose and muscular glycogen stores.

In portunid crabs, there are no differences in post-fight metabolites between winners and losers and fighting does not seem to lead to exhaustion (Thorpe *et al.*, 1995; Sneddon *et al.*, 1999). In hermit crabs, *P. bernhardus*, the two distinct roles adopted in fights appear to make their decisions on the basis of different physiological changes. In attackers there is no difference in post-fight metabolites between outcomes but defenders that resist being evicted from their shell have higher circulating glucose levels than evicted defenders (Briffa and Elwood, 2002). Thus, in shell fights, energy status does vary between outcomes but only for defenders. Rather than being caused by crossing a cost threshold, however, this difference appears to be the result of an early decision about the level of resistance to offer. Mobilization and depletion of energy reserves is one potential cost of fighting but intense encounters are likely to involve the additional cost of anaerobic respiration. Differences in aerobic capacity, as well as in energy reserves, are therefore one way that individuals may differ in the rate at which costs accumulate as envisaged by self-assessment based models. Indeed the use of "opercular displays" during contests in fish may have evolved as a means of demonstrating aerobic capacity. Raising the gill covers disrupts ventilation and Siamese fighting fish, *B. splendens*, show a severely reduced display rate under hypoxic conditions (Abrahams *et al.*, 2005). Anaerobic respiration appears to play a role in fights in a range of other vertebrates (e.g., cichlid fish Neat *et al.*, 1998; snakes Schuett and Grober, 2000) but appears to be particularly important in crustaceans. In *C. maenas*, there is a prolonged post-fight elevation in lactate (Sneddon *et al.*, 1999), which constrains the capacity for other activities until the oxygen debt has been repaid. This indicates that if the basis of giving up decisions is an individual threshold, these costs may persist after the fight has been resolved. Nevertheless, lactic acid accumulation also has the capacity to constrain agonistic behavior during the encounter, and has been investigated in relation to contest structure and outcomes. In the fiddler crab, *Uca lectea perplexa*, claw waving rates decline with increasing lactate as the contest progresses (Matsumasa and Murai, 2005). This is predicted by the E-WOA, which is well suited to examples such as this based on costly signals. De-escalation is also possible under the CAM but this model is more appropriate to situations where the opponents directly impose costs on one another. A similar result is seen during shell fights in hermit crabs. In both attackers and defenders, lactate increases as a result of fighting but there is only a difference between outcomes in the case of attackers

(Briffa and Elwood, 2001b, 2002). During this phase of the encounter, the pattern of rapping can either escalate or de-escalate with successful attackers increasing in intensity before evicting the defender and attackers that give up performing progressively fewer raps in each bout (see Figure 6.3). For attackers, the accumulated costs of fighting appear to drive changes in intensity as predicted by the E-WOA and CAM. However, these models cannot completely explain agonistic behavior in hermit crabs as they do not encompass contests with strongly asymmetric roles using different agonistic behaviors. Nevertheless, these studies on contest energies, in a range of study systems, show that energy metabolism is a clear source of the directly incurred costs assumed by theory to be associated with non-injurious fighting.

Mowles *et al.* (2009) investigated the links between circulating hemocyanin levels and fight outcomes in hermit crabs. This pigment is the main respiratory compound in crustaceans, and is involved in the transport of oxygen from the gills to the hemolymph-bathed muscle tissues via a semi-closed circulatory system. Hemocyanin levels vary between individuals and will strongly influence aerobic capacity – the rate at which oxygen can be transported to muscle tissues and therefore the amount of work that can be done aerobically before the onset of anaerobic respiration. In this sense, hemocyanin will be a key determinant of stamina, specifically endurance. Both attacking and defending hermit crabs that won the encounter had high hemocyanin levels compared to those that gave up. Thus, although the basis of their decision rules seems to be a different physiological variable, success in a fight appears to be influenced by this correlate of stamina for both roles. While hemocyanin is a physiological sample taken at the end of the encounter, it does not vary during the timescale of the fight (although it can vary seasonally and with medium term changes in the oxygen content of seawater). Therefore, as a correlate of RHP, hemacyanin is closer to the whole body performance capacities of interest to functional morphologists than to the dynamic measures of energetic status that are typically analyzed by evolutionary physiologists. Another approach is to obtain some direct measure of whole body performance in fighting animals.

WHOLE BODY PERFORMANCE AND AGONISTIC SIGNALS

Animals such as hermit crabs and red deer perform bouts of non-injurious but energy demanding activity during agonistic interactions. Such behaviors appear to function as costly signals of stamina, assumed to be a key component of RHP by models such as the E-WOA and the CAM. If this is true, an individual performing such activities is actually demonstrating a key whole body performance capacity to its rival (one might therefore argue that these agonistic signals are best thought of as *indices* rather than *handicap* signals but they do not appear to fit well with the idea that indices should be cost free). Stamina is assumed to be of central importance by current models, but other whole body performance capacities, for example, strength or speed, could also influence RHP. It is easy to imagine agonistic behaviors that could demonstrate these factors, particularly strength, as many contests do involve trials of strength with activities such as grappling in shore crabs (Huntingford et al., 1995) and antler wrestling in red deer (Clutton-Brock and Albon, 1979). Jump clashes in fallow deer (Jennings et al., 2005) may even depend on both strength and speed. Thus performance capacities that influence RHP can be demonstrated through the use of costly signals or activities such as wrestling but they could also be advertised through the use of conventional signals given by relatively static postural display of morphological traits. These may be used either in combination with costly signals (e.g., claw waving in fiddler crabs, Morrell et al., 2005) or during a particular phase of the agonistic encounter, which then escalates to a more costly phase (e.g., shore crabs). As signals of RHP are expected to be honest on average, there would need to be a positive correlation between the expression of the morphological trait and the performance capacity in question in order for such a signaling system to be evolutionarily stable. The approach of functional morphologists to the investigation of agonistic behavior has been to investigate the possibility of such correlations.

In some cases, morphological traits that could be used to advertise RHP might also be used as weapons to influence the outcome through direct physical effects on the opponent. In crustaceans, for example, chelar size correlates with pinch strength (Sneddon et al., 2000; Bywater et al., 2008). In male fig wasps where the males have a one-off chance of mating success the mandibles are used to decapitate rivals males (Hamilton, 1967) and fatality rates increase with average mandible size (Bean and Cook, 2001). Within the inflorescence of a fig, however, it is unlikely that opponents could assess one another's RHP using information about mandible size so the mandibles are probably not used for signaling activities in this example. In other examples, however, although mandibles can be formidable weapons, they are also used in flaring displays, as in various species of cricket (Hack, 1997; Hofmann and Schildberger, 2001). This is similar to the "meral spread" used during fights in crustaceans such as shore crabs, C. maenas and large weapons equate to high RHP in both shore crabs (Sneddon et al., 1997) and house crickets (Briffa, 2008). In these examples, there is still an obvious link between the ornaments used to advertise RHP and RHP itself because the ornaments are also weapons, which generally increase in effectiveness with size. But in other examples, ornaments which predict fighting success appear to have little obvious utility as weapons and it is therefore unclear how they could directly influence fight outcomes other than as a source of information on RHP. While both males and females may bear weapons (such as claws or powerful mandibles) such ornaments are more likely to be sexually selected, usually appearing in males only (although the opposite is of course true in species with sex-role reversed mating systems such as several species of pipe fish where it is the females that possess coloration and secondary sexual ornaments used in competition over access to males).

Examples of male ornaments that seem to be associated with winning contests include horns and elaborate mandibles in beetles and dewlaps in lizards. In several species of lizard of the genus Anolis, the dewlaps or throat fans are displayed during agonistic encounters. Vanhooydonck et al. (2005) demonstrated a positive correlation between bite force and throat fan area independent of body size. Thus there is the potential for throat fans to advertise a performance capacity that would contribute to RHP in escalated injurious fighting. On the other hand, further work across different species in this genus shows that the links between performance capacities and dewlap size may be complicated by inter-specific differences in the likelihood of escalation to actual biting in these encounters. Bite force predicted agonistic success to a greater extent and dewlap size correlated with bite force most convincingly in species where there was a high degree of sexual dimorphism compared to those where dimorphism was low. Despite the fact that one would therefore expect dewlap displays to feature to a greater extent in highly dimorphic species, the opposite is seen with highly dimorphic species more likely to use biting (Lailvaux and Irschick, 2007). Thus, the information content of such morphological signals may vary with differences in life history and ecology in ways that are difficult to predict. This possible lack of generality in the information contained in agonistic signals can be revealed by such comparative studies, a perspective that is currently absent from studies of costly agonistic signals. Other examples of contest behavior in lizards similarly underline the complexities of the links between conventional signals and whole body performance. Huyghe *et al.* (2005) found that males of the wall lizard *Gallotia galloti* are larger than females and various measures of male head size correlate with maximum bite force. This performance capacity is a clear predictor of the outcome of agonistic encounters, with winners capable of biting significantly harder than losers. In addition, males possess distinct patches of blue colored scales along their flanks and the degree of expression of this coloration increases with position in dominance hierarchies, a key resource over which males fight. However, there was no correlation between maximum bite force and the area covered by blue scales (Huyghe *et al.*, 2005). Recent work (Huyghe *et al.*, 2005) indicates that head morphology and hence bite force varies between color morphs of the lizard *Podarcis melisellensis*, perhaps due to differences in hormone levels. At present, however, the precise way in which patches of colored scales advertise RHP during aggression in lizards remains unclear.

Male beetles often bear ornaments formed by cuticular outgrowths projecting forward from the head or in some cases elaborate mandibles. They are used in a variety of ways during fights (reviewed by Snell-Rood and Moczek, 2013) including locking and grappling in a manner similar to antler locking in red deer. The weapons may therefore be used to demonstrate strength directly, but there are also effects of weapon size, independent of body size, on the likelihood of winning. In the horned dung beetles *Onthophagus acuminatus* and *Onthophagus taurus*, for example, horn size predicts victory when opponents are matched in body size (Emlen, 1997; Moczek and Emlen, 2000). These effects may be due to mechanical advantages associated with large horn size such as greater reach and leverage. However, the use of horns to throw, pin, or dislodge opponents will also be influenced by performance capacities such as strength, speed, and stamina. Therefore there is the potential for the expression of these traits to vary with RHP independently of their role as weapons, if there are correlations between measures of performance capacity obtained in a non-agonistic contest and measures of the size of these structures. In *E. intermedius*, horn size predicts fighting success more clearly than does overall body size (Pomfret and Knell, 2006) and Lailvaux *et al.* (2005) investigated the correlations between horn size and two performance capacities: pulling force, a measure of strength, and maximal distance achieved during forced locomotion, a measure of endurance. In this study, *E. intermedius* were found to be able to pull weights ranging from 228 times to an impressive 425 times their own body weight, a reminder of the incredible performance capacities of many arthropod species compared to members of other phyla such as chordates, for example. Although pulling force varied with body length, when corrected for body length, both pulling force and endurance increased with horn size. Furthermore, although these male–male contests occur within tunnels (dug within the dung exclusively by females) and involve tactics such as pulling, there is also the opportunity for opponents to make a visual assessment of one another's horn length (Lailvaux *et al.*, 2005). Thus, there is the clear potential for a link between whole body performance and a signal of RHP in these species.

Links between whole body performance and contests have also been studied in crustaceans. In the fiddler crab *Uca mjoebergi*. Males bear one claw that is markedly oversized, often exceeding carapace width in length. The resulting asymmetry in claw length is absent in females, where both claws are small and of equal size. The large claw possessed by males is not useful as a tool for feeding (Valiela *et al.*, 1974) and in some species makes males more visible to predators. During agonistic encounter, it is used in static postural displays and also in costly waving displays, which also involve using the walking legs to raise the carapace off the ground. Through these conventional and dynamic displays, RHP assessment plays a key role in determining the outcome of fiddler crab fights (Jennions and Backwell, 1996; Backwell *et al.*, 2000; Morrell *et al.*, 2005; Reaney *et al.*, 2007). In highly escalated encounters, the claws may be used as weapons when the crabs engage in activities such as grappling, pinching, and pulling. Lailvaux *et al.* (2009) investigated the links between whole body performance and the expression of claw length, the conventional component of the signal. Claw length increased independently of body size with two performance measures relevant to fighting, maximum pinch force, and the force required to dislodge the crab from a burrow. Thus there is a clear link between whole body performance and this conventional signal used in agonistic encounters. As described earlier, there is the potential in crustaceans for performance capacity to become decoupled (Lailvaux *et al.*, 2009) from signal magnitude due to the processes of post-moult hardening and regeneration of appendages that have been lost. Regenerated claws are not as strong as non-regenerated claws as evidenced by a reduction in both types of performance capacity in crabs with regenerated claws. Nevertheless, there are only slight differences in appearance between the two types of claws and individuals bearing regenerated claws are still able to win agonistic encounters by using these inferior appendages for agonistic signaling (Jennions and Backwell, 1996; Backwell *et al.*, 2000; Lailvaux *et al.*, 2009). Therefore, the approach of analyzing signal residuals applied to species where body size is assumed to be the key correlate of

RHP might also be useful where there is a relationship between a signal and whole body performance. This approach has already been used to analyze relationships between dynamic signal performance and specific physiological variables (Briffa, 2006).

A final example of links between whole body performance and agonistic signals is given by another study by Mowles *et al.* (2010) on shell fighting in hermit crabs. As already noted, an earlier study (Mowles *et al.*, 2009) shows how variation in a potential correlate of whole body performance (hemocyanin) correlates with success in contests. In this next study, the crabs were forced to exercise around a circular raceway. Then following recovery from this procedure, the crabs were allowed to engage in shell fights, as described above. No hermit crabs were exercised until exhaustion, but measures of average walking speed and the speed of the fastest lap were obtained. Successful attackers and successful defenders had greater average walking speed than the opponents they defeated. Thus, while analysis of specific metabolites shows that the mechanisms of the giving up decision vary between roles, analysis of whole body performance provides a different insight: whole body performance, a component of stamina, is important for both parties in the contest.

These examples demonstrate links between whole body performance capacities relative to agonistic behaviors and morphological traits that could advertise RHP. Such correlations are a pre-requisite for the possibility that morphological traits could be used in conventional agonistic signaling. The techniques adopted in this approach, however, are of further relevance to questions about the functions of agonistic signals. First, as noted already, the traits examined are often sexually selected so such studies have focused on male–male fights but there is no reason why the approach could not also be applied to other examples of contest behavior that are not a result of sexual selection (note equally that the body of theory developed around repeated agonistic signals could also be applied to sexual signals – see Mowles and Ord, 2012). Second, one performance capacity in particular, endurance, occupies a central role in current theories of agonistic signaling. Therefore the approach of investigating endurance capacity, would be directly useful for testing the assumptions of current threshold models such as the CAM, E-WOA. In this case, we would expect to see significant correlations between endurance capacity and the chance of victory and endurance capacity and the vigor with which repeated signals are performed. While several studies have analyzed the links between performance capacities and the expression of morphological traits used in signaling, it is curious that the approach has yet to be applied to the question of repeated signals; as these signals are assumed to advertise stamina it seems obvious that it would be useful to obtain measures of stamina independent of the signaling activity in question. Finally, functional morphologists bring a comparative approach to the study of agonistic behavior (e.g., between related species Lailvaux and Irschick, 2007; or color morphs Huyghe *et al.*, 2009) a perspective that is traditionally missing from studies where the focus is on signals based on behavioral activities rather than morphological displays. Nevertheless, the comparative approach has been successfully applied to other types of costly signal and could also be adopted for the study of repeated signals used in aggression when closely related species perform similar behaviors.

CONCLUSIONS

Two approaches to investigating the links between functions and underlying mechanisms of agonistic signals have to a certain extent run in parallel. One approach of investigating contest energetics has primarily focused on the use of costly signals and other demanding agonistic behavior whereas the second approach of investigating performance capacities has primarily focused on conventional signals given by relatively static postural displays. Clearly, the distinction between the two types of signal will not be clear cut and in some cases both types of signals have been described as ornaments (e.g., Kotiaho, 2001; Lailvaux and Irschick, 2006). Nevertheless a distinction may be useful, at least in defining the questions to be addressed and in identifying areas of useful communication between the two approaches. Both types of signals are costly in that resources must be allocated to their production but the difference lies in when these costs are incurred. If the agonistic signal involves demanding, often repeated performances of a behavioral activity such as shell rapping or vocalizing this will incur the accumulation of costs within the time-frame of the encounter. If the agonistic signal involves the display of an ornament or a weapon, such as a mandible, chelar, or dewlap flare, the cost will have been paid through the allocation of energy to the growth of the ornament during developmental time. Thus, the best approach might be determined by the type of signal under investigation but as noted earlier, agonistic encounters may involve the use of both types of signals of RHP.

Although both approaches have been primarily applied to one type of agonistic signal, each approach has the potential to bring new insights to the study of aggression if applied to a different type of signal, both in terms of the theoretical framework and the practicalities of investigating agonistic signal function. The performance capacity approach favored by functional morphologists has the potential to bring a developmental and comparative aspect to the study of costly signals, broadening the context in which they are usually discussed. While the main signal costs may be paid during the timescale of an agonistic encounter, the necessary muscle physiology may be influenced during development. Indeed, studies like Royle *et al.* (2005) on dominance in swordfish, *Xiphophorus helleri*, show how developmental experiences such as catch-up growth might influence both the development of morphological traits used in signaling and the performance of the display activity itself; in this case, development had no effect on ornament expression, but it did influence the chance of winning a fight. More directly, the performance capacity approach can provide ways of testing the key idea from theory that costly agonistic signals demonstrate stamina, in addition to testing alternative hypotheses that they could demonstrate other performance capacities that could influence RHP. In terms of studies of conventional signals, approaches usually applied to the study of costly signals could lead to a more in-depth consideration of how such signals relate to the main theoretical models of contest behavior. Indeed, attending to the predictions of the alternative models could help to solve outstanding questions about whether morphological features that correlate with performance capacities, such as horns, claws, or elaborate mandibles (and indeed, patches of coloration), are actually assessed by the opponent. Recent developments in distinguishing between mutual and self assessment (Taylor and Elwood, 2003; Gammell and Hardy, 2003; Briffa and Elwood, 2009; review in Arnott and Elwood, 2009) decision rules could be useful in distinguishing between traits that can be thought of as signals and traits that, although correlate with whole body performance, are not assessed during fights and are therefore best regarded as weapons or sexual signals rather than agonistic signals.

Figure 6.3 provides a schematic of how the two approaches could provide insights into the functions of costly and conventional signals and suggestions for the specific questions at which they could be targeted. Table 6.1 provides examples of studies that have applied these approaches to these questions; it can be seen that although several study systems have been investigated from both perspectives,

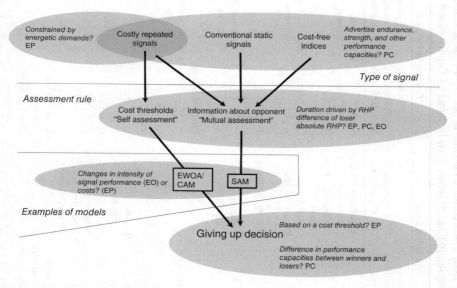

Figure 6.3. Schematic illustrating key components of agonistic signaling systems. Arrows represent the flow of information, which may be carried by a signal or relate to costs that have been accrued as a result of performing a signal. Questions that can give insights into the functions of agonistic signals are adjacent to each component that they apply to. They can be broadly grouped into questions about: the type of signal, the nature of the assessment rule, and how well the behavior fits a given theoretical model of fighting. A range of approaches can be used to study agonistic signals but the best approach may vary according to the component of the signaling system being studied: EP, evolutionary physiology (*sensu* Irschick *et al.*, 2007); PC, whole organism performance capacity; and EO, ethological observation.

TABLE 6.1. Selected Examples of Study Systems where the Links Between the Proximate Mechanisms and the Functional Significance of Agonistic Signals have been Investigated using Approaches of Whole Body Performance Capacities and Evolutionary Physiology

| Study System | Approaches Used | | Features Investigated | | Main Theoretical Questions Addressed | References |
	Whole Body Performance Capacities	Evolutionary Physiology	Conventional Postural or Static Signals	Dynamic Repeated Signals		
Red and fallow deer		Anecdotal reports of post-fight exhaustion in red deer	Antler morphology	Roaring/groaning	Assessment rules, signal honesty, sexual selection	Clutton-Brock and Albon (1979); Jennings et al. (2005)
Siamese fighting fish		Lactate	Opercular size	Opercular flaring	Signal honesty	Abrahams et al. (2005)
Crickets		Respiration rate		Stridulation	Assessment rules, sexual selection	Hack (1997); Hofmann and Schildberger (2001)
		Glucose, glycogen	Mandible asymmetry		Assessment rules	Briffa (2008)
Hermit crabs		Lactate, glucose, glycogen,	Cheliped displays	Pattern of shell rapping	Assessment rules	Briffa and Elwood (2001a, 2001b, 2002, 2004, 2005)
	Respiratory pigment			Pattern of shell rapping	Signal honesty	Briffa (2006)
	Locomotion			Pattern of shell rapping	Assessment rules, signal honesty	Mowles et al. (2009)
				Pattern of shell rapping	Assessment rules	Mowles et al. (2010)
Portunid crabs		Lactate, glucose, glycogen.			Assessment rules	Thorpe et al. (1995); Sneddon et al. (1999)
	Claw strength	Respiration and heartbeat rates	Claw size		Assessment rules	Sneddon et al. (1997, 2000)
Fiddler crabs	Claw closing force, claw pull resistance	Lactate, glucose	Claw size and weight	Waving	Assessment rules	Smith and Taylor (1993); Rovero et al. (2000)
					Signal honesty	Matsumasa and Murai (2005)
					Assessment rules	Lailvaux et al. (2009)
Wall lizards and Anolis lizards	Bite force, endurance		Dewlap size, blue patches		Signal honesty	Huyghe et al. (2005); Lailvaux and Irschick (2007)
Horned beetles	Endurance, strength		Horn morphology	Horn morphology	Sexual selection	Lailvaux et al. (2005)

there is relatively little overlap between the two approaches. A final potential benefit of applying both approaches to these questions is illustrated by the example of claw waving in male fiddler crabs. As noted earlier, enlarged ornaments such as claws can be used in both postural and dynamic displays. Thus, a focus on the constraints on dynamic performances during a contest might neglect the effects of developmental constraints on the structures used for the display. Similarly, focusing on links between whole body performance and the expression of a trait may not take into account how the trait is actually used during a contest. Such examples highlight the possibility that there may be trade-offs between investment in growth and performance capacities and between different performance capacities.

As yet there is little cross-fertilization between these two approaches to the study of agonistic signals. Although elements of both approaches have been applied to studies of crustaceans including portunid crabs, hermit crabs, and fiddler crabs they have yet to be integrated within the same studies (see Table 6.1). From a behavioral ecology perspective, the study of contest behavior has focused on testing the assumptions and predictions of models derived from evolutionary game theory; the aim of these models has been to propose functions – that is, fitness benefits – for various types of agonistic behavior including signals. Thus, the aim of behavioral ecologists investigating aggression has been to determine the adaptive value of agonistic behavioral patterns employed during contests. Although the various models of fighting can be distinguished through the behavioral differences they predict, they also assume different relationships between the costs of fighting, agonistic behavior, and contest outcomes so behavioral ecologists have also had recourse to physiological investigations of contest behavior. This has primarily been achieved through the assay of contest energetics. Functional morphologists are also interested in the adaptive value of agonistic signals and have also had recourse to physiological approaches. Here the focus has been on whole body performance capacities and conventional signals. Combining these two approaches has the potential to advance our understanding of the adaptive value of agonistic signals; this will require ethological, morphometric, metabolic, and whole body physiological approaches to address the hypotheses generated by current models of animal contests. In particular, studies looking at potential links between whole body performance and the performance of costly dynamic signals used during fights are currently lacking.

REFERENCES

Abrahams, M.V., Robb, T.L., and Hare, J.F. (2005) Effect of hypoxia on opercular displays: evidence for an honest signal? *Animal Behaviour*, **70**, 427–432.

Adams, E.S. and Caldwell, R.L. (1990) Deceptive communication in the stomatopod crustacean Gonodactylus bredini. *Animal Behaviour* **39**, 706–716.

Arnott, G. and Elwood, R.W. (2009) Assessment of fighting ability in animal contests. *Animal Behaviour*, **77**, 991–1004.

Arnott, G. and Elwood, R.W. (2010) Signal residuals and hermit crab displays: flaunt it if you have it!. *Animal Behaviour*, **79**, 137–143.

Backwell, P.R.Y., Christy, J.H., Telford, S.R., Jennions, M.D., and Passmore, N.I. (2000) Dishonest signalling in a fiddler crab. *Proceedings of the Royal Society of London, Series B: Biological Sciences*, **267**, 719–724.

Bean, D. and Cook, J.M. (2001) Male mating tactics and lethal combat in the non pollinating fig wasp *Sycoscapter australis*. *Animal Behaviour*, **66**, 535–542.

Blanchet, S., Paez, D.J., Bernatchez, L., and Dodson, J.J. (2008) An integrated comparison of captive-bred and wild Atlantic salmon (*Salmo salar*): implications for supportive breeding programs. *Biological Conservation*, **141**, 1989–1999.

Breithaupt, T. and Atema, J. (2000) The timing of chemical signaling with urine in dominance fights of male lobsters (*Homarus americanus*). *Behavioral Ecology and Sociobiology*, **49**, 67–78.

Briffa, M. (2006) Signal residuals during shell fighting in hermit crabs: can costly signals be used deceptively? *Behavioral Ecology*, **17**, 510–514.

Briffa, M. (2008) Decisions during fights in the house cricket, *Acheta domesticus*: mutual or self assessment of energy, weapons and size? *Animal Behaviour*, **75**, 1053–1062.

Briffa, M., Elwood, R.W., and Dick, J.T.A. (1998) Analysis of repeated signals during shell fights in the hermit crab *Pagurus bernhardus*. *Proceedings of the Royal Society of London, Series B: Biological Sciences*, **265**, 1467–1474.

Briffa, M. and Elwood, R.W. (2000a) Analysis of the finescale timing of repeated signals: does shell rapping in hermit crabs signal stamina? *Animal Behaviour*, **59**, 159–165.

Briffa, M. and Elwood, R.W. (2000b) The power of shell rapping influences rates of eviction in hermit crabs. *Behavioral Ecology*, **11**, 288–293.

Briffa, M. and Elwood, R.W. (2001a) Motivational change during shell fights in the hermit crab *Pagurus bernhardus*. *Animal Behaviour*, **62**, 505–510.

Briffa, M. and Elwood, R.W. (2001b) Decision rules, energy metabolism and vigour of hermit-crab fights. *Proceedings of the Royal Society of London, Series B: Biological Sciences*, **268**, 1841–1848.

Briffa, M. and Elwood, R.W. (2002) Power of shell-rapping signals influences physiological costs and subsequent decisions during hermit crab fights. *Proceedings of the Royal Society of London, Series B: Biological Sciences*, **269**, 2331–2336.

Briffa, M., Elwood, R.W., and Russ, J.M. (2003) Analysis of multiple aspects of a repeated signal: power and rate of rapping during shell fights in hermit crabs. *Behavioral Ecology*, **14**, 74–79.

Briffa, M. and Elwood, R.W. (2004) Use of energy reserves in fighting hermit crabs. *Proceedings of the Royal Society of London, Series B-Biological Sciences*, **271**(1537): 373–379.

Briffa, M. and Elwood, R.W. (2005) Rapid change in energy status in fighting animals: causes and effects of strategic decisions. *Animal Behaviour*, **70**, 119–124.

Briffa, M. and Elwood, R.W. (2009) Difficulties remain in distinguishing between mutual and self-assessment in animal contests. *Animal Behaviour*, **77** (3), 759–762.

Briffa, M. and Sneddon, L.U. (2010) Chapter 15 – Contest behaviour, in *Evolutionary Behavioral Ecology* (eds D.F. Westneat and C.W. Fox), Oxford University Press, New York.

Bywater, C.L., Angilletta, M.J. Jr., and Wilson, R.S. (2008) Weapon size is a reliable indicator of strength and social dominance in female slender crayfish (*Cherax dispar*). *Functional Ecology*, **22**, 311–316.

Clutton-Brock, T.H. and Albon, S.D. (1979) The roaring of red deer and the evolution of honest advertisement. *Behaviour*, **69**, 145–170.

Davis, N., Schaffner, C.M., and Wehnelt, S. (2009) Patterns of injury in zoo-housed spider monkeys: a problem with males? *Applied Animal Behaviour Science*, **116**, 250–259.

DeCarvalho, T.N., Watson, P.J., and Field, S.A. (2004) Costs increase as ritualized fighting progresses within and between phases in the sierra dome spider, *Neriene litigiosa*. *Animal Behaviour*, **68**, 473–482.

Dugatin, L.E. and Reeve, H.K. (1998) *Game Theory and Animal Behaviour*, Oxford University Press, New York.

Elwood, R.W., Wood, K.E., Gallagher, M.B., and Dick, J.T.A. (1998) Probing motivational state during agonistic encounters in animals. *Nature*, **393**, 66–68.

Emlen, D.J. (1997) Alternative reproductive tactics and male-dimorphism in the horned beetle *Onthophagus acuminatus* (Coleoptera: Scarabaeidae). *Behavioral Ecology and Sociobiology*, **41**, 335–341.

Enquist, M. and Jakobsson, S. (1986) Decision-making and assessment in the fighting behavior of *Nannacara anomala* (Cichlidae, Pisces). *Ethology*, **72**, 143–153.

Enquist, M. and Leimar, O. (1983) Evolution of fighting behaviour: decision rules and assessment of relative strength. *Journal of Theoretical Biology*, **102**, 387–410.

Enquist, M., Leimar, O., Ljungberg, T., Mallner, Y., and Segerdahl, N. (1990) A test of the sequential assessment game – fighting in the cichlid fish *Nannacara anomala*. *Animal Behaviour*, **40**, 1–14.

FitzGibbon, C.D. and Fanshawe, J.H. (1988) Stotting in Thomson's gazelles: an honest signal of condition. *Behavioral Ecology and Sociobiology*, **23**, 69–74.

Gammell, M.P. and Hardy, I.C.W. (2003) Contest duration: sizing up the opposition? *Trends in Ecology and Evolution*, **18**, 491–493.

Grafen, A. (1990) Biological signals as handicaps. *Journal of Theoretical Biology*, **144**, 517–546.

Haley, M.P. (1994) Resource-holding power asymmetries, the prior residence effect, and reproductive payoffs in male northern elephant seal fights. *Behavioral Ecology and Sociobiology*, **34**, 427–434.

Hack, M.A. (1997) Assessment strategies in the contests of male crickets, *Acheta domesticus* (L.). *Animal Behaviour*, **53**, 733–747.

Hamilton, W.D. (1967) Extraordinary sex ratios. *Science*, **156**, 477–488.

Hasson, O. (1997) Towards a general theory of biological signalling. *Journal of Theoretical Biology*, **185**, 139–156.

Hofmann, H.A. and Schildberger, K. (2001) Assessment of strength and willingness to fight during aggressive encounters in crickets. *Animal Behaviour*, **62**, 337–348.

Hsu, Y., Lee, S.-P., Chen, M.-H., Yang, S.-Y., and Cheng, K.-C. (2008) Switching assessment strategy during a contest: fighting in killifish *Kryptolebias marmoratus*. *Animal Behaviour*, **75**, 1641–1649.

Huber, R., Smith, K., Delago, A., Isaksson, K., and Kravitz, E.A. (1997) Serotonin and aggressive motivation in crustaceans: altering the decision to retreat. *Proceedings of the National Academy of Sciences of the United States of America*, **94**, 5939–5942.

Hughes, M. (2000) Deception with honest signals: signal residuals and signal function in snapping shrimp. *Behavioral Ecology*, **11**, 614–623.

Huntingford, F.A., Taylor, A.C., Smith, I.P., and Thorpe, K.E. (1995) Behavioural and physiological studies of aggression in swimming crabs. *Journal of Experimental Marine Biology and Ecology*, **193**, 21–39.

Huntingford, F. and Adams, C. (2005) Behavioural syndromes in farmed fish: implications for production and welfare. *Behaviour*, **142**, 1207–1221.

Hurd, P.L. (1997) Is signalling of fighting ability costlier for weaker individuals? *Journal of Theoretical Biology*, **184**, 83–88.

Huyghe, K., Vanhooydonck, B., Scheers, H., Molina-Borja, M., and Van Damme, R. (2005) Morphology, performance and fighting capacity in male lizards, *Gallotia galloti*. *Functional Ecology*, **19**, 800–807.

Huyghe, K., Herrell, A., Adriaens, D., Tadic, Z., and Van Damme, M. (2009) It is all in the head: morphological basis for differences in bite force among colour morphs of the Dalmatian wall lizard. *Biological Journal of the Linnean Society*, **96**, 13–22.

Irschick, D.J. and Garland, T. Jr. (2001) Integrating function and ecology in studies of adaptation: studies of locomotor capacity as a model system. *Annual Reviews of Ecology and Systematics*, **32**, 367–396.

Irschick, D.J., Herrel, A.Y., Vanhooydonck, B., and Van Damme, R. (2007) A functional approach to sexual selection. *Functional Ecology*, **21**, 621–626.

Jennings, D.J., Gammell, M.P., Carlin, C.M., and Hayden, T.J. (2005) Win, lose or draw: a comparison of fight structure based on fight conclusion in the fallow deer. *Behaviour*, **142**, 423–429.

Jennions, M.D. and Backwell, P.R.Y. (1996) Residency and size affect fight duration and outcome in the fiddler crab, *Uca annulipes*. *Biological Journal of the Linnean Society*, **57**, 293–306.

Kemp, D.J. (2002) Butterfly contests and flight physiology: why do older males fight harder? *Behavioral Ecology*, **13**, 456–461.

Kokko, H. (2013) Dyadic contests: modelling fights between two individuals, in *Animal Contests* (eds I.C.W. Hardy and M. Briffa), Cambridge University Press, Cambridge.

Kotiaho, J. (2001) Costs of sexual traits: a mismatch between theoretical considerations and empirical evidence. *Biological Reviews*, **76**, 365–376.

Krebs, J.R. and Davies, N.B. (1993) *An Introduction to Behavioural Ecology*, 3rd edn, Blackwell, Oxford.

Lailvaux, S., Pomfret, J., Hathway, J., and Knell, R.J. (2005) Horn size predicts physical performance in the beetle *Euoniticellus intermedius*. *Functional Ecology*, **19**, 632–639.

Lailvaux, S. and Irschick, D.J. (2006) A functional perspective on sexual selection: insights and future prospects. *Animal Behaviour*, **72**, 263–273.

Lailvaux, S. and Irschick, D.J. (2007) The evolution of performance-based male fighting ability in Caribbean *Anolis* lizards. *The American Naturalist*, **170**, 573–586.

Lailvaux, S.P., Reaney, L.T., and Backwell, P.R.Y. (2009) Regenerated claws signal performance and fighting ability in the fiddler crab *Uca mjoebergi*. *Functional Ecology*, **23**, 359–366.

Maynard Smith, J. and Parker, G. (1976) The logic of asymmetric contests. *Animal Behaviour*, **24**, 159–175.

Matsumasa, M. and Murai, M. (2005) Changes in blood glucose and lactate levels of male fiddler crabs: effects of aggression and claw waving. *Animal Behaviour*, **69**, 569–577.

McGregor, P.K. (2005) *Communication*, in *The Behaviour of Animals* (eds L.-A. Giraldeau and J.J. Bolhuis), Blackwell, Oxford.

Snell-Rood, E. and Moczek, A.P. (2013) Horns and the role of development in the evolution of beetle contests (eds I.C.W. Hardy and M. Briffa), Cambridge University Press, Cambridge.

Moczek, A.P. and Emlen, D.J. (2000) Male horn dimorphism in the scarab beetle *Onthophagus taurus*: do alternative reproductive tactics favor alternative phenotypes? *Animal Behaviour*, **59**, 459–466.

Morrell, L.J., Backwell, P.R.Y., and Metcalfe, N.B. (2005) Fighting in fiddler crabs *Uca mjoebergi*: what determines duration? *Animal Behaviour*, **70**, 653–662.

Mowles, S.L., Cotton, P.A., and Briffa, M. (2009) Aerobic capacity influences giving up decisions in fighting hermit crabs: does stamina constrain contests? *Animal Behaviour*, **78**, 735–740.

Mowles, S.L., Cotton, P.A., and Briffa, M. (2010) Whole-organism performance capacity predicts resource holding potential in the hermit crab *Pagurus bernhardus*. *Animal Behaviour*, **80**, 277–282.

Mowles, S.L. and Ord, T.J. (2012) Repetitive signals and mate choice: insights from contest theory. *Animal Behaviour*, **84**, 295–304.

Neat, F.C., Taylor, A.C., and Huntingford, F.A. (1998) Proximate costs of fighting in male cichlid fish: the role of injuries and energy metabolism. *Animal Behaviour*, **55**, 875–882.

Parker, G.A. (1970) The reproductive behaviour and the nature of sexual selection in *Scatophaga stercoraria* L. (Diptera: Scatophagidae). II. The fertilization rate and the spatial and temporal relationships of each sex around the site of mating and oviposition. *Journal of Animal Ecology*, **39**, 205–228.

Parker, G.A. (1974) Assessment strategy and the evolution of fighting behaviour. *Journal of Theoretical Biology*, **47**, 223–243.

Payne, R.J.H. and Pagel, M. (1996a) Escalation and time costs in displays of endurance. *Journal of Theoretical Biology*, **183**, 185–193.

Payne, R.J.H. and Pagel, M. (1996b) When is false modesty a false economy? An optimality model of escalating signals. *Proceedings of the Royal Society of London, Series B: Biological Sciences*, **263**, 1545–1550.

Payne, R.J.H. (1998) Gradually escalating fights and displays: the cumulative assessment model. *Animal Behaviour*, **56**, 651–662.

Payne, R.J.H. and Pagel, M. (1997) Why do animals repeat displays? *Animal Behaviour*, **54**, 109–119.

Pomfret, J.C. and Knell, R.J. (2006) Sexual selection and horn allometry in the dung beetle *Euoniticellus intermedius*. *Animal Behaviour*, **71**, 567–576.

Reaney, L., Milner, R., Detto, T., and Backwell, P.R.Y. (2007) The effects of claw regeneration on territory ownership and mating success in the fiddler crab *Uca mjoebergi*. *Animal Behaviour*, **75**, 1437–1478.

Royle, N.J., Lindstrom, J., and Metcalfe, N.B. (2005) A poor start in life negatively affects dominance status in adulthood independent of body size in green swordtails *Xiphophorus helleri*. *Proceedings of the Royal Society of London, Series B: Biological Sciences*, **272**, 1917–1922.

Ryan, M.J. (1988) Energy, calling, and selection. *American Zoologist*, **28**, 885–898.

Rovero, F., Hughes, R.N., Whiteley, N.M., and Chelazzi, G. (2000) Estimating the energetic cost of fighting in shore crabs by noninvasive monitoring of heartbeat rate. *Animal Behaviour*, **59**, 705–713.

Schuett, G.W. and Grober, M.S. (2000) Post-fight levels of plasma lactate and corticosterone in male copperheads, *Agkistrodon contortrix* (Serpentes, Viperidae): differences between winners and losers. *Physiology & Behavior*, **71**, 335–341.

Smith, I.P. and Taylor, A.C. (1993) The energetic cost of agonistic behavior in the velvet swimming crab, *Necora* (=*Liocarcinus*) *puber* (L.). *Animal Behaviour*, **45**, 375–391.

Sneddon, L.U., Huntingford, F.A., and Taylor, A.C. (1997) Weapon size versus body size as a predictor of winning in fights between shore crabs, *Carcinus maenas* (L.). *Behavioral Ecology and Sociobiology*, **41**, 237–242.

Sneddon, L.U., Taylor, A.C., and Huntingford, F.A. (1999) Metabolic consequences of agonistic behaviour: crab fights in declining oxygen tensions. *Animal Behaviour*, **57**, 353–363.

Sneddon, L.U., Huntingford, F.A., Taylor, A.C., and Orr, J.F. (2000) Weapon strength and competitive success in the fights of shore crabs (*Carcinus maenas*). *Journal of Zoology*, **250**, 397–403.

Szamado, S. (2000) Cheating as a mixed strategy in a simple model of aggressive communication. *Animal Behaviour*, **59**, 221–230.

Szamado, S. (2003) Threat displays are not handicaps. *Journal of Theoretical Biology*, **221**, 327–348.

Tanner, C.J. (2008) Aggressive group behaviour in the ant *Formica xerophila* is coordinated by direct nestmate contact. *Animal Behaviour*, **76**, 1335–1341.

Taylor, P.W. and Elwood, R.W. (2003) The mis-measure of animal contests. *Animal Behaviour*, **65**, 1195–1202.

Taylor, P.W., Hasson, O., and Clark, D.L. (2000) Body posture and patterns as amplifiers of physical condition. *Proceedings of the Royal Society of London, Series B: Biological Sciences*, **267**, 917–922.

ten Cate, C., Slabbekoorn, H., and Ballintijn, M.R. (2002) Birdsong and male–male competition: causes and consequences of vocal variability in the collared dove (*Streptopelia decaocto*). *Advances in the Study of Behavior*, **31**, 31–75.

Thorpe, K.E., Taylor, A.C., and Huntingford, F.A. (1995) How costly is fighting? Physiological effects of sustained exercise and fighting in swimming crabs, *Necora puber* (L.) (Brachyura, Portunidae). *Animal Behaviour*, **50**, 1657–1666.

Tinbergen, N. (1963) On aims and methods in ethology. *Zeitschrift für Tierpsychologie*, **20**, 410–433.

Turner, G.F. and Huntingford, F.A. (1986) A problem for game-theory analysis – assessment and intention in male mouthbrooder contests. *Animal Behaviour*, **34**, 961–970.

Valiela, I., Babiec, D.F., Atherton, W., Seitzinger, S., and Krebs, C. (1974) The consequences of sexual dimorphism: feeding in male and female fiddler crabs, *Uca pugnax* (Smith). *Biological Bulletin*, **147**, 652–660.

Vanhooydonck, B., Herrel, A.Y., Van Damme, R., and Irschick, D.J. (2005) Does dewlap size predict male bite performance in Jamaican *Anolis* lizards? *Functional Ecology*, **19**, 38–42.

Vanhooydonck, B., Van Damme, R., Herrel, A.Y., and Irschick, D.J. (2007) A performance based approach to distinguish indices from handicaps in sexual selection studies. *Functional Ecology*, **21**, 645–652.

Weary, D.M., Lambrechts, M.M., and Krebs, J.R. (1991) Does signalling exhaust male great tits? *Animal Behaviour*, **41**, 540–542.

Wilson, D.S. (1975) A theory of group selection. *Proceedings of the National Academy of Sciences of the United States of America*, **72**, 143–146.

Zahavi, A. (1977) The cost of honesty (further remarks on the handicap principle). *Journal of Theoretical Biology*, **53**, 205–214.

Zahavi, A. (1975) Mate selection – a selection for a handicap. *Journal of Theoretical Biology*, **67**, 603–605.

Zeil, J., Hemmi, J., and Backwell, P.R.Y. (2006) Fiddler crabs. *Current Biology*, **16**, R40–R41.

7

ACOUSTIC SIGNAL EVOLUTION: BIOMECHANICS, SIZE, AND PERFORMANCE

Jeffrey Podos[1] and S.N. Patek[1,2]

[1]*Department of Biology, Organismic and Evolutionary Biology Program,
University of Massachusetts at Amherst, Amherst, MA, USA*
[2]*Department of Biology, Duke University, Durham, NC, USA*

INTRODUCTION

Acoustic signals are a prominent mode of communication in many animal taxa, including insects, crustaceans, fishes, anurans, birds, and mammals. In recent decades, acoustic (and other) signals have been studied mainly from the perspective of signal function, with much emphasis on the hypothesis that signals should provide reliable indicators of signaler attributes, and should thus be costly (e.g., Johnstone, 1997; Maynard Smith and Harper, 2003; Searcy and Nowicki, 2005; Seyfarth *et al.*, 2010; see also Husak *et al.*, chapter, this volume). Yet there is also considerable potential value for studies of signal evolution in applying knowledge about signal mechanics – that is, the nuts and bolts of how animals actually go about signaling. As a general rule, proximate mechanisms of signaling may shape signal structure evolution in its origins, maintenance, and patterns of diversification. Additionally, evolutionary history or context can dictate the kinds of mechanisms available in the first place to animals for signaling. The theme of this chapter is that evolutionary and proximate perspectives on acoustic signaling are mutually informative and that a comprehensive understanding of acoustic signaling benefits from attention to both.

Animal Signaling and Function: An Integrative Approach, First Edition.
Edited by Duncan J. Irschick, Mark Briffa, and Jeffrey Podos.

As an opening illustration of the mutually informative nature of evolutionary and mechanistic perspectives, we describe a series of recent papers by Marlene Zuk and collaborators about field crickets, *Teleogryllus oceanicus*. Field crickets comprise a fascinating model system for quantifying the competing pressures of mate selection and predation on acoustic calling behavior. As in many taxa, calls are produced by males only, and serve primarily as a mating signal. Calls help males avoid direct conflict with male rivals, and help females orient to and locate males, and to choose from among multiple prospective mates (e.g., Rebar *et al.*, 2009; Logue *et al.*, 2010). Calls may also serve as reliable indicators of male attributes, thus enabling females to choose optimal mating partners (e.g., Tolle and Wagner, 2011). Acoustic signaling is thus generally considered a fundamental feature of field cricket social and sexual behavior.

It is in this context that a recent, fascinating discovery has emerged: In an introduced population of field crickets on the Hawaiian island of Kauai, mating now occurs in virtual silence, with only a small percentage of males calling at all (Zuk *et al.*, 2006). Moreover, field observations suggest that the evolutionary processes driving this virtual loss of calling behavior have acted *in situ* and with great speed, over only about 4 years (~20 generations). This situation is very surprising. How could a field cricket population persist in the near absence of the main advertisement signal that enables mate localization and selection? What conditions could have possibly enabled this rapid and unexpected evolutionary change?

One proposed explanation for the silence of the crickets focuses on *Ormia ochracea*, a parasitoid fly that locates male crickets by sound. This fly, present on Kauai, is known for killing crickets in brutal fashion: female flies deposit their larvae on a host cricket, which burrow into it and consume the host cricket from the inside out (Cade, 1975). Silent males evade detection by *Ormia* and thus stand a much better chance of not being parasitized. But, while being silent may help male crickets to survive, what good is being silent if calls are necessary for mate localization and attraction? How is it that silence as a phenotype has not been expunged in short order from the population, because of a presumed associated lack of reproductive success for its bearers? Several explanations have now been put forth for how silent Kauai males may still succeed in the dual challenges of mate localization and choice.

Regarding mate localization, field observations indicate that Kauai females do not approach silent males, as expected. However, Kauai males make up for their lost sexual magnetism by engaging in an alternative behavioral tactic, in which they position themselves near callers (as satellites) and then intercept approaching females (Tinghitella *et al.*, 2009). Regarding mate choice, laboratory trials reveal that island populations of female field crickets are, contrary to all expectations, actually ready and willing to accept silent males as mates (Tinghitella and Zuk, 2009). This willingness stands in contrast to the crickets' more finicky continental counterparts, who only mate with males that call (Tinghitella and Zuk, 2009). Kauai females' initial willingness to accept silent males, together with the presence of an acoustically orienting parasitoid, may indeed favor males who are silent.

A central question raised by this example, to our eyes, is this: Are Kauai males silent by necessity or by choice? In other words, do the silent crickets have the ability as individuals to modulate their calling behavior as necessary? This possibility would be consistent with many observations in the field of behavioral ecology, in which animals are seen to opt for behavioral outcomes that maximize fitness gains and minimize fitness costs (Krebs and Davies, 1993). Under such circumstances, the "decision" whether to call and when to call would be mediated by a neural control system that evaluates environmental circumstances, and that inhibits the calling motor pattern when silence is beneficial to the sender. Kauai crickets would, for instance, refrain from calling when faced with an abundance of parasitic flies, and perhaps continue calling if flies were suddenly absent, or if the satellite mating tactic was no longer feasible. Further investigation into the evolutionary loss of calling would thus require delving into the neuroethology of calling decisions.

However, in the case of the Kauai crickets, available evidence indicates that silence has evolved by a simple mutation that has modified the morphology and function of the calling mechanism itself. As a result, it appears that silent males actually do not have the option to call at all. To step back, field crickets (like many other orthopterans) produce sound via stridulation, in which a file on one wing is rubbed rapidly on a scraper on the other wing. File and scraper are both morphological modifications to wing veins, expressed in most males, yet not at all in females. In the Kauai population, most males now develop wings that resemble female wings, insofar as they completely lack files and scrapers. Males with these so-called "flatwing" mutations are actually incapable of making sound by stridulation and so their silence is obligatory. This observation points future studies on Kauai cricket evolution away from neuroethology *per se*, and more directly into the realms of population genetics, developmental genetics, and developmental morphology. This observation also specifies the forms of behavioral plasticity that have enabled the evolution of this fascinating phenomenon – in this case, alternative mating strategies by males, and liberal mate selection by females.

This brief opening example is intended to illustrate how an understanding of signal evolution can benefit from knowledge of signal mechanics. The example also illustrates that the interface of acoustic mechanics and evolution can be complex and multi-faceted, bridging topics such as natural and sexual selection, behavior and communication, perception, biomechanics, development, physiology, and genetics. Is there a useful and general framework for studying the interface of signal mechanics and evolution, as a complement to more traditional research programs that focus on external selective factors that shape signal evolution, such as natural and sexual selection for signal exaggeration, reliability, and efficacy (reviewed by Maynard Smith and Harper, 2003; Searcy and Nowicki, 2005)? This leads us to a three-part framework for our chapter. The first considers the biomechanics of acoustic signal generation, and how an evolutionary approach can inform our understanding of signal structural diversity. The second addresses the role of size in acoustic signal generation, also ideally addressed from an evolutionary perspective. The third links signal mechanics with performance – an organism's capacity to achieve certain behavioral outcomes. Our ultimate goal in this chapter is to show how characterizing these three components – biomechanics, size, and performance – together help define the range of acoustic signal pheno- types that can be expressed and thus may evolve. More specifically, as we hope becomes apparent, we regard the study of biomechanics as a fundamental first step in understanding acoustic signal structure, thus holding special relevance in comparative studies. In turn, body size and performance hold particular relevance for understanding the expression of signaling variations within species, thus linking more directly back to questions about signal function and reliability.

BIOMECHANICS

To begin, we consider the biomechanics of sound production, which can be defined most simply as the physical processes by which acoustic signals are produced. The biomechanics of acoustic systems have fascinated biologists for centuries and range from the remarkable elastic efficiency of cicadas (Patek *et al.*, 2011; Young and Bennet-Clark, 1995) to the complex mechanics of syringeal vibrations in birds (Goller and Larsen, 1997). Basic discoveries about mechanical sound production systems continue apace: recent discoveries of acoustic mechanisms include muscle-generated infrasound in mantis shrimp (Patek and Caldwell, 2006; Staaterman *et al.*, 2011), stick-slip sound production in spiny lobsters (Patek, 2001, 2002; Patek and Baio, 2007), and stridulating feathers in manikins (Bostwick and Prum, 2005). While there is a tradition of acoustic biomechanics in which the importance of biomechanics in explaining signal diversity is recognized, relatively few studies have examined the influence of mechanics on macroevolutionary diversification, or have used phylogenetic approaches to inform interpretations of biomechanical diversity (Patek and Oakley, 2003). In this section, we explore as a focal example the biomechanics and evolution of acoustic signals in katydids – an informative system given recent advances and general perspectives in the areas of biomechanics and evolution. Through this particular example, we show how biomechanical systems can influence signal features in surprising ways, how signaling biomechanics can potentially impose trade-offs with other mating behaviors, and how macroevolutionary patterns of mechanical and acoustic systems do not necessarily meet standard expectations, for example, following a "progression" from broadband noisy signals to tonal calls.

Like the crickets mentioned earlier, katydids produce sound via stridulation. Katydids rub one forewing, the "scraper," over a series of bumps or teeth (the "file") on the other forewing to produce sound (Gwynne, 2001). In its simplest case, the katydid's sound-producing mechanism works like a fingernail rubbing over the tines of a comb, or a stick rubbing across a washboard. A major defining feature of a stridulatory sound is its fundamental frequency, which is equivalent to pulse rate (number of strikes per second). According to the standard model for stridulation, pulse rate is determined by tooth strike rate, which in turn is determined by (i) the speed with which the scraper rubs over the file and (ii) the spacing of teeth along the file (Elliott and Koch, 1985). For a given scraper speed, files with teeth separated by a greater distance should produce lower pulse rates than files with more closely spaced teeth.

Katydids and crickets have augmented this simple washboard-like system with a small, thin membrane on the scraper forewing, called the *mirror*, which radiates sound bursts produced by scraper–file interactions (Montealegre-Z and Postles, 2010). The mirror is analogous to a speaker cone in how it radiates source vibrations with efficiency. Traditionally, much attention in studies of stridulation has been paid to the radiating mirror and its acoustic properties. Like other sound-generating structures, radiating mirrors have resonant vibration frequencies, which help determine the acoustic properties (especially frequency and amplitude) of calls produced. Katydid calls vary widely by species, yet can be classified into three categories – broadband, sonic, and ultrasonic (Figure 7.1). Broadband calls are wide bandwidth signals featuring dominant frequencies spanning 10–100 kHz. Tonal signals fall within narrow ranges of frequencies and are either sonic (<50 kHz) or ultrasonic (>50 kHz).

Based on the standard model of stridulation coupled with a mirror-resonance system, one might expect tonal katydids to have highly regular tooth spacing to produce uniformly tonal signals, and one might also expect ultrasonic katydids to have the most closely spaced teeth in order to produce the highest pulse rates and associated fundamental frequencies. Yet surprisingly, katydids that sing tonally at the highest, ultrasonic frequencies do not necessarily exhibit the most closely set teeth, evenly spaced teeth, or even the fastest forewing translations during

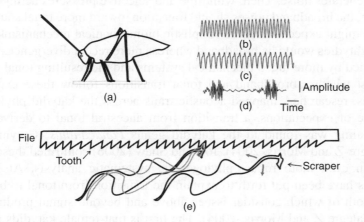

Figure 7.1. Katydids (a) are insects that produce sound by rubbing one forewing across the other with a washboard-like mechanism. They can produce tonal signals (b,c) or broadband signals (d) as represented by oscillograms. (From Montealegre-Z and Morris (2004).). (e) One unique feature of some ultrasonic katydids is that they use deflection (orange) and subsequent vibrations of the scraper (blue) to generate ultrasonic vibrations, analogous to the "twanging" of a thumb harp. This deflection of the scraper is also used to elastically preload the system such that subsequent tooth impacts can happen at a much higher rate than the driving wing movement. (Images adapted from Montealegre-Z et al. (2006).) (*See insert for color representation of this figure.*)

sound production (Montealegre-Z *et al.*, 2006). Closer examination provides several intriguing explanations for this finding. First, carrier frequencies are generated through resonances not only of the mirror, but also of the scraper (Figure 7.1). Thus, the rate at which the scraper plucks the teeth can be decoupled from the inherent resonances of the vibrating scraper. More simply, this acoustic mechanism is analogous to a thumb harp, in which metal strips are tuned to particular frequencies and excited by "twanging" with a finger; in the katydid's case, the scraper is elastically loaded and then rapidly released at various points along the file. Furthermore, bending and elastic energy stored in the scraper allow subsequent tooth impacts to occur at much higher rates than the translational forewing speed would suggest. The greater tooth spacing, and possibly its irregularity, in some ultrasonic katydids may actually be adaptive in terms of the efficiency of call production. Larger tooth spacing likely allows the scraper to generate more momentum in between tooth-strikes, such that the scraper can hit the next tooth with more energy and thus activate vibrations of enhanced amplitude (Montealegre-Z, 2009). In sum, ultrasonic katydids incorporate scraper resonances and elastic energy storage to circumvent frequency limits of the standard mode of stridulation (Montealegre-Z *et al.*, 2006).

Taking these surprising biomechanical discoveries into a macroevolutionary framework, one might ask how the transitions between tonal and broadband signaling have occurred in this clade. In a theoretical acoustic system, one might imagine that the initial stages of sound production would consist of a simple, translating system that generates noise. Then, with time and selective pressures acting to "tune" the system, the broadband noises should transition toward more tonal sounds. Similarly, one might expect that more simplistic biomechanical mechanisms found in ancestral katydids would be broadband whereas more recent divergences would be accompanied by more tuned mechanical systems and the resulting tonal calls.

To test whether broadband and tonal transitions follow these expectations in katydids, researchers mapped acoustic traits across the katydid phylogeny. In the reverse of expectations, a transition from ancestral tonal to derived broadband signaling was found in the katydid genus *Panacanthus* (Gwynne, 2001; Montealegre-Z and Morris, 2004). Jost and Shaw (2006) confirmed these findings through an even more robust phylogenetic systematic analysis. At least two hypotheses have been put forth to explain the transition from tonal to broadband singing, both of which consider issues above and beyond signal production *per se* (Montealegre-Z and Morris, 2004). The first is that female katydids may have broadband hearing sensitivity, thus broadband signals that maximize acoustic stimulation of the female auditory system are more effective than tonal signals in attracting females. Second, katydids may sing in noisy environments, and broadband signals in such environments would be less likely than tonal signals to be masked during transmission.

Focusing on the particular signal features of tonal calls, to what extent do scraper and file morphology predict call features? Studies addressing this question have been conceptually broad, taking a macroevolutionary approach, integrating a consideration of mating behavior, signaling energetics, and acoustic mechanics (Del Castillo and Gwynne, 2007; Montealegre-Z, 2009). To begin, one can look at the most fundamental mechanical and morphological parameters of the katytid's acoustic system – body size, file length, and number of teeth on the file – and how they influence katydid call acoustics. At one level, relationships among these parameters can be examined at the individual or within-species level, which establishes the mechanical correlations as we discussed above. However, moving to a broader, comparative framework, the network of mechanical and acoustic correlations can be examined across species to both test and generate hypotheses about the variation, function, and evolution of this system.

This was the context for Monteleagre-Z's (2009) investigations of the relationships between morphology and acoustics across 58 katydid species that produce either sonic or ultrasonic pure tone calls (thus excluding species that produce broadband calls). The researcher focused on morphological features that may explain variations in carrier frequency and pulse structure, and tested three specific hypotheses: (i) body size should correlate positively with mirror and file size, because of simple scaling considerations; (ii) file length should correlate positively with tooth size or tooth spacing, again because of scaling; and (iii) tooth density should correlate positively with carrier frequency, and correlate negatively with pulse duration.

As expected, larger katydids were found to have larger mirrors and longer files (hypothesis 1), and to have files with greater tooth size and spacing (hypothesis 2). Moreover, relationships between body size × tooth number and density scaled isometrically, whereas relationships between body size × mirror size and file length scaled allometrically. A lack of isometry in body size × acoustic morphology is explained in part by the observation that in some ultrasonic-calling species, males of some species are large yet have disproportionately small mirrors. This makes sense because small mirrors are better suited for radiating calls at ultrasonic frequencies. Deviations from isometry for body size × acoustic morphology relationships might also be explained by constraints imposed by the multiple costs of sexually selected features – for example, the costs of operating the stridulatory apparatus versus other investments in mating such as nuptial gifts, as we will discuss below.

How do these scaling variables relate to variation in the acoustic signals across the group (hypothesis 3)? The phylogenetically corrected results both follow and diverge from expectations. As expected for tonal calls, pulse duration correlated positively with file length, frequency, and number of teeth. Given the same closing speed, a larger file system presumably allows greater time between impacts and thus a greater pulse duration. Moreover, carrier frequency correlated negatively with file length and body size, consistent with expectations that smaller animals should radiate higher frequency signals. Yet in contrast to expectations, pulse duration was not correlated with tooth density, even though one would expect more tightly spaced teeth to yield lower pulse duration. Moreover, carrier frequency was not correlated with tooth number or density.

These last results suggest a partial decoupling between the morphological arrangement of teeth and the production of tonal signals. This decoupling between morphology and acoustics points to some new, testable hypotheses about katydid call evolution (Montealegre-Z, 2009). To illustrate, the observation that carrier frequency does not correlate with tooth density suggests the presence of alternate mechanisms for controlling carrier frequency. One possible mechanism parallels that described earlier for ultrasonic katydids, in which impacts generate ultrasonic vibrations of the scraper. In addition, the scraper stores elastic energy and releases it via a much more rapid translation across the file teeth than the wing movement would suggest. Another interesting possibility is that katydids use a mechanism analogous to the so-called "escapement" mechanism found in crickets. Crickets use the emergent mechanics of the scraper–tooth interactions to drive the pulse rate independently of the speed of the scraper (Elliott and Koch, 1985; Koch et al., 1988). Although katydids are not thought to use this type of mechanism, the results of the comparative analyses certainly suggest a broader decoupling than previously thought.

These discussions are also worth considering within the context of developmental constraints. Katydids are known to have a fixed number of teeth throughout their development, which means that larger katydids cannot or do not add teeth as they grow. This lack of plasticity in tooth number may drive the decoupling of morphology and acoustics out of necessity. Similarly, larger katydids with constant tooth number must geometrically have lower tooth density. Thus, to maintain consistent call parameters across development, it becomes necessary to optimize or control factors other than impact rate – particularly through the momentum hypothesis discussed earlier.

As a final example of how acoustic mechanics in a comparative perspective can be used to illuminate fundamental questions about the evolution of acoustic systems, researchers examined the connection in katydids between energetics of acoustic mechanics for mate attraction and its potential trade-offs with other relevant mating behaviors. Specifically, Del Castillo and Gwynne (2007) tested for the presence of trade-offs between the energetic costs of producing a signal with a high carrier frequency (and the associated higher impact rate, driven by greater scraper velocity) and the size of the male's spermatophore, a nuptial gift given to females. In their analysis of calls, the authors appear to have restricted their results to species with carrier frequencies below 40 kHz, thus circumventing the lack of correlation between scraper velocity and carrier frequency observed in ultrasonic signalers (as described above). For species included in the analysis, carrier frequency was assumed to correlate positively with scraper velocity, thus implying that higher carrier frequencies would be more expensive or costly to produce. Looking across 58 katydid species, and using independent contrasts analysis (to account for phylogenetic non-independence of study species, Felsenstein, 1985), the researchers identified a significant negative correlation between carrier frequency and spermatophore size. This finding supports the hypothesis that males do indeed face a trade-off between the energetic costs of signaling and the costs of providing a nuptial gift. Thus the evolution of carrier frequencies within tonal or broadband call categories may be constrained by physiological or metabolic limitations – a significant and important issue that ties directly back to understanding the acoustic mechanics of a signaling system.

To summarize this section, our discussion of katydid acoustic biomechanics points to a fascinating tension between predictive and stochastic factors in call evolution. The calls of individual katydid species can be predicted to some extent by the morphology of their acoustic apparatus and by assumptions about the biomechanical function of this morphology. File length, size, mirror size, body size, and so on, all predict to some extent the duration and frequency profiles of the calls of individual katydid species. However, at the same time, a comparative approach suggests that katydid call evolution has been somewhat haphazard, featuring events such as the decoupling of file and call properties, the evolution of broadband signals from tonal signals, and the evolution of ultrasonic capabilities via modifications in scraper use and function.

Returning briefly to our general framework, we suggest that analysis of biomechanics is most useful for gaining insights into signal diversity across species and over macroevolutionary scales. Biomechanical principles determine the kinds of acoustic signals animals can and cannot produce. The remaining two factors to be discussed, body size and performance, also describe constraints on signal mechanics and production. However, they are more relevant at finer, within-species scales, essentially addressing minor variations of signaling apparati and their potential for producing sound. This is why, in contrast to biomechanics, the topics of body size and performance have already been, to some extent, integrated into the literature on signal reliability (e.g., Searcy and Nowicki, 2005).

BODY SIZE

Body size influences the production and evolution of many acoustic signal features, including signaling amplitude, frequency, duration, dexterity, and vigor. For example, larger animals may show greater signaling vigor because they typically have more robust metabolic reserves that should enable sustained calling, although they likely also expend more energy per unit time signaling (e.g., Wells *et al.*, 1995; Prestwich *et al.*, 1989). The relationship between energy expenditure and metabolic resource depletion during acoustic signaling remains virtually unexplored. For specific acoustic signaling events, larger animals typically produce sound at greater amplitudes, because the larger size of their production sources enables increased power. Correlations between body size and amplitude can be weak, however, because signal amplitude is also highly dependent on other factors not related to morphology, such as the resonance and transmission properties of the signaling environment (e.g., Bennet-Clark, 1987). In the temporal domain, larger animals may be able to sustain individual calls for longer duration, for example, in phonation as a result of larger tidal volumes (e.g., Hartley and Suthers, 1989). On the other hand, larger animals may possess limited mechanical dexterity, thus obstructing the production of rapid or temporally complex acoustic patterns. For the remainder of this section, we focus on a correlate of body size already discussed in the katydid example: acoustic frequency. The relationship between body size and frequency serves as a particularly useful focal example for examining the interplay of signal mechanics, morphology, and evolution.

Across a wide range of taxa, body size is predicted to correlate with acoustic frequencies because of simple scaling considerations. Body size typically scales with the size or mass of the sound production mechanism, which in turn dictates ranges of acoustic frequencies animals can effectively produce. More specifically, larger and more massive acoustic source tissues resonate most efficiently at lower frequencies, whereas smaller and lighter tissues resonate most effectively at higher frequencies. Moreover, the size of the sound source sets absolute constraints on the production of low frequencies, by delimiting maximum wavelengths and thus minimum frequencies that can be attained (Bradbury and Vehrencamp, 2011). Thus, animals with diminutive production sources are effectively limited to high frequencies, whereas animals with larger production sources can produce broader ranges of acoustic frequencies.

Some of the most widely cited empirical studies on the effects of body size on acoustic frequency have focused on vocal production in birds, and both support and expand upon the predictions outlined above. Wallschläger (1980) analyzed the songs of 90 European passerine species, and found that body size explained 59% of variance in song frequency. Consistent with these findings, Ryan and Brenowitz (1985) found significant negative relationships between body size and emphasized vocal frequency, in analyses of 159 bird species from Panama. When dividing the Panama sample into three phylogenetically distinct groups – nonpassserines, suboscine, and oscine passerines – significant negative relationships also emerged, but regression values differed among groups, suggesting an influence of divergent evolutionary histories on song frequency evolution (Ryan and Brenowitz, 1985). Recent comparative studies of body size × frequency scaling have achieved greater precision by attending to the lack of statistical independence among species due to phylogenetic relatedness, using independent contrasts analysis. In most of these studies, patterns still hold between body size or mass and song/call frequency, for example, across ~200 species of antbirds inhabiting rainforests in the Neotropics (Seddon, 2005), 47 species of tinamou, a primitive group of ground dwelling birds in Central and South America (Bertelli and Tubaro, 2002), and 28 species of crows and ravens spanning their worldwide range (Laiolo and Rolando, 2003). In contrast to these comparative studies, within-species analyses in birds have not always revealed tight relationships between body size and vocal frequency (e.g., Logue et al., 2007; Cardoso et al., 2008; Patel et al., 2010 but see Hall et al., 2013).

While studies that correlate body size and acoustic frequency represent a good starting point, more comprehensive analyses of size by vocal frequency relationships would ideally account for the size and function of the syrinx, rather than treating it as a black box. This is because scatter in body size by vocal frequency correlations could emerge from a lack of precision in how syrinx dimensions scale with either body size or vocal frequency. Perhaps our best comparative data along these lines comes from morphological studies of Galapagos finches, which revealed a positive and nearly isometric relationship between body mass and syrinx (sound source) volume (Cutler, 1970), with larger-bodied species tending to sing at lower vocal frequencies (Bowman, 1983). For a group such as the Darwin's finches with near isometric scaling of the syrinx, scatter in body size by frequency relationships can thus be best attributed to factors besides syrinx size or function, such as adaptation to divergent acoustic environments or divergent trajectories of cultural evolution (e.g., Bowman, 1983; Goodale and Podos, 2010).

Several studies in birds also identify a role for the beak in shaping vocal frequency production. The beak, trachea, and associated structures form the vocal tract, which is now known to influence song production by filtering harmonic frequencies produced at the sound source (Nowicki, 1987; see also Fitch, 1999). Vocal tract volume influences the frequencies of vocalizations that can be filtered effectively, such that large-volume vocal tracts are most effective as resonance filters of vocalizations of low frequency (Nowicki, 1987; Nowicki and Marler, 1988). The corresponding prediction, that beak volume should scale negatively with vocal frequencies, has been supported in two comparative studies that employed independent contrast analysis: across 39 species of Neotropical woodcreepers (Palacios and Tubaro, 2000); and across 15 species of corvids (Laiolo and Rolando, 2003). This prediction has also been supported by a study within a single Darwin's finch species (Huber and Podos, 2006). In all of these studies, acoustic frequencies regressed negatively onto beak size residuals, that is, measures of beak size that were standardized to body size. The effects of beak size *per se* on vocal frequencies could thus be isolated independently of body size variation (Figure 7.2).

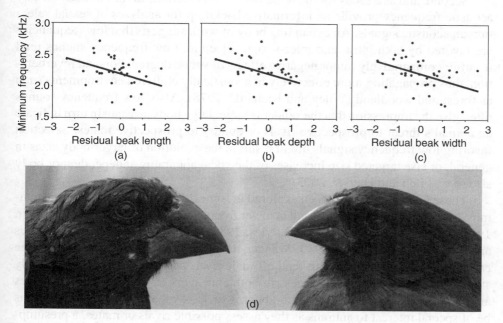

Figure 7.2. (a–c) Minimum vocal frequency as a function of residual beak dimensions (beak length, depth, and width, respectively, calculated relative to body size), for the medium ground finch *Geospiza fortis* at El Garrapatero, Santa Cruz Island (Huber and Podos, 2006). The negative relationships are in accord with the prediction that birds with greater volume vocal tracts relative to their body sizes should sing at lower vocal frequencies. (d) Representative large and small beak morphs of this population. A similar pattern is shown by morph: after factoring out body size, large-beaked morphs sing lower-frequency songs (Huber and Podos, 2006). (*See insert for color representation of this figure.*)

Studies such as those just reviewed for birds provide insight into acoustic signal evolution in at least two ways. First, they suggest that acoustic frequency evolution could evolve as an incidental byproduct of the evolution of body size variation. Body size is known to evolve in response to a wide array of environmental factors such as temperature, aridity, and food availability (e.g., Schmidt-Nielsen, 1984; Graves, 1991), and as a result of selection on life-history traits such as fecundity, reproductive rate, and dispersal (e.g., Roff, 1992; Blanckenhorn, 2000). As body size evolves, we would predict, as a null hypothesis, corresponding (isometric) changes in the size of the acoustic apparatus and thus in acoustic frequencies. These changes may be incidental and not affect signal function, so long as the perception of and response to evolving signals change in concert. But body size evolution can have evolutionary consequences in other ways. For example, species in a given radiation that diverge in body size may experience incidental, parallel divergence in acoustic frequencies, which may in turn enhance mating isolation among diverging populations, at least in taxa in which vocal frequencies are assessed as part of mate choice (Podos and Hendry, 2006).

Second, and as a focus for the remainder of this section, studies of body size by acoustic frequency provide an informative backdrop for analyses of sexual selection on acoustic signals. An expanding body of work suggests that low frequencies are favored by both intra- and inter-sexual selection. Low frequency sounds tend to suffer comparatively minor negative effects of sound degradation and reverberation, thus propagating more effectively in acoustically cluttered environments such as forest and woodland (Wiley and Richards, 1978). Also, low frequency sounds often give the impression that the sender is large; large body size can in turn indicate an animal's threat to prospective rivals or quality to prospective mates. Selection favoring low frequency signals might result in the evolution of larger body sizes in general, or else targeted size increases of the vocal apparatus, independent of body size (e.g., Fitch and Hauser, 2003).

These possibilities have been explored in some detail for vocal behavior in anurans. Larger individuals tend to call at lower frequencies, both across species (Blair, 1964; Martin, 1972; Duellman and Pyles, 1983) and within species (e.g., Davies and Halliday, 1978; Ryan, 1980; Wagner, 1989; Howard and Young, 1998; Bee et al., 1999). Males call to both fend off male rivals and attract females, and body size generally predicts male success in both endeavors (e.g., Wells, 1977; Davies and Halliday, 1979). Indicators of body size such as vocal frequency should thus be of special interest to anurans as they assess possible rivals or mates, a presumption supported by ample experimental evidence (e.g., Davies and Halliday, 1978; Arak, 1983; Wagner, 1989; Morris and Yoon, 1989; Ryan et al., 1992). Yet, when assessed statistically, correlations between body size and vocal frequency in anurans tend to be weak. For example, in a review of 12 studies across 11 species, Searcy and Nowicki (2005) note that on average, body size only explained about 36% of variation in call frequency. This diminishes the value of vocal frequency as a reliable indicator of body size. Could adaptations targeted to the sound-producing apparatus, and a resulting lack of isometric scaling of the vocal apparatus to body size, account for some of this observed scatter? This possibility is supported by a study of vocal mechanics in 33 *Bufo* toads; in this case, the mass of vocal cords correlates tightly with the resultant resonant frequency (correlation coefficient for kilohertz × vocal chord mass index of 0.835; Martin, 1971). Extrapolating to additional anuran species, we might conclude that scatter in body size by frequency correlations in *Bufo* toads results reflects instead from a loose correspondence between body size and sound source mass, and that anurans circumvent body size constraints by augmenting larynx mass independently of body size. The decoupling of body size and larynx morphology is nicely illustrated in recent studies on call evolution in Túngara frogs, *Physalaemus pustulosus*. These frogs are well known for producing calls that feature tonal frequency sweeps (whines) followed by lower-frequency elements (chucks). The number of chucks included in calls is augmented by sexual selection favoring signal complexity, yet diminished by natural selection favoring crypsis (Ryan, 1985). An experimental study of calling anatomy has now revealed that the capacity to produce chucks derived from the evolution of a fibrous mass on the vocal folds, which presumably enabled the associated vocal chords to vibrate effectively at lower frequencies. Removal of the fibrous mass resulted in frogs that could no longer produce the low frequency chuck, whereas production of the whine was unaffected (Gridi-Papp *et al.*, 2006).

These studies together lead to several general conclusions. First, acoustic parameters are best evaluated with reference to the sound-producing apparatus rather than the size of the entire body. Indeed, as seen in the Túngara frogs, some organisms actively modify the mass of the sound-producing apparatus to circumvent body size limitations. Second, the relevant structures involved in modifying frequency can extend beyond the fundamental vibrating structure. This was seen in the example of bird vocalizations, in which the beak and vocal tract play as much of a role in frequency modulation as does the syrinx. Finally, and perhaps of most interest, is the finding that deviations of the sound-producing apparatus from body size scaling rules hold important clues to other selective factors at play in acoustic signal evolution. For example, finches show deviations from isometric scaling in their beaks, which holds important information relevant to correlates of environmental or cultural selection on the signals. Thus, while body size can provide a rough guide in predicting the sounds an animal will produce, deviations from these rules may hold the most interesting and surprising insights into modifications of acoustic frequencies across taxa.

PERFORMANCE

A third factor that shapes acoustic signal evolution concerns performance, defined most generally as an organism's capacity to achieve behavioral outcomes that hold functional relevance (e.g., Irschick, 2003). As with other behavioral phenotypes, acoustic signals typically vary in how they are performed across renditions, individuals, species, and often according to circumstances such as time of year or social context (e.g., Catchpole and Slater, 2008; Bradbury and Vehrencamp, 2011). Some signal renditions may be produced easily, yet others may challenge signalers' production capacities. The concept of performance has received significant attention in recent years in the realms of feeding and locomotion, and is providing new insights into the interface of behavior, mechanics, ecology, and fitness (e.g., Arnold, 1983; Irschick, 2003; Wainwright, 1994). A recurring theme in these studies is that maximal performance events, in which behavioral capacities are pushed to the limits of possibility, have particular bearing on function and evolution. The recent literature on lizard locomotion well-illustrates this point. In the laboratory, researchers have been able to quantify maximal sprint speeds, using track or treadmill tests (Garland and Losos, 1994; Irschick and Garland, 2001). In nature, maximal speeds are achieved only rarely, during predator evasion (Irschick, 2003). By contrast, under typical contexts, locomotion rarely approaches maximum speed, for example, in *Anolis* lizards normally achieving only 10–40% of the maximum (Irschick, 2003). Maximal, and not typical performance has been found to be a main locus of natural selection. While all lizards can move slowly, the ones that can reach the highest speeds are favored by natural selection for predator avoidance (Irschick *et al.*, 2008).

What can analyses of performance tell us about acoustic signal production and evolution? First, studies of performance can help us to understand how behavioral mechanics may enable or constrain certain behavioral outcomes. For instance, male frogs produce a diversity of call types, but within the limits of production capacities. Males cannot call continuously, because of energy limitations on call output (Taigen and Wells, 1985). They face trade-offs in the expression of temporal call traits such as call rate and call duration (Klump and Gerhardt, 1987) and also are limited in call frequency and amplitude by size limitations of the vocal apparatus (Martin, 1971). Such performance constraints can counterbalance selection for exaggerated or elaborate signal phenotypes, and thus help to explain observed phenotypic variation in signal structure. Second, studies of performance can help us to understand the origins of structural trade-offs in signal production. When animals face performance challenges in a certain acoustic feature, such as call rate, they might be able to further maximize the production of that feature only by diminishing performance in another feature, for example, call duration (Klump and Gerhardt, 1987). Third, as discussed throughout this book, maximal signal performance is a likely locus of sexual selection, because signals at maximal performances are well-suited to provide reliable information about signaler quality.

In this section, we focus on the causes and consequences of acoustic performance variation, specifically the production of trilled vocalizations by songbirds. With this example, we illustrate how acoustic production mechanics can drive trade-offs between temporal and frequency song features, how selection on morphological components of the vocal apparatus might indirectly alter vocal performance capacities, and how performance-limited features may influence the evolution of signals as well as the animals that produce those signals (see also Podos *et al.*, 2009). Bird vocalizations are a good place to look for performance constraints, because their production involves a mechanism that is remarkably complex and intricate (Suthers, 2004; Podos and Nowicki, 2004).

Birds generate sound at the syrinx, a specialized bilateral sound source located at the juncture of the bronchi and trachea (Greenewalt, 1968; Nottebohm, 1971). Membranes or labia on the syrinx vibrate when activated by outward respiratory airflow, with left and right sides of the syrinx sometimes producing sound independently of each other (Suthers, 1990; Goller and Larsen, 1997). Birds breathe in tight synchrony with vocal output, typically only producing sound during expiration and remaining silent during inspiration, and often coordinating the opening of one syrinx side with the closure of the other (Suthers, 2004). As mentioned in the prior section, in recent decades we have learned that song production additionally involves the vocal tract (the trachea, beak, and associated structures), which acts as a resonance filter for sounds produced at the syrinx, selectively filtering harmonic overtones, and thus enabling the production of songs that are highly pure-tonal or whistle-like (Nowicki, 1987). Moreover, singing songbirds often rapidly adjust the configuration of their vocal tracts, so that vocal tract volume tracks shift in vocal frequency (e.g., Westneat *et al.*, 1993; Riede *et al.*, 2006). These adjustments allow birds to maintain the effectiveness of their vocal tract filter across shifting source frequencies (Hoese *et al.*, 2000). The syrinx, respiratory system, and vocal tract are all activated and coordinated by a complex, hierarchically arranged neural control system (Brainard and Doupe, 2002; Jarvis, 2004).

Mechanical constraints would seem particularly likely for song features that involve rapid temporal modulations, which should be difficult to achieve because they require correspondingly rapid mechanical reconfiguration and maintained coordination of vocal apparatus components (e.g., Hartley and Suthers, 1989; Podos *et al.*, 1995). Rapid temporal modulations are often found in acoustic trills, a feature of the songs of many bird species in which notes or note clusters are repeated in succession (e.g., Thorpe and Lade, 1961; Podos, 1997). A good place to begin a discussion of motor constraints on trill production is with a study conducted on swamp sparrows (*Melospiza georgiana*), a species that sings simple trills and that is known to copy song structure from adults through imitative learning (Marler and Peters, 1977). In this study, young male swamp sparrows were trained with song models modified to have artificially elevated trill rates, achieved by digitizing song recordings and reducing the durations of pauses that separate notes and syllables (Podos, 1996). It was hypothesized that young males would attempt to memorize and reproduce the models presented, yet encounter performance limitations in their subsequent attempts to reproduce the rapid models accurately. Consistent with this hypothesis, birds proved able to memorize the rapid models, but unable to reproduce the models with accuracy; copy errors emerged that suggested an impact of motor constraints. In particular, birds reproduced the rapid models at slowed trill rates, or with notes omitted, or with "broken" syntax, in which groups of syllables were interspersed with brief periods of silence. All of these outcomes made it possible for birds to produce songs that otherwise would not have been mechanically possible. A similar demonstration of motor constraints was reported more recently for mockingbirds reared with the songs of canaries; mockingbirds attempting to replicate rapid canary trills reproduced those songs with broken syntax parallel to that seen in the swamp sparrow study (Zollinger and Suthers, 2004).

Evidence for performance constraints on trill production also comes from descriptive studies. One particular focus has been a trade-off in the expression of two vocal parameters: trill rate (the number of syllables delivered per unit time) and frequency bandwidth (the range of frequencies a trill spans). This trade-off was first described for 34 species of emberizid sparrows, which show a triangular distribution in a trill rate × frequency bandwidth scatterplot, circumscribed by a declining upper-bound regression (Podos, 1997). This distribution contains many songs with both low trill rates and low frequency bandwidths, and some songs with either high trill rates or high frequency bandwidth, but no songs with both high trill rates and wide frequency bandwidths. This trade-off is thought to emerge as a result of performance limitations, as follows: to produce songs with fast trill rates, the vocal apparatus must undergo rapid modulations, which necessarily limit the time span during which individual modulations can be performed. These rapid modulations in turn limit realized frequency bandwidth, because wide bandwidth songs would require broad (and thus time-consuming) mechanical modulations. Similarly, the requirement for broad mechanical modulations in the production of wide frequency bandwidth necessarily limits cycle rates and thus realized trill rates. Trade-offs between trill rate and frequency bandwidth have since been described in greater detail for a wide range of songbird species, including swamp sparrows (Ballentine *et al.*, 2004; Liu *et al.*, 2008), dark-eyed juncos (Cardoso *et al.*, 2007), Lincoln's sparrows (Sockman, 2009), yellow warblers (Beebee, 2004), black-throated blue warblers (Colbeck *et al.*, 2010), banded wrens (Illes *et al.*, 2006), and red-winged blackbirds (Cramer and Price, 2007). Similar trade-offs have also been described outside of songbirds, including in two seabirds, the brown skua (Janicke *et al.*, 2008) and the great frigatebird (Juola and Searcy, 2011), in a rodent, the neotropical singing mouse (Pasch *et al.*, 2011), and in bat echolocation calls (Schumm *et al.*, 1991; but see Schmeider *et al.*, 2010). The mechanical origins for the observed trade-offs in these other taxa are probably not related to vocal tract filtering, as they are in songbirds (see Cardoso, 2008; Janicke and Hahn, 2008; Pasch *et al.*, 2011). Instead they may arise as a more general feature of the production of repetitive vocal units, perhaps as a result of a constrained relationship between modulation rate and breadth in respiration.

Returning to songbirds, a performance constraint hypothesis on trill production is supported by other lines of descriptive data, including observations of variation in the expression or consistency of notes and note syllables across different trill types (Lambrechts, 1997) or by age (de Kort *et al.*, 2009b; Ballentine, 2009), and by a negative relationship between trill rate and the relative amplitude of notes (Forstmeier *et al.*, 2002). Taxonomically, broad constraints on trill structure such as these are of particular interest because they run counter to the marked plasticity that characterizes other aspects of songbird song evolution related to imitative learning and cultural transmission (e.g., Podos and Warren, 2007).

Constraints on song production in songbirds are also of interest in illuminating connections between the evolution of song and feeding behavior. This connection results from the fact that feeding and singing make use of a common morphological component, the beak. Birds are well known for radiating into divergent feeding niches, resulting in pronounced divergence in beak size, form, and function. Given the additional role of the beak and vocal tract in song production, feeding adaptations might impose secondary consequences on vocal performance and thus song evolution (Nowicki et al., 1992). One specific prediction along these lines is that the adaptations for increased force application (to crack hard food items) should diminish birds' vocal performance capabilities. This is because force by speed biomechanical trade-offs should render hard-biting birds unable to adjust beak gape rapidly when singing, thus leading to trills that are low in trill rate and/or frequency bandwidth (Podos and Nowicki, 2004). Several comparative studies support this hypothesis. Across eight species of Darwin's finches on Santa Cruz Island in central Galapagos, species with large beaks were found to produce songs with relatively slow trill rates and/or frequency bandwidth (Podos, 2001). A similar negative relationship between beak size (bill width) and trill rate was described by Seddon (2005) for the Neotropical antbird family Thamnophilidae. In Darwin's finches, field measures of beak gape velocity and bite force support the proposed mechanism of beak-imposed constraints on singing: birds that bite harder (measured directly from biting birds) were found to modulate their gapes slowly while singing (measured via high-speed videography), whereas birds that bite softly were found to adjust their gapes rapidly (Herrel et al., 2009). Other studies have focused within species, and have led to mixed results. Beak morphology does not correlate with song structure in African seedcrackers (Slabbekoorn and Smith, 2000), but does predict trill performance in swamp sparrows (Ballentine, 2006) and in one Darwin's finch species with unusually wide variation in beak morphology (Podos, 2001; Huber and Podos, 2006; Figure 7.3).

These and other emerging data on vocal trill performance bear not just on song evolution but also song function. Like other mating signals, bird songs are presumed to provide reliable indicators of signalers' social, behavioral, or physiological status (Zahavi, 1975; Grafen, 1990; Johnstone, 1997; Searcy and Nowicki, 2005). This is because signals that allow cheating should come to be ignored by signal receivers, and thus be dropped from usage, whereas signals that provide consistently reliable information should be favored in evolution because of their utility to receivers (Searcy and Nowicki, 2005). Reliability might be ensured by receiver-dependent costs such as predation or retaliation, but might also be ensured because of proximate limits on display mechanics. With regard to the latter possibility, animals might encounter limits in their ability to develop and maintain a display's morphological and neuromuscular underpinnings (e.g., Rome et al., 1996; Patek, 2001; Bostwick and Prum, 2005), or physical constraints in actual display performance (e.g., Gibson, 1996; Patricelli and Krakauer, 2009). Reliability in both scenarios should emerge because only the highest quality signalers should be positioned to meet challenges inherent in display execution, especially when displays are exaggerated or physically challenging (Byers et al., 2010). As such, displays can achieve reliability by being uncheatable.

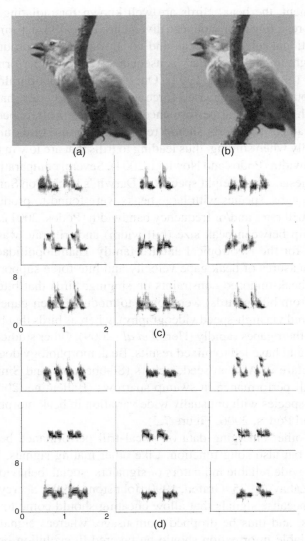

Figure 7.3.   (a,b) Video images of a singing male Darwin's finch, *Camarhynchus psittacula*. Panel (a) shows the bird during the production of a low frequency note, Panel (b) during a high frequency note. Notable changes in thoracic volume and beak gape are apparent, and correlate with changing vocal frequencies such that the vocal tract's resonance filtering function is retained across varying source frequencies. (After Podos *et al.* (2004).) (c,d) Spectrograms of songs produced by small and large morphs (Panels c and d respectively) of another Darwin's finch species, *Geospiza fortis*. The songs of the two morphs differ in frequency bandwidth, in a manner predicted by a vocal tract constraint hypothesis. (After Huber and Podos (2006).) *X*-axis = 2 seconds, *Y*-axis spans 0–8 kHz. (*See insert for color representation of this figure.*)

Returning to bird song, do songs provide reliable indicators of the status of singers? Performance limitations on vocal production could help ensure signal reliability, in that birds of varying quality or status might diverge in their ability to produce challenging trill phenotypes. Two ways to test whether trills offer reliable indicators of signaler quality or status are to evaluate the contexts in which trills are produced, and to evaluate how other birds respond to those trills (the "context" and "response" criteria, respectively, for evaluating signal function; see Searcy and Beecher, 2009). With regard to the context criterion, a number of species have been documented singing trills more often, or with higher performance levels, during aggressive contexts than during non-aggressive contexts (Trillo and Vehrencamp, 2005; Kunc et al., 2006; Schmidt et al., 2006; DuBois et al., 2009; Sprau et al., 2010). Moreover, the ability to produce high-performance or consistent trills has been shown to correlate with success in extra-pair paternity (Forstmeier et al., 2002; Janicke et al., 2008). With regard to the response criterion, numerous species have been shown to respond differentially to songs of high versus low performance. Males in natural conditions respond more strongly to song bouts that include greater proportions of trills (Schmidt et al., 2008), and differentially to trills that express high versus low performance levels (Illes et al., 2006; Cramer and Price, 2007; de Kort et al., 2009a). Females in captivity have also been shown to differentiate songs of varying performance levels, typically reserving their strongest responses for high-performance songs (Draganoiu et al., 2002; Ballentine et al., 2004; Vallet and Kreutzer 1995; Vallet et al., 1998). Of particular note is a recent hypothesis that suggests an interaction between vocal performance and song type matching, a phenomenon in which countersinging birds match each other to songtype. According to this hypothesis, song type matching provides listeners a uniquely useful opportunity for assessing and comparing vocal performance among multiple individuals (Logue and Forstmeier 2008).

Given song's role in communication, performance limits on song production may hold implications for patterns of reproductive isolation and speciation. In short, adaptive evolution of the beak in distinct populations is predicted to drive divergence in vocal performance capacities, which could in turn drive vocal divergence and increased probabilities of reproductive isolation among diverging populations. This possibility has been explored in a series of recent studies on the medium ground finch at El Garrpatero, Santa Cruz Island, Galapagos. This species exhibits two distinct forms – small and large "morphs" (Hendry et al., 2006) – which appear favored by disruptive selection (Hendry et al., 2009). Consistent with the performance constraint hypothesis, the morphs were found to differ in vocal performance (Herrel et al., 2009) and vocal phenotypes, such that small-morph songs tend to show greater frequency bandwidth (Huber and Podos, 2006). Birds are able to discriminate songs by morph (Podos, 2010), a capacity that may help to explain observed patterns of assortative pairing and genetic divergence by morph (Huber et al., 2007). The two morphs also differ in a key visual signal, beak size, so that visual and acoustic signals offer redundant information about the identity of prospective rivals and mates. This redundancy likely offers birds greater certainty in assortative mating, and on a larger scale may help to explain rapid rates of speciation in Galapagos finches (Podos, 2001; Podos and Nowicki, 2004).

In summary, as with sprinting in lizards, acoustic systems may experience trade-offs that relate directly to constraints on behavioral performance. The evolutionary boundaries on trill rate and frequency bandwidth are replicated at developmental levels when birds attempt to learn "challenging" songs. When multiple levels of acoustic control are considered in studies of performance, such as the integrated role of the syrinx and beak in modulating songs, even broader implications become evident. For example, natural selection for particular beak features, such as greater mass and strength for crushing large seeds, can have direct impacts on acoustic signal evolution. These and the other findings featured in this section ultimately point to the behavioral role of acoustic signals in communication systems. An individual's performance can provide essential, reliable information about his or her quality, thus bringing the role of acoustic mechanics full circle, to encompass both proximate and ultimate levels of communication.

CONCLUDING REMARKS

Biomechanics, size, and performance can help define patterns of acoustic signal variation within and among species. Evolutionary and developmental variation, as well as constraints, determine the acoustic phenotype sets that can be expressed, and therefore available for evolution and selection. Notably, for every apparent production constraint in a group of animals, multiple evolutionary or behavioral outcomes emerge. In katydids, stridulation frequencies appear at first glance to be limited by mirror resonances or forewing speed, yet some taxa substitute mirror for scraper resonances and have introduced spring-loaded, rapid scraper translation that far exceeds the underlying forewing translation speeds. In the case of body size limitations, frogs appear to exhibit vocal frequencies that are body size-limited, yet some species circumvent this limitation by scaling their vocal chords allometrically, or even just by adding tissue masses to their vocal chords to decrease frequencies independently from body size scaling rules. Lastly, trill rate maxima in songbirds are determined by performance limits, yet some songbirds can instead sing with "broken" syntax in order to achieve maximal trill rates with minimal performance costs.

In these and other examples, integrating mechanistic analyses with developmental, ecological, or evolutionary perspectives provides a window into the parameters essential to the production of particular signals. Looking at the mechanics of just one species or population might yield a straightforward relationship between structure and signal, but the broader comparative approach often uncovers surprising and informative discoveries about the ways that organisms circumvent limitations and achieve the performances that characterize the fabulous diversity of acoustic signals in nature. In addition, a more precise accounting of within-species determinants of signal diversity will ultimately provide a useful complement to more traditional emphases on signal function and reliability.

ACKNOWLEDGMENTS

We are grateful to Duncan Irschick and Bill Searcy for their helpful feedback on an earlier version of this manuscript. Supported by National Science Foundation grants IOS-1028964 (JP) and IOS-1014573 (SNP).

REFERENCES

Arak, A. (1983) Sexual selection by male–male competition in natterjack toad choruses. *Nature*, **306**, 261–262.

Arnold, S.J. (1983) Morphology, performance, and fitness. *American Zoologist*, **23**, 347–361.

Ballentine, B. (2006) Morphological adaptation influences the evolution of a mating signal. *Evolution*, **60**, 1936–1944.

Ballentine, B. (2009) The ability to perform physically challenging songs predicts age and size in male swamp sparrows, *Melospiza georgiana*. *Animal Behaviour*, **77**, 973–978.

Ballentine, B., Hyman, J., and Nowicki, S. (2004) Vocal performance influences female response to male bird song: an experimental test. *Behavioral Ecology*, **15**, 163–168.

Bee, M.A., Perrill, S.A., and Owen, P.C. (1999) Size assessment in simulated territorial encounters between male green frogs (*Rana clamitans*). *Behavioral Ecology and Sociobiology*, **45**, 177–184.

Beebee, M.D. (2004) Variation in vocal performance in the songs of a wood warbler: evidence for the function of distinct singing modes. *Ethology*, **110**, 531–542.

Bennet-Clark, H.C. (1987) The tuned singing burrow of mole crickets. *Journal of Experimental Biology*, **128**, 383–409.

Bertelli, S. and Tubaro, P.L. (2002) Body mass and habitat correlates of song structure in a primitive group of birds. *Biological Journal of the Linnean Society*, **77**, 423–430.

Blair, W.F. (1964) Evolution at populational and interpopulational levels; isolating mechanisms and interspecies interactions in anuran amphibians. *Quarterly Review of Biology*, **39**, 333–344.

Blanckenhorn, W.U. (2000) The evolution of body size: what keeps organisms small? *Quarterly Review of Biology*, **75**, 385–407.

Bostwick, K.S. and Prum, R.O. (2005) Courting bird sings with stridulating wing feathers. *Science*, **309**, 736.

Bowman, R.I. (1983) The evolution of song in Darwin's finches, in *Patterns of Evolution in Galápagos Organisms* (eds R.I. Bowman, M. Berson, and A.E. Leviton), American Association for the Advancement of Science, Pacific Division, San Francisco, CA, pp. 237–537.

Bradbury, J.W. and Vehrencamp, S.L. (2011) Principles of Animal Communication, 2nd edn, Sinauer Associates, Sunderland, MA.

Brainard, M.S. and Doupe, A.J. (2002) What songbirds teach us about learning. *Nature*, **417**, 351–358.

Byers, J.A., Hebets, E., and Podos, J. (2010) Female mate choice based upon male motor performance. *Animal Behaviour*, **79**, 771–778.

Cade, W. (1975) Acoustically orienting parasites: fly phonotaxis to cricket song. *Science*, **190**, 1312–1313.

Cardoso, G.C. (2008) On the performance of brown skua, *Catharacta antarctica*, vocalizations. *Animal Behaviour*, **76**, e1–e2.

Cardoso, G.C., Atwell, J.W., Ketterson, E.D., and Price, T.D. (2007) Inferring performance in the songs of dark-eyed juncos (*Junco hyemalis*). *Behavioral Ecology*, **18**, 1051–1057.

Cardoso, G.C., Mamede, A.T., Atwell, J.W., Mota, P.G., Ketterson, E.D., and Price, T.D. (2008) Song frequency does not reflect differences in body size among males in two oscine species. *Ethology*, **114**, 1084–1093.

Catchpole, C.K. and Slater, P.J.B. (2008) Bird Song: Biological Themes and Variations, 2nd edn, Cambridge University Press, Cambridge, UK.

Colbeck, G.J., Sillett, T.S., and Webster, M.S. (2010) Asymmetric discrimination of geographical variation in song in a migratory passerine. *Animal Behaviour*, **80**, 311–318.

Cramer, E.R.A. and Price, J.J. (2007) Red-winged blackbirds *Ageliaus phoeniceus* respond differently to song types with different performance levels. *Journal of Avian Biology*, **38**, 122–127.

Cutler, B. (1970) Anatomical Studies of the Syrinx of Darwin's Finches, San Francisco State University, San Francisco, p. 272.

Davies, N.B. and Halliday, T.R. (1978) Deep croaks and fighting assessment in toads: *Bufo bufo*. *Nature*, **274**, 683–685.

Davies, N.B. and Halliday, T.R. (1979) Competitive mate searching in male common toads, *Bufo bufo*. *Animal Behaviour*, **27**, 1253–1267.

de Kort, S.R., Eldermire, E.R.B., Cramer, E.R.A., and Vehrencamp, S.L. (2009a) The deterrent effect of bird song in territory defense. *Behavioral Ecology*, **20**, 200–206.

de Kort, S.R., Eldermire, E.R., Valderrama, S., Botero, C.A., and Vehrencamp, S.L. (2009b) Song consistency reflects age in banded wrens. *Proceedings of the Royal Society of London, Series B: Biological Sciences*, **276**, 2315–2321.

Del Castillo, R.C. and Gwynne, D.T. (2007) Increase in song frequency decreases spermatophore size: correlative evidence of a macroevolutionary trade-off in katydids (Orthoptera: Tettigoniidae). *Journal of Evolutionary Biology*, **20**, 1028–1036.

Draganoiu, T.I., Nagle, L., and Kreutzer, M. (2002) Directional female preference for an exaggerated male trait in canary (*Serinus canaria*) song. *Proceedings of the Royal Society of London, Series B: Biological Sciences*, **269**, 2525–2531.

DuBois, A.L., Nowicki, S., and Searcy, W.A. (2009) Swamp sparrows modulate vocal performance in an aggressive context. *Biology Letters*, **5**, 163–165.

Duellman, W.E. and Pyles, R.A. (1983) Acoustic resource partitioning in anuran communities. *Copeia*, **1983**, 941–953.

Elliott, C.J.H. and Koch, U.T. (1985) The clockwork cricket. *Naturwissenchaften*, **72**, 150–153.

Felsenstein, J. (1985) Phylogenies and the comparative method. *American Naturalist*, **125**, 1–15.

Fitch, W.T. (1999) Acoustic exaggeration of size in birds via tracheal elongation: comparative and theoretical analyses. *Journal of Zoology*, **248**, 31–48.

Fitch, W.T. and Hauser, M.D. (2003) Unpacking "honesty": vertebrate vocal production and the evolution of acoustic signals, in *Acoustic Communication* (eds A.M. Simmons, A.N. Popper, and R.R. Fay), Springer, New York, pp. 65–137.

Forstmeier, W., Kempenaers, B., and Meyer, A. (2002) A novel song parameter correlates with extra-pair paternity and reflects male longevity. *Proceedings of the Royal Society of London, Series B: Biological Sciences*, **269**, 1479–1485.

Garland, T. Jr. and Losos, J.B. (1994) Ecological morphology of locomotor performance in squamate reptiles, in *Ecological Morphology: Integrative Organismal Biology* (eds P.C. Wainwright and S.M. Reilly), University of Chicago Press, Chicago, IL, pp. 240–302.

Gibson, R.M. (1996) Female choice in sage grouse: the roles of attraction and active comparison. *Behavioral Ecology and Sociobiology*, **18**, 117–123.

Goller, F. and Larsen, O.N. (1997) A new mechanism of sound generation in songbirds. *Proceedings of the National Academy of Sciences of the United States of America*, **94**, 14787–14791.

Goodale, E. and Podos, J. (2010) Persistence of song types in Darwin's finches, *Geospiza fortis*, over four decades. *Biology Letters*, **6**, 589–592.

Grafen, A. (1990) Biological signals as handicaps. *Journal of Theoretical Biology*, **144**, 517–546.

Graves, G.R. (1991) Bergmann's rule near the equator: latitudinal clines in body size of an Andean passerine bird. *Proceedings of the National Academy of Sciences of the United States of America*, **88**, 2322–2325.

Greenewalt, C. (1968) *Bird Song: Acoustics and Physiology*, Smithsonian Institution Press, Washington, DC.

Gridi-Papp, M., Rand, A.S., and Ryan, M.J. (2006) Complex call production in the Túngara frog. *Nature*, **441**, 38.

Gwynne, D.T. (2001) *Katydids and Bush-Crickets: Reproductive Behavior and Evolution of the Tettigoniidae*, Cornell University Press, Ithaca, NY.

Hall, M., Kingma, S.A., and Peters, A. (2013) Male songbird indicates body size with low-pitched advertising songs. PLoS ONE, **8**, e56717.

Hartley, R.S. and Suthers, R.A. (1989) Airflow and pressure during canary song: evidence for mini-breaths. *Journal of Comparative Physiology A*, **165**, 15–26.

Hendry, A.P., Grant, P.R., Grant, B.R., Ford, H.A., Brewer, M.J., and Podos, J. (2006) Possible human impacts on adaptive radiation: beak size bimodality in Darwin's finches. *Proceedings of the Royal Society of London, Series B: Biological Sciences*, **273**, 1887–1894.

Hendry, A.P., Huber, S.K., de Leon, L.F., Herrel, A., and Podos, J. (2009) Disruptive selection in a bimodal population of Darwin's finches. *Proceedings of the Royal Society of London, Series B: Biological Sciences*, **276**, 753–759.

Herrel, A., Podos, J., Vanhooydonck, B., and Hendry, A.P. (2009) Force-velocity trade-off in Darwin's finch jaw function: a biomechanical basis for ecological speciation? *Functional Ecology*, **23**, 119–125.

Hoese, W.J., Podos, J., Boetticher, N.C., and Nowicki, S. (2000) Vocal tract function in birdsong production: experimental manipulation of beak movements. *Journal of Experimental Biology*, **203**, 1845–1855.

Howard, R.D. and Young, J.R. (1998) Individual variation in male vocal traits and female mating preferences in *Bufo americanus*. *Animal Behaviour*, **55**, 1165–1179.

Huber, S.K. and Podos, J. (2006) Beak morphology and song features covary in a population of Darwin's finches (*Geospiza fortis*). *Biological Journal of the Linnean Society*, **88**, 489–498.

Huber, S.K., de Leon, L.F., Hendry, A.P., Bermingham, E., and Podos, J. (2007) Reproductive isolation of sympatric morphs in a population of Darwin's finches. *Proceedings of the Royal Society of London, Series B: Biological Sciences*, **274**, 1709–1714.

Illes, A.E., Hall, M.L., and Vehrencamp, S.L. (2006) Vocal performance influences male receiver response in the banded wren. *Proceedings of the Royal Society of London, Series B: Biological Sciences*, **273**, 1907–1912.

Irschick, D.J. (2003) Measuring performance in nature: implications for studies of fitness within populations. *Integrative and Comparative Biology*, **43**, 396–407.

Irschick, D.J. and Garland, T. Jr. (2001) Integrating function and ecology in studies of adaptation: investigations of locomotor capacity as a model system. *Annual Review of Ecology, Evolution, and Systematics*, **32**, 367–396.

Irschick, D.J., Meyers, J.J., and Le Galliard, J.F. (2008) How does selection operate on whole-organism functional performance capacities? A review and synthesis. *Evolutionary Ecology Research*, **10**, 177–196.

Janicke, T. and Hahn, S. (2008) On the performance of brown skua, *Catharacta antarctica*, vocalizations: reply. *Animal Behaviour*, **76**, e3–e5.

Janicke, T., Hahn, S., Ritz, M.S., and Peter, H.-U. (2008) Vocal performance reflects individual quality in a nonpasserine. *Animal Behaviour*, **75**, 91–98.

Jarvis, E.D. (2004) Brains and birdsong, in *Nature's Music: the Science of Birdsong* (eds P. Marler and H. Slabbekoorn), Elsevier Academic Press, San Diego, CA, pp. 226–271.

Johnstone, R.A. (1997) The evolution of animal signals, in *Behavioural Ecology* (eds J.R. Krebs and N.B. Davies), Blackwell Publishing, Oxford, UK, pp. 155–178.

Jost, M.C. and Shaw, K.L. (2006) Phylogeny of Ensifera (Hexapoda: Orthoptera) using three ribosomal loci, with implications for the evolution of acoustic communication. *Molecular Phylogenetics and Evolution*, **38**, 510–530.

Juola, F.A. and Searcy, W.A. (2011) Vocalizations reveal body condition and are associated with visual display traits in great frigatebirds (*Fregata minor*). *Behavioral Ecology and Sociobiology*, **65**, 2297–2303.

Klump, G.M. and Gerhardt, H.C. (1987) Use of non-arbitrary acoustic criteria in mate choice by female gray tree frogs. *Nature*, **326**, 286–288.

Koch, U.T., Elliot, C.J.H., Schaffner, K., and Kleindienst, H.-U. (1988) The mechanics of stridulation of the cricket Gryllus campestris. *Journal of Comparative Physiology A*, **162**, 213–223.

Krebs, J.R. and Davies, N.B. (1993) An Introduction to Behavioral Ecology, 4th edition, Oxford University Press, Oxford.

Kunc, H.P., Amrhein, V., and Naguib, M. (2006) Vocal interactions in nightingales (*Luscinia megarhynchos*): more aggressive males have higher pairing success. *Animal Behaviour*, **72**, 25–30.

Lambrechts, M.M. (1997) Song frequency plasticity and composition of phrase versions in great tits *Parus major*. *Ardea*, **85**, 99–109.

Laiolo, P. and Rolando, A. (2003) Comparative analysis of the rattle calls in *Corvus* and *Nucifraga*: the effect of body size, bill size, and phylogeny. *Condor*, **105**, 139–144.

Liu, I.A., Lohr, B., Olsen, B., and Greenberg, R. (2008) Macrogeographic vocal variation in subspecies of swamp sparrow. *Condor*, **110**, 102–109.

Logue, D.M., Abiola, I.O., Rains, D., Bailey, N.W., Zuk, M., and Cade, W.H. (2010) Does signalling mitigate the cost of agonistic interactions? A test in a cricket that has lost its song. *Proceedings of the Royal Society of London, Series B: Biological Sciences*, **277**, 2571–2575.

Logue, D.M., Droessler, E.E., Roscoe, D.W., Vokey, J.R., Rendall, D., and Kunimoto, R.M. (2007) Sexually antithetical song structure in a duet singing wren. *Behaviour*, **144**, 331–350.

Logue, D.M. and Forstmeier, W. (2008). Constrained performance in a communication network: implications for the function of song-type matching and for the evolution of multiple ornaments. *American Naturalist*, **172**, 34–41.

Martin, W.F. (1971) Mechanics of sound production in toads of the genus *Bufo*: passive elements. *Journal of Experimental Zoology*, **176**, 273–294.

Martin, W.F. (1972) Evolution of vocalization in the toad genus *Bufo*, in *Evolution in the Genus Bufo* (ed W.F. Blair), University of Texas Press, Austin, Texas, pp. 279–309.

Marler, P. and Peters, S. (1977) Selective vocal learning in a sparrow. *Science*, **198**, 519–521.

Maynard Smith, J. and Harper, D. (2003) *Animal Signals*, Oxford University Press, Oxford.

Montealegre-Z, F. (2009) Scale effects and constraints for sound production in katydids (Orthoptera: Tettigoniidae): correlated evolution between morphology and signal parameters. *Journal of Evolutionary Biology*, **22**, 355–366.

Montealegre-Z, F. and Morris, G.K. (2004) The spiny devil katydids, *Panacanthus* Walker (Orthoptera: Tettigoniidae): an evolutionary study of acoustic behavior and morphological traits. *Systematic Entomology*, **29**, 21–57.

Montealegre-Z, F., Morris, G.K., and Mason, A.C. (2006) Generation of extreme ultrasonics in rainforest katydids. *Journal of Experimental Biology*, **209**, 4923–4937.

Montealegre-Z, F. and Postles, M. (2010) Resonant sound production in *Copiphora gorgonensis* (Tettigoniidae: Copiphorini), an endemic species from Parque Nacional Natural Gorgona, Colombia. *Journal of Orthoptera Research*, **19**, 347–355.

Morris, M.R. and Yoon, S.L. (1989) A mechanism for female choice of large males in the treefrog *Hyla chrysoscelis*. *Behavioral Ecology and Sociobiology*, **25**, 65–71.

Nottebohm, F. (1971) Neural lateralization of vocal control in a passerine bird. I. Song. *Journal of Experimental Zoology*, **177**, 229–262.

Nowicki, S. (1987) Vocal tract resonances in oscine bird sound production: evidence from birdsongs in a helium atmosphere. *Nature*, **325**, 53–55.

Nowicki, S. and Marler, P. (1988) How do birds sing? *Music Perception*, **5**, 391–426.

Nowicki, S., Westneat, M.W., and Hoese, W.J. (1992) Birdsong: motor function and the evolution of communication. *Seminars in Neuroscience*, **4**, 385–390.

Palacios, M.G. and Tubaro, P.L. (2000) Does beak size affect acoustic frequencies in woodcreepers? *Condor*, **102**, 553–560.

Pasch, B., George, A.S., Campbell, P., and Phelps, S.M. (2011) Androgen-dependent male vocal performance influences female preference in Neotropical singing mice. *Animal Behaviour*, **82**, 177–183.

Patek, S.N. (2001) Spiny lobsters stick and slip to make sound. *Nature*, **411**, 153–154.

Patek, S.N. (2002) Squeaking with a sliding joint: mechanics and motor control of sound production in palinurid lobsters. *Journal of Experimental Biology*, **205**, 2375–2385.

Patek, S.N. and Baio, J.E. (2007) The acoustic mechanics of stick-slip friction in the California spiny lobster (*Panulirus interruptus*). *Journal of Experimental Biology*, **210**, 3538–3546.

Patek, S.N. and Caldwell, R.L. (2006) The stomatopod rumble: sound production in *Hemisquilla californiensis*. *Marine and Freshwater Behaviour and Physiology*, **39**, 99–111.

Patek, S.N., Dudek, D.M., and Rosario, M.V. (2011) From bouncy legs to poisoned arrows: elastic movements in invertebrates. *The Journal of Experimental Biology*, **214**, 1973–1980.

Patek, S.N. and Oakley, T.H. (2003) Comparative tests of evolutionary tradeoffs in a palinurid lobster acoustic system. *Evolution*, **57**, 2082–2100.

Patel, R., Mulder, R.A., and Cardoso, G.C. (2010) What makes vocalisation frequency an unreliable signal of body size in birds? A study on black swans. *Ethology*, **116**, 554–563.

Patricelli, G.L. and Krakauer, A.H. (2009) Tactical allocation of effort among multiple signals in sage grouse: an experiment with a robotic female. *Behavioral Ecology*, **21**, 97–106.

Podos, J. (1996) Motor constraints on vocal development in a songbird. *Animal Behaviour*, **51**, 1061–1070.

Podos, J. (1997) A performance constraint on the evolution of trilled vocalization in a songbird family (Passeriformes: Emberizidae). *Evolution*, **51**, 537–551.

Podos, J. (2001) Correlated evolution of morphology and vocal signal structure in Darwin's finches. *Nature*, **409**, 185–188.

Podos, J. (2010) Acoustic discrimination of sympatric morphs in Darwin's finches: a behavioral mechanism for assortative mating? *Philosophical Transactions of the Royal Society B*, **365**, 1031–1039.

Podos, J. and Hendry, A.P. (2006) The biomechanics of ecological speciation, in *Ecology and Biomechanics: a Mechanical Approach to the Ecology of Animals and Plants* (eds A. Herrel, T. Speck, and N. Rowe), CRC Press, Boca Raton, FL, pp. 301–321.

Podos, J. and Nowicki, S. (2004) Performance limits on birdsong, in *Nature's Music: the Science of Birdsong* (eds P. Marler and H. Slabbekoorn), Elsevier Academic Press, San Diego, CA, pp. 318–342.

Podos, J., Lahti, D.C., and Moseley, D.L. (2009) Vocal performance and sensorimotor learning in songbirds. *Advances in the Study of Behaviour*, **40**, 159–195.

Podos, J., Sherer, J.K., Peters, S., and Nowicki, S. (1995) Ontogeny of vocal tract movements during song production in song sparrows. *Animal Behaviour*, **50**, 1287–1296.

Podos, J., Southall, J.A., and Rossi-Santos, M.R. (2004) Vocal mechanics in Darwin's finches: correlation of beak gape and song frequency. *Journal of Experimental Biology*, **207**, 607–619.

Podos, J. and Warren, P.S. (2007) The evolution of geographic variation in birdsong. *Advances in the Study of Behaviour*, **37**, 403–458.

Prestwich, K.N., Brugger, K.E., and Topping, M. (1989) Energy and communication in three species of hylid frogs: power output and efficiency. *Journal of Experimental Biology*, **144**, 53–80.

Rebar, D., Bailey, N.W., and Zuk, M. (2009) Courtship song's role during female mate choice in the field cricket *Teleogryllus oceanicus*. *Behavioral Ecology*, **20**, 1307–1314.

Riede, T., Suthers, R.A., Fletcher, N.H., and Blevins, W.E. (2006) Songbirds tune their vocal tract to the fundamental frequency of their song. *Proceedings of the National Academy of Sciences of the United States of America*, **103**, 5543–5548.

Roff, D.A. (1992) *The Evolution of Life Histories: Theory and Analysis*, Chapman and Hall, New York.

Rome, L.C., Syme, D.A., Hollingworth, S., Lindstedt, S.L., and Baylor, S.M. (1996) The whistle and the rattle: the design of sound producing muscles. *Proceedings of the National Academy of Sciences of the United States of America*, **93**, 8095–8100.

Ryan, M.J. (1980) Female choice in a neotropical frog. *Science*, **209**, 523–525.

Ryan, M.J. (1985) *The Túngara Frog*, University of Chicago Press, Chicago.

Ryan, M.J. and Brenowitz, E.A. (1985) The role of body size, phylogeny, and ambient noise in the evolution of bird song. *American Naturalist*, **126**, 87–100.

Ryan, M.J., Perrill, S.A., and Wilczynski, W. (1992) Auditory tuning and call frequency predict population-based mating preferences in the cricket frog *Acris crepitans*. *American Naturalist*, **139**, 1370–1383.

Schmeider, D.A., Kingston, T., Hashim, R., and Siemers, B.M. (2010) Breaking the trade-off: rainforest bats maximize bandwidth and repetition rate of echolocation calls as they approach prey. *Biology Letters*, **6**, 604–609.

Schmidt, R., Kunc, H.P., Amrhein, V., and Naguib, M. (2006) Responses to interactive playback predict future mating status in nightingales. *Animal Behaviour*, **72**, 1355–1362.

Schmidt, R., Kunc, H.P., Amrhein, V., and Naguib, M. (2008) Aggressive responses to broadband trills are related to subsequent pairing success in nightingales. *Behavioral Ecology*, **19**, 635–641.

Schmidt-Nielsen, K.S. (1984) *Scaling: Why is Animal Size so Important?*, Cambridge University Press, Cambridge.

Schumm, A., Krull, D., and Neuweiler, G. (1991) Echolocation in the notch eared bat, *Myotis emarginatus*. *Behavioral Ecology and Sociobiology*, **28**, 255–261.

Searcy, W.A. and Beecher, M.D. (2009) Song as an aggressive signal in songbirds. *Animal Behaviour*, **78**, 1281–1292.

Searcy, W.A. and Nowicki, S. (2005) *The Evolution of Animal Communication: Reliability and Deception in Signalling Systems*, Princeton University Press, Princeton, NJ.

Seddon, N. (2005) Ecological adaptation and species recognition drives vocal evolution in Neotropical suboscine birds. *Evolution*, **59**, 200–215.

Seyfarth, R.M., Cheney, D.L., Bergman, T., Fischer, J., Zuberbuhler, K., and Hammerschmidt, K. (2010) The central importance of information in studies of animal communication. *Animal Behaviour*, **80**, 3–8.

Slabbekoorn, H. and Smith, T.B. (2000) Does bill size polymorphism affect courtship song characteristics in the African finch *Pyrenestes ostrinus*? *Biological Journal of the Linnean Society*, **71**, 737–753.

Sockman, K.W. (2009) Annual variation in vocal performance and its relationship with bill morphology in Lincoln's sparrows, *Melospiza lincolnii*. *Animal Behaviour*, **77**, 663–671.

Sprau, P., Schmidt, R., Roth, T., Amrhein, V., and Naguib, M. (2010) Effects of rapid broadband trills on responses to song overlapping in nightingales. *Ethology*, **116**, 300–308.

Staaterman, E.R., Clark, C.W., Gallagher, A.J., deVries, M.S., Claverie, T., and Patek, S.N. (2011) Rumbling in the benthos: the acoustic ecology of the California mantis shrimp. *Aquatic Biology*, **13**, 97–105.

Suthers, R.A. (1990) Contributions to birdsong from the left and right sides of the intact syrinx. *Nature*, **347**, 473–477.

Suthers, R.A. (2004) How birds sing and why it matters, in *Nature's Music: the Science of Birdsong* (eds P. Marler and H. Slabbekoorn), Elsevier Academic Press, San Diego, CA, pp. 272–295.

Taigen, T.L. and Wells, K.D. (1985) Energetics of vocalization by an anuran amphibian, *Hyla versicolor*. *Journal of Comparative Physiology B*, **155**, 163–170.

Thorpe, W.H. and Lade, B.I. (1961) The songs of some families of the Passeriformes: II. The songs of the buntings (Emberizidae). *Ibis*, **103**, 246–259.

Tinghitella, R.M. and Zuk, M. (2009) Asymmetric mating preferences accommodated the rapid evolutionary loss of a sexual signal. *Evolution*, **63**, 2087–2098.

Tinghitella, R.M., Wang, J.M., and Zuk, M. (2009) Preexisting behavior renders a mutation adaptive: flexibility in male phonotaxis behavior and the loss of singing ability in the field cricket *Teleogryllus oceanicus*. *Behavioral Ecology*, **20**, 722–728.

Tolle, A.E. and Wagner, W.E. (2011) Costly signals in a field cricket can indicate high or low-quality direct benefits depending on the environment. *Evolution*, **65**, 283–294.

Trillo, P.A. and Vehrencamp, S.L. (2005) Song types and their structural features are associated with specific contexts in the banded wren. *Animal Behaviour*, **70**, 921–935.

Vallet, E., Beme, I., and Kreutzer, M. (1998) Two-note syllables in canary songs elicit high levels of sexual display. *Animal Behaviour*, **55**, 291–297.

Vallet, E. and Kreutzer, M. (1995) Female canaries are sexually responsive to special song phrases. *Animal Behaviour*, **49**, 1603–1610.

Wagner, W.E. (1989) Fighting, assessment, and frequency alteration in Blanchard's cricket frog. *Behavioral Ecology and Sociobiology*, **25**, 429–436.

Wainwright, P.C. (1994) Functional morphology as a tool in ecological research, in *Ecological Morphology: Integrative Organismal Biology* (eds P.C. Wainwright and S.M. Reilly), University of Chicago Press, Chicago, IL, pp. 42–59.

Wallschläger, D. (1980) Correlation of song frequency and body weight in passerine birds. *Experientia*, **36**, 69–94.

Wells, K.D. (1977) Territoriality and male mating success in the green frog (*Rana clamitans*). *Ecology*, **58**, 750–762.

Wells, K.D., Taigen, T.L., Rusch, S.W., and Robb, C.C. (1995) Seasonal and nightly variation in glycogen reserves of calling grey treefrogs (*Hyla versicolor*). *Herpetologica*, **51**, 359–368.

Westneat, M.W., Long, J.H., Hoese, W., and Nowicki, S. (1993) Kinematics of birdsong: functional correlation of cranial movements and acoustic features in sparrows. *Journal of Experimental Biology*, **182**, 147–171.

Wiley, R.H. and Richards, D.G. (1978) Physical constraints on acoustic communication in the atmosphere: implications for the evolution of animal vocalizations. *Behavioral Ecology and Sociobiology*, **3**, 69–94.

Young, D. and Bennet-Clark, H. (1995) The role of the tymbal in cicada sound production. *Journal of Experimental Biology*, **198**, 1001–1020.

Zahavi, A. (1975) Mate selection – selection for a handicap. *Journal of Theoretical Biology*, **53**, 205–214.

Zollinger, S.A. and Suthers, R.A. (2004) Motor mechanisms of a vocal mimic: implications for birdsong production. *Proceedings of the Royal Society of London, Series B: Biological Sciences*, **271**, 483–491.

Zuk, M., Rotenberry, J.T., and Tinghitella, R.M. (2006) Silent night: adaptive disappearance of a sexual signal in a parasitized population of field crickets. *Biology Letters*, **2**, 521–524.

8

DISHONEST SIGNALING DURING AGGRESSIVE INTERACTIONS: THEORY AND EMPIRICAL EVIDENCE

Robbie S. Wilson[1] and Michael J. Angilletta Jr.[2]

[1]*School of Biological Sciences, The University of Queensland, St Lucia, QLD, Australia*
[2]*School of Life Sciences, Arizona State University, Tempe, AZ, USA*

Fool me once, shame on you. Fool me twice, shame on me.
—*Montgomery "Scotty" Scott, Star Trek*

Fool me once, shame on — shame on you. Fool me, you can't get fooled again.
—*George W. Bush, 43rd President of the United States of America*

INTRODUCTION

Humans are inherently fascinated with dishonesty. Who has not told a white lie? Who has not cheated during a game of cards or sport? If you are thinking "Not me," you likely reside in a very small minority. Whether at home, work, or play, humans use subtle forms of exaggeration to gain resources and to manipulate the behavior of others. But how often does dishonesty really pay off? After all, no one believes a perpetual liar. Indeed, people are often taught that the best way to hide a lie is between two truths. Cheating and dishonesty have undoubtedly played an important role in the evolution of humans and other primates (Lachmann *et al.*, 2001). In fact,

Animal Signaling and Function: An Integrative Approach, First Edition.
Edited by Duncan J. Irschick, Mark Briffa, and Jeffrey Podos.
© 2015 John Wiley & Sons, Inc. Published 2015 by John Wiley & Sons, Inc.

the expansion of the primate neocortex seems associated with increased tendencies for deception (Byrne and Corp, 2004). However, should we also expect dishonest communication to be routinely used by other species?

Dishonesty has been one of the most controversial topics in animal communication (Maynard Smith and Harper, 2003; Searcy and Nowicki, 2005). The central issue has always been whether individuals actually convey reliable information to receivers (Searcy and Nowicki, 2005). Communication between individuals of the same species often involves a conflict of interest, and the role that deception plays in this struggle not only fascinates researchers but also potentially explains much about human behavior. In this chapter, we pursue the subject of dishonesty and discuss its importance during animal communication when interests oppose. In particular, we focus on what functional studies of animal performance can offer researchers interested in dishonest communication. We briefly review the relevant theoretical work on dishonest signaling and the empirical studies that support its existence in natural populations. Although mounting evidence supports the existence of dishonest communication, we critically evaluate this evidence and discuss whether dishonesty provides the only explanation. We finish by suggesting ways to link functional studies of animal performance to evolutionary analyses of dishonest signaling during agonistic encounters.

THE EVOLUTION OF SIGNALING

A signal is a behavior or structure that alters the actions of other organisms, commonly referred to as *receivers* (Maynard Smith and Harper, 2003). Krebs and Dawkins (1984) describe animal signaling as an arms race between "manipulators" (signalers) and "mind-readers" (receivers). A signaler has private information about some aspect of its quality or intention and selects a signal to send to a receiver. The receiver responds by selecting an action and the benefit (or cost) depends upon the actions of both individuals. Signals can be acoustic, visual, or chemical, with individuals often combining these different forms of signals when displaying to receivers (Berglund *et al.*, 1996; Johnstone, 1997). Such signals portray resource requirements to parents, fighting ability to competitors, or genetic quality to mates (Backwell *et al.*, 2000; Searcy and Nowicki, 2005; Lappin *et al.*, 2006).

A central tenet of signaling theory is that, at an evolutionary equilibrium, both signalers and receivers must benefit from the exchange of information. If receivers did not benefit from their responses, selection would favor receivers that ignore the signal. This requirement implies that signals must convey reliable information about the intentions, quality, or needs of the signaler. In other words, signalers should be honest. This idea has been the source of controversy surrounding theoretical and empirical studies of animal communication (Enquist, 1985; Grafen, 1990; Számadó, 2000). To understand this controversy, we must define what we mean by dishonest signaling. Searcy and Nowicki (2005) provided a formal definition of dishonest signaling. Dishonesty occurs when a receiver registers X from a signaler and responds in a way that not only benefits the signaler but would also benefit the receiver if X means Y, however, Y is false. In other words, dishonest communication occurs when a signal becomes disassociated from the signaler's quality, ability, or need. This definition highlights the benefit of dishonesty to the signaler and begs a fundamental question: Why would any organism signal honestly to their opponents when dishonesty potentially delivers significant rewards? If a weak individual signaled as if it was strong, then this weak individual might convince other weak individuals and some strong individuals to yield resources (Maynard Smith, 1974). Because of this benefit, the dishonest signaler's genes might spread within the population. Ultimately, the success of dishonest individuals would destroy the entire system of communication. As the frequency of dishonest signaling increased, natural selection would begin to favor receivers that ignored signals of strength over those that regarded them. Once receivers ignored the signal, natural selection would no longer favor the structure or behavior required for signaling. As the population approached an evolutionary equilibrium, signaling would disappear altogether. Thus, the possibility that signalers might benefit from dishonesty poses a serious problem for the evolutionary stability of animal communication.

Two mechanisms can maintain honest communication when interests oppose: (i) constraints on the production of signals (often called *indices*) and (ii) costs of developing or displaying signals (reviewed by Searcy and Nowicki, 2005; Husak *et al.*, 2012). Constraints exist when certain individuals (i.e., low-quality) cannot produce a signal, because the signal is mechanistically linked to the phenotype being advertised (Rohwer and Ewald, 1981; Enquist, 1985; FitzGibbon and Fanshawe, 1988). For example, male collared lizards signal their strength to opponents by opening their mouths, revealing the jaw muscles that generate force during biting (Lappin *et al.*, 2006). Thus, this display of jaw muscles signals the potential for an individual to inflict a costly wound on an opponent, and is a signal that cannot be faked (Lappin *et al.*, 2006). Costs of developing or displaying signals (including handicaps) can also maintain honest communication (Zahavi, 1977; Yachi, 1995; Kotiaho, 2000). Many types of costs can occur, ranging from the loss of energetic resources needed to develop signaling structures to the risk of injury resulting from signaling behaviors (Moller, 1989; Barbosa and Moller, 1999a, 1999b; Kotiaho, 2001; Basolo and Alcaraz, 2003). Such costs decrease the survival or fecundity of individuals bearing these signals and it is expected that only high-quality individuals can afford to produce high-quality signals, and thus bear these costs. Even signals that are inexpensive to produce can impose a cost of expression. The most notable examples are the "badges of status" that occur in many species of birds. The size of a melanin-based patch appears to honestly signal the aggressiveness of an individual during social interactions. The patch requires little energy to produce, but imposes potentially large social costs that are metered out to cheaters by honest signalers; birds with experimentally altered badges receive more aggression from higher-quality opponents than honest individuals do (for review, see Senar, 1999). Substantial social costs of dishonest signaling have also been demonstrated among female paper wasps (*Polistes dominulus*) using staged encounters between pairs of unfamiliar wasps (Tibbetts and Dale, 2004). In these encounters, subordinate wasps with experimentally altered facial features that displayed dominant signals (cheaters) received considerably more aggression from "real" dominant wasps than did control wasps. These social costs seemingly outweigh the benefit of dishonest signaling. Together, constraints on the production of signals and costs of developing or displaying signals might ensure that natural selection favors signals that convey reliable information about the qualities, abilities, or needs of individuals (Maynard Smith and Harper, 1995).

THE THEORY OF DISHONESTY

Costs and constraints are best considered in the context of an evolutionary game, in which one determines whether a system of signaling can persist as an evolutionarily stable strategy (Maynard Smith, 1982) and whether this evolutionarily stable strategy includes some proportion of dishonest behavior. Enquist (1985) analyzed a simple evolutionary game between strong and weak individuals who were able to signal either honestly or dishonestly. The benefit of signaling was access to resources without fighting. Fighting was assumed to impose a cost that depended on the relative strength of the opponents; a weak individual paid a greater cost when fighting a strong individual than it did when fighting another weak individual. Under this condition, dishonest signaling by weak individuals cannot invade a population of honest signalers. Similar analyses have shown that a system of signaling can be evolutionarily stable when weaker individuals suffer greater costs to express signals of strength (Adams and Mesterton-Gibbons, 1995; Számadó, 2000, 2003).

Should we then expect to observe any dishonest signaling within a population that has reached an evolutionarily equilibrium? As stated earlier, dishonesty lessens the value of the signal to receivers, potentially favoring genotypes that ignore signals altogether. Despite this consideration, dishonest signaling might evolve under certain conditions. Very weak individuals should signal dishonestly when fighting imposes an extremely high cost (Adams and Mesterton-Gibbons, 1995) or when contested resources are extremely valuable (Számadó, 2000). In fact, dishonest signaling could be a relatively general feature of evolutionarily stable systems of communication. Using adaptive dynamics, Szalai and Számadó (2009) found that about 35–40% of simulations to an equilibrium resulted in populations that included weak individuals that signaled dishonestly. Nevertheless, these individuals generally constituted less than 20% of the population, suggesting that these evolutionarily stable systems of signaling tended to convey reliable information (on average). In the rare case that weak individuals can de-escalate an interaction and escape without fighting, they pay little cost of signaling dishonestly (assuming these social interactions are the main pathway for costs). In such cases, dishonesty can be part of an evolutionarily stable system of signaling (Számadó, 2008). But if dishonest signalers can evade fights too easily, the system of signaling would break down entirely (Számadó, 2008). Therefore, we can expect to observe dishonest signaling within populations, as long as signals are honest "on average." This condition does not mean that dishonest signaling must be rare; indeed, dishonest signaling can be more common than honest signaling when receivers pay great costs to avoid deception or small costs to accept deception (Searcy and Nowicki, 2005).

Dishonest signaling might also be particularly common among individuals that cannot accurately determine their own abilities. Botero and colleagues (2010) discovered this phenomenon when modeling the adaptive dynamics of signaling between individuals that compete for resources. These researchers assumed that individuals produce a signal that affects their probability of winning resources without fighting. The size of the signal was a function of the individual's perception of its quality (i.e., fighting ability). Receivers responded to this signal according to its size and their perception of their own quality. If encounters escalated to fighting, the stronger individual was more likely to acquire resources from the weaker individual. In addition to losing resources needed for reproduction, the loser paid a cost that presumably stemmed from injury. By producing a larger signal, an individual could increase the chances of procuring resources without fighting; however, producing a larger signal reduced survival during development. When individuals accurately perceived their own quality, evolution of the population resulted in an honest system of communication, as characterized by two key features: (i) the size of a signal was positively related to the quality of an individual and (ii) receivers were less likely to fight an individual that signaled a higher quality. When individuals were unable to accurately assess their own quality, a very different and somewhat surprising system of signaling evolved, in which the quality of the individual had little bearing on the size of its signal or its propensity to fight. With modest errors of self-assessment, two distinct sets of personalities evolved: one set of individuals produced large signals and attacked frequently, whereas the other set produced small signals and attacked infrequently. With larger errors of self-assessment, more sets of personalities evolved but the surprising outcome of the model remained intact: many weak individuals produced dishonest signals that enabled them to acquire and defend resources from stronger individuals.

Importantly, this phenomenon only emerged when asexual reproduction prevented strategies of signaling and receiving from segregating independently. When these signals were allowed to recombine during sexual reproduction, an honest system of signaling emerged, as was the case when individuals perceived their own strength without error. In sexual populations, linkage or pleiotropy of signaling and receiving strategies could provide the genetic correlation needed for dishonest signaling to evolve. These genetic constraints coupled with poor self-assessment would favor the evolution of dishonest signaling.

DISHONEST SIGNALING IN AGGRESSIVE INTERACTIONS BETWEEN CONSPECIFICS

Theory tells us that dishonesty can evolve under certain circumstances but how common is dishonest signaling in real populations? To answer this question, we must be able to recognize and quantify dishonesty. Furthermore, we must be able to rule out alternative explanations for apparently dishonest signals. Only then can we address key questions about the evolution of dishonesty. Are signalers aware of their dishonesty? Do receivers fall for dishonest signals? Does the frequency of dishonest signaling vary according to environmental conditions? These questions bear directly on our theoretical understanding of animal communication.

In the remainder of this chapter, we explore these empirical issues while focusing on dishonest signaling during aggressive interactions. During these circumstances, animals commonly signal their fighting ability to conspecifics, presumably as a means of settling disputes before they escalate to physical combat (Figure 8.1). In this way, costs of fighting can be avoided when they likely outweigh the benefits. We focus on aggressive interactions between male arthropods, mainly because some of the best evidence of dishonesty come from studies of these animals. Arthropods such as stomatopods and decopods possess enlarged appendages that act as weapons during aggressive interactions. These weapons are often displayed ritualistically when encountering other males, presumably to demonstrate size and strength. As we shall see, weapons do not always provide accurate information about the fighting ability of their bearers. Therefore, arthropods make excellent models for investigating the causes and consequences of dishonest signaling.

How Do We Recognize and Compare Dishonesty?

Regression has been the most common approach to quantify the reliability of signals. First, one estimates the quality (or ability) of individuals and the magnitudes of their signals. Then, one regresses one of these variables onto the other (the order is arbitrary). Finally, one examines the residuals for information about potential dishonesty of the signal. For example, let us assume we regressed the magnitude of the signal onto the quality of the individual (Figure 8.2). In this case, the signal residual measures the magnitude of a signal produced by an individual relative to the average signal for the population. A positive signal residual might mean that an individual has exaggerated its quality, whereas a negative signal residual might mean that an individual has understated its quality. Thus, the proportion of unexplained variation in the regression model provides a measure of the unreliability of the signal. This unreliability has been interpreted as dishonest signaling.

Hughes (2000) was the first to use signal residuals to quantify dishonesty. She investigated aggressive interactions between males of the big-clawed snapping shrimp, *Alpheus heterochaelis*. In this species, fights between rivals are usually won by the larger individual; small animals only win contests with similar-sized individuals (Hughes, 1996a, 1996b). Therefore, body size constitutes a useful proxy for competitive ability. Males display their chelae to mutually assess competitive ability, even though chela size does not contribute to fighting success once one controls for body size (Hughes, 1996a, 1996b). Despite a strong relationship between body size and chela size ($r^2 = 0.82$; Hughes, 1996a, 1996b), signal residuals suggest that some males produce relatively large chela for their size, and vice versa (Hughes,

Figure 8.1. Examples of some of the diversity of signals used by animals to display their physical prowess to rivals. (a) *Rhinoceros* beetles use their enormous horns during both displays and fights (Photo: Phil Matthews.), (b) Koalas communicate their size by bellowing during territorial bouts (Photo: William Ellis.), (c) Crabs display their strength to opponents using their enlarged front claws (Photo: Daniel Hancox.), (d) Male frogs use their calls to signal their size to attract females and intimidate opponents (Photo: Ben Barth.), and (e) *Euastacus* crayfish signal their potential to injure competitors using their claws. (Photo: Ben Barth.) (*See insert for color representation of this figure.*)

2000). These individuals are, in effect, displaying a signal that exaggerates their competitive ability.

Importantly, signal residuals reflect only *potential* dishonesty, because alternative hypotheses can account for residual variation (Hughes, 2000). First, the magnitude of an individual's signal or its ability could deviate from the expected

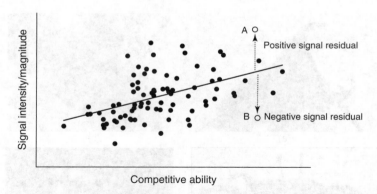

Figure 8.2. Relationship between competitive ability and signal intensity for an animal signaling to an opponent. This variation in signal size that is independent of competitive ability is referred to as *signal residuals*.

because of errors associated with assessment, production, or measurement. Errors in assessment arise when the production of a signal becomes temporally uncoupled from the development of quality: either an individual cannot predict its future quality when producing the signal, or the individual's quality changes more rapidly than the signal can change. Production errors arise when plastic signals, such as a vocal or visual display, cannot be produced at the same magnitude repeatedly. In such cases, poor sampling within individuals inevitably leads to residual variation in the relationship between individual ability and signal strength. Finally, measurement errors result from an inability to accurately observe either the magnitude of the signal or the quality of the individual.

Measurement error seems a particularly important cause of signal residuals, because researchers usually do not measure the quality of an individual directly. Rather, most researchers rely on an index of quality that plausibly correlates with true quality. For example, imagine that we do not know the ability of a signaler to overcome an opponent during conflict. We might assume that this ability relates to an easily measured trait, such as body size. Thus, body size becomes a proxy for quality. When researchers use proxies of quality, residual variation can reflect an inadequate proxy more than dishonest signaling. In other words, the signal may be highly correlated with quality but only weakly correlated with the chosen proxy. In which case, signal residuals reveal additional and accurate information about an individual's competitive ability.

More direct measures of quality enable one to minimize the errors associated with proxies of competitive ability. Quality is easier to estimate when the signal plays a direct role in combat. For example, a male fiddler crab (*Uca* spp.) possesses a greatly enlarged chela that serves as a weapon against conspecifics (Jordao and Oliveira, 2001). Because the chela serves as both a signal and a weapon, one can use the relationship between the structure and function of this weapon to assess the honesty of signaling. Using this approach, Bywater and Wilson (2012) measured the size and strength of the major chela in more than 100 males of *Uca vomeris*. After

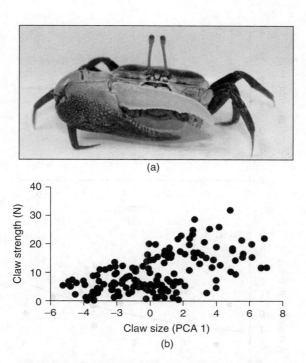

Figure 8.3. (a) Male two-toned fiddler crabs (*Uca vomeris*) possess an enlarged front claw. (b) The relationship between claw size and strength for 120 male *U. vomeris*. (Adapted from Bywater and Wilson, 2012.). Maximum claw strength increases with claw size but strength is highly variable and difficult to predict based on the size of an individual's claw alone (unreliable signal). Interestingly, a ceiling on maximal strength exists over a given size range, presumably due to biomechanical constraints. (*See insert for color representation of this figure.*)

regressing strength onto size, they found substantial residual variation in strength (Figure 8.3); in fact, the strength of this weapon seems virtually independent of size! Because the strength of the chela reflects its effectiveness during combat, size seems a deceptive signal of fighting ability.

Despite the advantage of using strength residuals to quantify dishonesty, even these can represent the error resulting from methodological limitations or motivational bias. For example, relatively weak individuals could merely lack motivation to perform during assays of strength. This concern can be addressed by establishing the repeatability of strength within individuals. Bywater *et al.* (2008) measured the maximal force generated by the chelae of slender crayfish (*Cherax dispar*) on two consecutive days. Force on day 1 was highly correlated with force of day 2, giving some assurance that strength residuals reflect real variation in fighting ability rather than a transient lack of motivation. Similarly, Angilletta and Wilson (2012) found that strength residuals for crayfish at two different temperatures were highly correlated. This finding indicates repeatability between different environmental contexts as well as repeatability between times. Besides establishing repeatability, one can

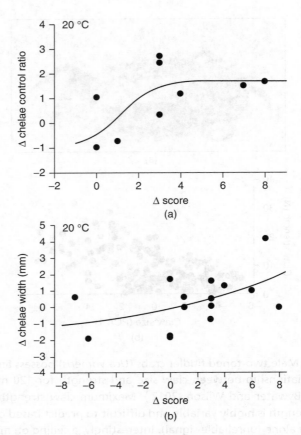

Figure 8.4. Differences in dominance score from bouts between paired competing common yabbies (*Cherax destructor*) and its relationship with differences in (a) respiratory control ratio in their chela muscle, and (b) chela size. (Taken from Seebacher and Wilson, 2007.)

also examine the mechanistic basis of variation in strength to determine whether strength residuals reflect ability or motivation. For example, Seebacher and Wilson (2006) found significant relationships between the respiratory control ratio of chela muscle and dominance in the freshwater crayfish, *Cherax destructor* (Figure 8.4). Ultimately, functional studies of this kind are necessary to establish the relationship between weapon size and competitive ability, which might confirm the presence of dishonest signaling (also see Husak *et al.*, 2012).

Are Signalers Aware of their Dishonesty?

Botero and colleagues showed that dishonest signaling can evolve when individuals cannot accurately assess their own ability when producing the signal. For crustaceans, one can reasonably assume that errors in assessment occur. The size of a chela is set during the process of molting, after which their size remains fixed until the next moult. Changes in ability between moults would not be reflected in the size of the chelae. If signal residuals represent error more than dishonesty, the

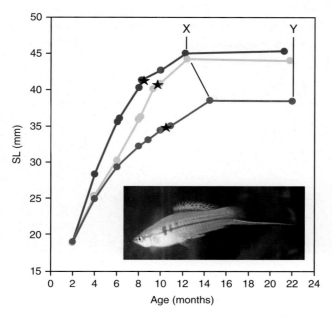

Figure 2.2. Growth trajectories (standard length) of male green swordtail fish that experienced different resource availability during development. The green line shows the average trajectory of males that received *ad lib* resources, whereas the red line shows the average trajectory of males that received a restricted diet throughout development. The yellow line shows males that were released from a restricted diet at 6 months of age and subsequently re-aligned growth (growth compensation). Following maturation (indicated by stars), males compensated for their poor start in life by accelerating growth of both body size and sword length, so that there was no difference in body size or sword length compared to *ad lib* fed males by the time body size reached an asymptote. As a result, the window of reproductive opportunity (difference between points X and Y) was similar for compensating and *ad lib* fed males. Males that remained on a restricted diet, however, had reduced mature body size and a smaller window of reproductive opportunity, putting them at a double disadvantage (see Section "Nutrition during growth and the expression of sexual signals as an adult").

Animal Signaling and Function: An Integrative Approach, First Edition.
Edited by Duncan J. Irschick, Mark Briffa, and Jeffrey Podos.
© 2015 John Wiley & Sons, Inc. Published 2015 by John Wiley & Sons, Inc.

Figure 4.1. A wall of sheep vertebrae is positioned as a backdrop for males displaying in front of the bower.

Figure 4.2. The bower of a great bowerbird near Mareena, Queensland. This bower is unusual because of its large size, the way the male had positioned it within a set of thick branches, and the uneven slope on which it is built. The pattern of decorations seen here and most other great bowers we have seen does not show a size gradient of decorations necessary to create the forced perspective illusion. In this case, decorations arrayed at the front of the bower are approximately the same size. Moreover, because these decorations are on a downward slope, they cannot be seen by a female being courted in the bower avenue to produce a forced perspective illusion even if there were a size gradient. Thus, the decorations appear to function in attracting females to the bower before they arrive in the bower avenue.

Figure 4.3. Shiny quartz rocks are positioned in the central avenue of a spotted bowerbird bower.

Figure 4.4. A spotted bowerbird bower with a sunken central stick bowl where preferred decorations are placed.

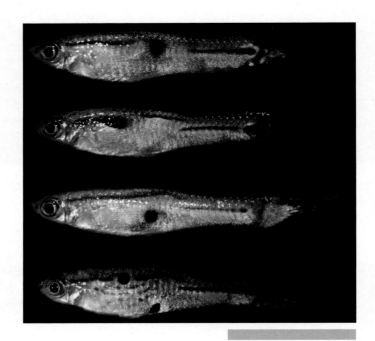

Figure 5.1. Four male guppies (*Poecilia reticulata*) from the Guanapo catchment of Trinidad, showing the complex mosaics of black, orange, blue, green, yellowish-bronze and silver markings that comprise the ornamental color pattern of this species. There is tremendous diversity among streams, populations, and individuals regarding the extent, placement, and spectral properties of each color element. The scale bar (lower right) indicates ~8 mm.

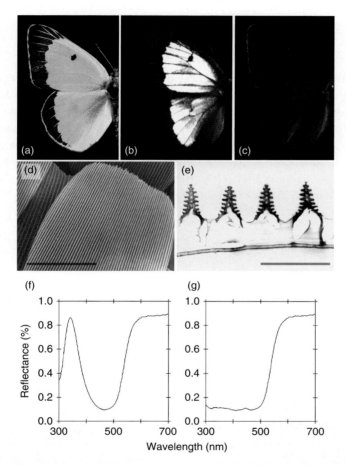

Box 5.1. The Multicomponent Nature of *Colias* Butterfly Coloration. Image reprinted in part from Kemp and Rutowski (2007) with permission (Rightslink License 3197970783626).

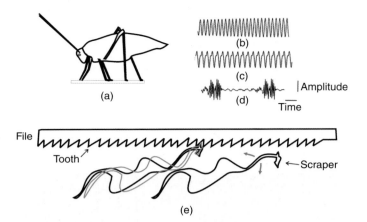

Figure 7.1. Katydids (a) are insects that produce sound by rubbing one forewing across the other with a washboard-like mechanism. They can produce tonal signals (b,c) or broadband signals (d) as represented by oscillograms. (From Montealegre-Z and Morris (2004).). (e) One unique feature of some ultrasonic katydids is that they use deflection (orange) and subsequent vibrations of the scraper (blue) to generate ultrasonic vibrations, analogous to the "twanging" of a thumb harp. This deflection of the scraper is also used to elastically preload the system such that subsequent tooth impacts can happen at a much higher rate than the driving wing movement. (Images adapted from Montealegre-Z *et al.* (2006).)

Figure 7.2. (a–c) Minimum vocal frequency as a function of residual beak dimensions (beak length, depth, and width, respectively, calculated relative to body size), for the medium ground finch *Geospiza fortis* at El Garrapatero, Santa Cruz Island (Huber and Podos, 2006). The negative relationships are in accord with the prediction that birds with greater volume vocal tracts relative to their body sizes should sing at lower vocal frequencies. (d) Representative large and small beak morphs of this population. A similar pattern is shown by morph: after factoring out body size, large-beaked morphs sing lower-frequency songs (Huber and Podos, 2006).

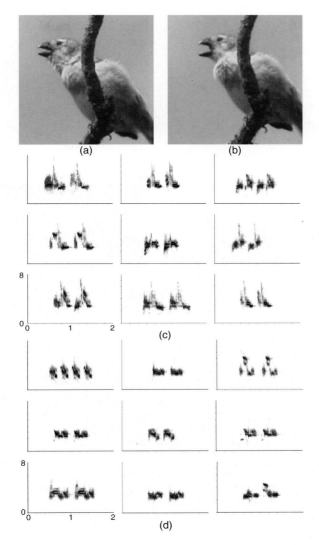

Figure 7.3. (a,b) Video images of a singing male Darwin's finch, *Camarhynchus psittacula*. Panel a shows the bird during the production of a low frequency note, Panel b during a high frequency note. Notable changes in thoracic volume and beak gape are apparent, and correlate with changing vocal frequencies such that the vocal tract's resonance filtering function is retained across varying source frequencies. (After Podos *et al*. (2004).) (c,d) Spectrograms of songs produced by large and small morphs (Panels c and d respectively) of another Darwin's finch species, *Geospiza fortis*. The songs of the two morphs differ in frequency bandwidth, in a manner predicted by a vocal tract constraint hypothesis. (After Huber and Podos (2006).) *X*-axis = 2 seconds, *Y*-axis spans 0–8 kHz.

Figure 8.1. Examples of some of the diversity of signals used by animals to display their physical prowess to rivals. (a) *Rhinoceros* beetles use their enormous horns during both displays and fights (Photo: Phil Matthews.), (b) Koalas communicate their size by bellowing during territorial bouts (Photo: William Ellis.), (c) Crabs display their strength to opponents using their enlarged front claws (Photo: Daniel Hancox.), (d) Male frogs use their calls to signal their size to attract females and intimidate opponents (Photo: Ben Barth.), and (e) *Euastacus* crayfish signal their potential to injure competitors using their claws. (Photo: Ben Barth.)

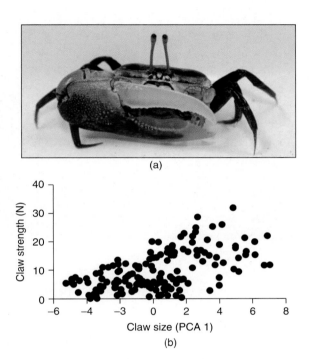

(a)

(b)

Figure 8.3. (a) Male two-toned fiddler crabs (*Uca vomeris*) possess an enlarged front claw. (b) The relationship between claw size and strength for 120 male *U. vomeris*. (Adapted from Bywater and Wilson, 2012.). Maximum claw strength increases with claw size but strength is highly variable and difficult to predict based on the size of an individual's claw alone (unreliable signal). Interestingly, a ceiling on maximal strength exists over a given size range, presumably due to biomechanical constraints.

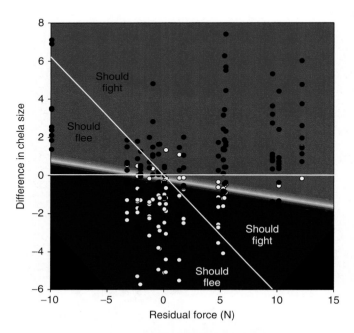

Figure 8.5. The probability that a crayfish engaged in a fight depended on the relative size of its chelae (the focal crayfish's mean chela size minus its opponents mean chela size) and the degree of deceptive signaling (residual of maximal chela force regressed onto chela size). Black and white circles represent staged encounters where the focal crayfish escalated or withdrew, respectively. Blue and red regions indicate phenotypes associated with 0% and 100% chances of fighting, respectively, in a generalized linear model of the data. White lines separate regions where a focal crayfish should fight or flee according to its strength relative to that of its opponent. Generally, crayfish fought opponents that had smaller chelae. However, honest signalers (positive residuals) fought crayfish with slightly larger chelae and dishonest signalers (negative residuals) only fought against opponents with slightly smaller chelae than their own. Despite this distinction, crayfish escalated aggressive encounters as if they were mostly ignorant of their chela strength, as indicated by the black circles in the region marked "should flee" and the white circles in the region marked "should fight."

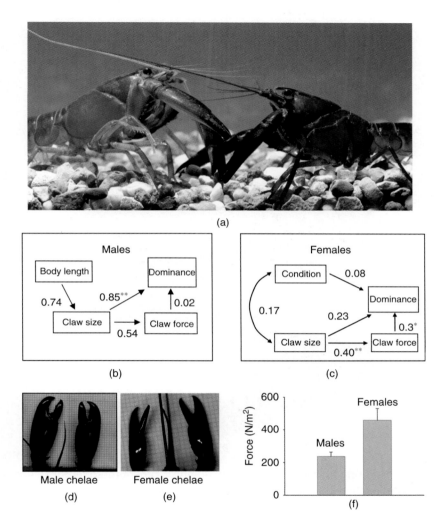

Figure 8.6. (a) Two male slender crayfish (*Cherax dispar*) signaling during a territorial dispute. Best path models explaining morphological and performance relationships with social dominance for (b) males and (c) females of *C. dispar*. Dominance was largely explained by claw size for males but claw strength for females (Significance at the level of *$P < 0.05$ and **$P < 0.001$). Shape of chelae for (d) males and (e) females of *C. dispar*. (f) Chela muscle from male crayfish produced only half of the strain (force per cross-sectional area of muscle) that was produced by chela muscle from female crayfish. (Taken from Wilson *et al*. (2007) and Bywater *et al*. (2008).)

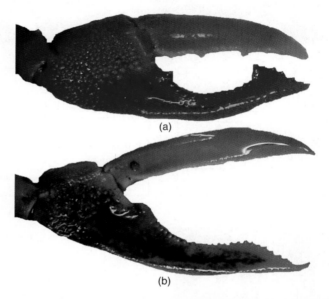

(a)

(b)

Figure 8.7. An original brachychelous claw (a) and a regenerated leptochelous claw (b) from males of the two-toned fiddler crab (*Uca vomeris*). (Photos: Daniel Hancox.)

signaler could be unaware of the discrepancy between its advertised ability and its true ability. Consequently, the behaviors of signaler would be unrelated to the values of signal or strength residuals. In other words, an individual with a positive residual would behave in a similar manner as would an individual with no residual or a negative residual. On the other hand, if signal or strength residuals truly reflect a strategy of dishonesty, these residuals should relate to behavior during aggressive encounters. Thus, one way to confirm that signal residuals represent degrees of dishonesty is to show that individuals with larger residuals behave differently than those with smaller residuals. This kind of evidence for dishonest signaling has been presented for big-clawed snapping shrimp (*A. heterochaelis*) and European hermit crabs (*Pagurus bernhardus*) (Hughes, 2000; Elwood *et al.*, 2006).

Recall that the competitive ability of a male snapping shrimp depends on its body size. Shrimp use chela size to signal their ability, despite the fact that some individuals produce relatively large chelae for their body size (Hughes, 2000). Males with disproportionate chelae are, in effect, exaggerating their competitive ability (i.e., positive signal residuals). Assuming that these residuals represented dishonest signaling, Hughes reasoned that individuals with large positive signal residuals should display their chelae more frequently than those with zero or negative signal residuals. Hughes (2000) found displays of chelae to males with larger bodies depended on signal residuals. Thus, individuals that had larger chela for their body size displayed their chelae more frequently to competitors of large body size. In addition, interactions with shrimps having larger signal residuals were longer and more escalated. Presumably, these behaviors enabled dishonest individuals to exaggerate their own body size.

Similarly, European hermit crabs behave in a manner that depends on the relative size or strength of their appendages (Elwood *et al.*, 2006). Hermit crabs commonly use two types of displays involving their chelipeds: presentation and extension. During the presentation display, both males hold their chelipeds in a stationary position that is perpendicular to the substrate. This appears to allow an accurate, mutual assessment of signals. Larger individuals were more likely to offer a "cheliped presentation" for a longer period and become the aggressor and win a subsequent shell fight (Elwood *et al.*, 2006). By contrast, the "cheliped extension" display, where they hold their chela horizontally and thrust them toward their opponents, does not appear to easily afford mutual assessment of signals, and there appears to be an element of bluff in this behavior. Interestingly, smaller individuals use the cheliped extension more commonly. Arnott and Elwood (2010) found that contestants with larger chelipeds for their body size (positive signal residuals) spent more time presenting their chelipeds. In addition, cheliped size residuals influenced the escalation of encounters to more aggressive events. The smaller of the competitors in shell fights (a form of escalated aggression) had larger cheliped size residuals than those competitors that did not elect to escalate to that level. These data suggest that individuals with larger signals relative to their body size tend to exaggerate their competitive ability by displaying these signals more frequently.

Comparisons of display rates among shrimp and crabs with different signal residuals suggest that dishonest signalers are aware of the mismatch between their signal and their ability. However, alternative hypotheses should not be dismissed without further investigation. For example, Hughes (2000) hypothesized that body mass imperfectly reflects the competitive ability of snapping shrimp, and thus positive signal residuals reflect the variation in competitive ability that does not covary with body mass. Unfortunately, chela strength has not been assessed for either snapping shrimp or hermit crabs, making interpretation of the functional importance of the chela display complex. Substantial variation in the strength of this weapon could occur among individuals, which could have important consequences for the outcomes of disputes between individuals of similar size. Signaling should evolve to reduce the potential costs associated with fighting, and the ability to inflict a cost upon an opponent should be mediated through the strength of a weapon rather than its size. By assessing the strength of the chelae, researchers can more directly estimate the ability to inflict injuries on an opponent.

Our analyses of strength residuals and aggressive behaviors of slender crayfish (*C. dispar*) support a similar conclusion about whether individuals were aware of their dishonesty (Angilletta and Wilson, unpublished). As with many species of freshwater crayfish slender crayfish use their enlarged front chelae to signal during aggressive encounters with conspecifics. During such disputes, each crayfish presents its chelae to its opponent, allowing the opponent to rub the chelae up and down the lengths of their own chelae. This mutually reciprocated, highly ritualistic interaction proceeds until one of the rivals backs down or engages the other in combat. Despite the importance of chelae size in these encounters (Wilson *et al.*, 2007), the force generated by large chelae varies among individuals by nearly an order of magnitude. We hypothesized that an individual's propensity to escalate an interaction to fighting would depend on its strength residual: relatively strong crayfish (positive strength residuals) should be more likely to fight an individual of equal or greater chela size than should relatively weak crayfish (negative strength residuals). We found that male crayfish tuned their aggression according to their own strength, such that males who signal dishonestly (i.e., weak for their size) escalated disputes less often (Wilson and Angilletta, unpublished data). Still, the lower aggression of weak males only partially reflected their true weakness, suggesting that deceptive crayfish were largely unaware of the magnitude of their deception. Our results supported such a prediction, as the magnitude of the strength residuals covaried with the propensity of a crayfish to engage with opponents (Figure 8.5). This conclusion held over a diverse combination of sizes for focal individuals and their opponents, suggesting that crayfish had at least some knowledge of their strength relative to conspecifics.

Do Receivers Fall for Dishonest Signals?

If we believe that signal or strength residuals reflect dishonest signals of competitive ability, we should immediately wonder whether such signals confer significant benefits to the signaler. Early models showed that dishonest signaling could select for receivers that ignore the signal, leading to a breakdown in communication (Maynard Smith and Harper, 2003; Searcy and Nowicki, 2005). The stability of a

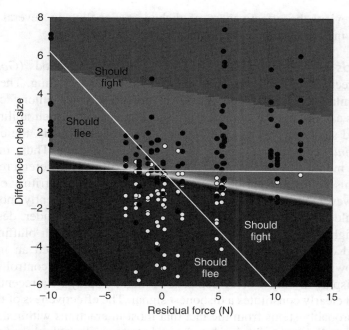

Figure 8.5. The probability that a crayfish engaged in a fight depended on the relative size of its chelae (the focal crayfish's mean chela size minus its opponents mean chela size) and the degree of deceptive signaling (residual of maximal chela force regressed onto chela size). Black and white circles represent staged encounters where the focal crayfish escalated or withdrew, respectively. Blue and red regions indicate phenotypes associated with 0% and 100% chances of fighting, respectively, in a generalized linear model of the data. White lines separate regions where a focal crayfish should fight or flee according to its strength relative to that of its opponent. Generally, crayfish fought opponents that had smaller chelae. However, honest signalers (positive residuals) fought crayfish with slightly larger chelae and dishonest signalers (negative residuals) only fought against opponents with slightly smaller chelae than their own. Despite this distinction, crayfish escalated aggressive encounters as if they were mostly ignorant of their chela strength, as indicated by the black circles in the region marked "should flee" and the white circles in the region marked "should fight." (*See insert for color representation of this figure.*)

signaling system that includes dishonesty depends on the abilities of receivers to distinguish honest and dishonest signals, as well as the cost of making an erroneous distinction (Maynard Smith and Harper, 2003; Searcy and Nowicki, 2005). The majority of signaling in an evolutionarily stable system could be dishonest when incorrectly identifying a signal as dishonest imposes a severe cost. The strength of chelae, and other appendages with an exoskeleton, would seem difficult to detect without engaging in some form of physical interaction. Therefore, an individual must put itself at risk to determine the honesty of a signal. Given these criteria, we should not be surprised that dishonest signaling occurs prevalently in arthropods.

Here, we review the effectiveness of dishonest signaling in several groups of arthropods, including stomatopods, wasps, crabs, and crayfish.

Maxillipeds of Stomatopods Recently molted stomatopods (*Gonodactylus bredini*) effectively defend territories through dishonest signaling. These animals use their enlarged, second maxillipeds to defend cavities among corals. The appendages can inflict considerable damage, as indicated by their ability to smash the calcified shells of prey. As with other species of arthropods, the exoskeleton of *G. bredini* becomes soft and pliable following molting. Thus, immediately following a moult, a stomatopod cannot use its maxillipeds to defend resources by fighting. This vulnerable period increases the risk of losing territory or incurring injury. However, Steger and Caldwell (1983) found that recently moulted individuals could hold their cavities when challenged by an intruder, despite being unable to fight. These stomatopods held their cavities through bluffing displays that warned off opponents. Recently moulted individuals used an intimidating display known as a *meral spread* in 15 of 17 contests, while control individuals used this display in only 4 of 19 contests. A meral display by a recently moulted stomatopod clearly constitutes a dishonest signal. The effectiveness of this dishonest signal probably stems from the fact that most interactions within a population involve two individuals with hardened exoskeletons. Thus, the benefit of calling an occasional bluff would be offset by the cost of injury associated with most encounters.

Mandibles of Fig Wasps Moore *et al.* (2009) provided compelling evidence of dishonest signaling during aggressive encounters between males of the non-pollinating fig wasp, *Philotrypesis* sp. ex *Ficus rubiginosa*. These males fight aggressively for females, making use of their potentially dangerous mandibles. Competitive ability depends on body length and mandible length, and males use the gape widths of their opponents' mandibles to assess their abilities (Pereira and do Prado, 2005a, 2005b). Moore *et al.* (2009) discovered that about 18% of males were larger than usual; atypically large males also possessed relatively long mandibles for their body size. When combat occurred between the two kinds of males, an atypical male incurred more injuries and was less likely to win than was a typical male. However, atypical males engaged in fewer fights and reproduced more successfully than did typical males of the same body size. Moore *et al.* (2009) concluded that atypical males were signaling dishonestly, which provided them with a greater chance of winning a dispute during the early phase of mutual assessment, before escalation into physical fighting.

Chelae of Freshwater Crayfish Males of the slender crayfish (*C. dispar*) routinely use dishonest signals during aggressive interactions (Wilson *et al.*, 2007; Bywater *et al.*, 2008) (Figure 8.6). Both males and females assess chela size, rather than body size, when determining whether or not to fight. On this basis, greater than 80% of disputes are resolved before they escalate into physical

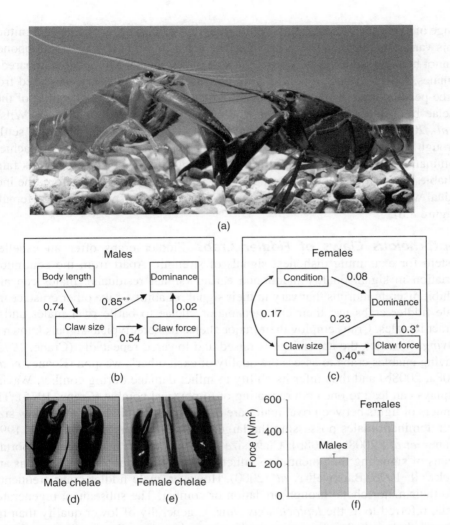

Figure 8.6. (a) Two male slender crayfish (*Cherax dispar*) signaling during a territorial dispute. Best path models explaining morphological and performance relationships with social dominance for (b) males and (c) females of *C. dispar*. Dominance was largely explained by claw size for males but claw strength for females (Significance at the level of *P* < 0.05 and **P* < 0.001). Shape of chelae for (d) males and (e) females of *C. dispar*. (f) Chela muscle from male crayfish produced only half of the strain (force per cross-sectional area of muscle) that was produced by chela muscle from female crayfish. (Taken from Wilson *et al.* (2007) and Bywater *et al.* (2008).) (*See insert for color representation of this figure.*)

contests (Wilson *et al.*, 2007); only encounters between individuals with similarly sized chelae escalate to combat. When these crayfish fight, the individual with the stronger chelae generally prevail (Wilson *et al.*, 2007). For males, however, the size of a male's chela provides very unreliable information about its strength; for a range of chela sizes, the strength of this weapon can vary by an order of magnitude. This variation creates a scenario where the actual fighting ability of an opponent cannot be assessed accurately from the size of his chelae. Moreover, compared to females, males produce a poor quality of muscle in the chelae (as assessed from force per unit area of muscle), indicating that males exaggerate the sizes of their chelae by spending less energy on the quantity and quality of muscle (Wilson *et al.*, 2007). Consequently, a large number of disputes between males are settled through dishonest signaling, such that weaker males with larger weapons achieve dominance (Figure 8.6). By contrast, the size of a female's chela provides fairly reliable information about its strength (Bywater *et al.*, 2008). Therefore, the individual with the larger and stronger chelae usually wins disputes between females (Figure 8.6).

Leptochelous Claws of Fiddler Crabs Fiddler crabs offer an excellent system for examining dishonest signals of strength. Apart from the continuous variation in signal quality mentioned earlier (signal residuals), males can also exhibit discrete morphs that vary in their signal reliability. Like other crustaceans, male fiddler crabs use their enlarged major chelae to batlle rival males and to attract females. Crabs employ their major chelae in stereotypical signals known as *waving*, in which the appendage is raised and lowered repeatedly (Crane, 1975). Waving enables crabs to assess the quality of each other's weapon (Reaney *et al.*, 2008a, 2008b) and thus infer its ability to inflict damage during conflict. Waving displays can lead to one crab retreating or to physical combat (Crane, 1975). The winners of fights between rival males are determined mainly by relative claw size, with dominant males possessing the larger claws (Jennions and Backwell, 1996; Reaney *et al.*, 2008a, 2008b). Claw size is considered to be the most important means of signaling the potential to inflict costs on their opponents (Jennions and Backwell, 1996; Backwell *et al.*, 2000). However, male fiddler crabs frequently lose their major chela through predation or combat. The subsequent regenerated chela, referred to as the *leptochelous claw*, is generally of lower quality than the original chela, which is known as the *brachychelous claw* (Backwell *et al.*, 2000). Although the length of a leptochelous claw roughly equals that of a brachychelous claw, leptochelous claws contain less muscle (Backwell *et al.*, 2000) and produce less force than do brachychelous claws (Lailvaux *et al.*, 2009). No matter how much the animal grows, a regenerated claw never develops to the same point as the original claw (Figure 8.7). For *Uca annulipes*, males with leptochelous claws are just as successful at attracting females as those with brachychelous claws, which suggests that honest signaling provides no advantage for mating. However, Backwell *et al.* (2000) found that males with leptochelous claws were inferior in combat to those with brachychelous claws; the former were defeated in more than 60% of disputes over territories. Males with leptochelous claws were unable to defend burrows against males with brachychelous claws. In fact, disputes between these two types of males tended to be short in duration, because males with

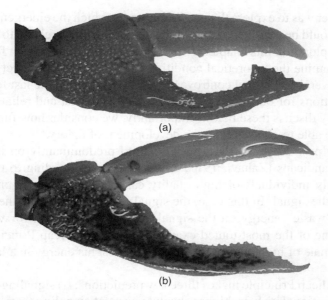

Figure 8.7. An original brachychelous claw (a) and a regenerated leptochelous claw (b) from males of the two-toned fiddler crab (*Uca vomeris*). (Photos: Daniel Hancox.) (*See insert for color representation of this figure.*)

leptochelous claws were unlikely to escalate aggressive encounters. However, despite the disadvantage of a leptochelous claw, male crabs seem unable to visually distinguish between those with leptochelous or brachychelous claws. Some leptochelous males were even able to fool some brachychelous males into retreating (Backwell *et al.*, 2000). The role of female choice in mediating the benefits and costs of the different claw-morphs also seems to be a fascinating complication. As females apparently cannot distinguish among the males with the different claw types, those males displaying a weaker claw should be able to fool females. But the offspring of males with regenerated claws do not inherit the weaker claw, rather they initially grow the original-type claw. In fact, males displaying a regenerated claw to a female may actually be displaying their high quality – as they are individuals that have been able to survive and grow a large replacement claw despite experiencing the earlier traumatic loss of their original claw. In any case, the production and use of leptochelous claws by male fiddler crabs represents some of the most convincing evidence for dishonest signaling in a natural system.

Empirical Analyses of Theoretical Predictions: Where Next for Functional Studies?

The future directions for functional studies of dishonest signaling should be guided by our theoretical understanding of the maintenance of signal reliability. A key goal for this chapter was to explore the conditions under which the emergence of dishonest signals should be expected. Some of the empirical issues related to this question have been explored in this chapter. However, it is equally important for functional studies to examine the theoretical conditions under which we expect reliability in signaling. Several theoretical controversies have emerged in the last few years that have implications for our understanding of both dishonest and reliable signaling, and we briefly discuss these below. Importantly, we consider how functional biologists will be able to contribute to the development of theory.

Over the past 20 years, theory has focused predominantly on the costs and benefits of handicaps (Zahavi, 1977; Grafen, 1990). According to the Handicap Principle, only individuals of high quality can bear the costs of producing and maintaining the signal. In this case, the signal actually represents the individual's capacity to "waste" energy on the signal, rather than using it for viability. The peacock is one of the most-touted examples of the Handicap Principle: presumably, only a male of high quality can afford to invest his energy in a larger train of feathers.

The Handicap Principle makes three key predictions: (i) signaling is costly; (ii) higher-quality signalers have higher absolute costs of signaling than lower-quality signalers – meaning that bigger signals cost more; and (iii) higher-quality signalers have lower relative costs of signaling than lower-quality signalers – meaning that the cost of the signal is lower relative to the entire energy budget. According to the Handicap Principle, honesty is maintained because only high-quality individuals can afford to produce the signal.

But recently, this view of the Handicap Principle has come under challenge (Getty, 1998; Számadó, 2008, 2011). New theory reveals that handicap models have incorrectly used *additive* functions to calculate viability and fecundity, rather than the correct *multiplicative* functions. When the appropriate fitness functions are substituted into the handicap models, the empirical predictions change dramatically: the absolute and relative costs of signaling *may or may not differ* with individuals' quality. This transforms our entire view of signaling in animal communication, indicating that discovering generalizations may be more appropriate than testing unidirectional predictions (Murai *et al.*, 2009) and alternatives to the Handicap Principle should be considered (Getty, 1998; Számadó, 2011).

In short, recent theoretical models show that (i) the inclusion of appropriate fitness functions changes the predictions that arise from handicap models; (ii) although handicaps may occur in nature, they are probably less common than previously thought; and (iii) testing whether a signal is unambiguously a handicap is very difficult. Demonstrating whether a signal is a handicap is difficult because the costs required to unambiguously communicate a message to a receiver (efficacy costs) and the costs of producing an honest signal, over and above the basic requirements to communicate (handicap costs), must be isolated and measured. The difficulty of quantifying handicap costs independent of efficacy costs prevents researchers from accurately assessing the role of handicaps in maintaining signal reliability.

The same mathematical models that challenge the Handicap Principle suggest that receiver-imposed costs could maintain honesty (Lachmann *et al.*, 2001;

Getty, 1998; Számadó, 2008, 2011). For example, honest signaling of strength during aggressive interactions are not maintained by handicaps but via the potential costs imposed on cheaters by their opponents. Imagine that a male creates a signal that makes him appear of higher-quality than he is – if his bluff gets called by an opponent, he will likely lose resources or mating potential and he may even sustain bodily damage. This exemplifies how receiver-imposed costs (punishment) could outweigh the benefits of dishonest signaling (Barbosa and Moller, 1999a, 1999b; Healy and Olsson, 2009; Tibbetts and Izzo, 2010). The mechanisms that can provide such receiver-imposed costs include individual recognition, proximity risk, and social punishment (Számadó, 2011). Functional biologists again seem well placed to explore the importance of these receiver-imposed costs on cheaters. For example, manipulation of signal quality can allow analysis of the social costs imposed on those dishonest and honest signalers. The severity of such costs should indicate the disadvantages of high-quality signals produced by low-quality individuals. Studies by Tibbetts and Dale (2004) on the paper wasps provide good evidence for the severity of such costs.

CONCLUSIONS

Our inherent fascination with questions about dishonest behavior will ensure that this topic will attract continued theoretical and empirical attention. Recent studies indicate dishonest signaling is far more common in natural populations than initially recognized. Functional biologists are uniquely placed to play a central role in testing theoretical models of dishonest signaling, because their methodologies allow accurate assessments of organismal performances. This means functional biologists can directly quantify an individual's fighting ability and associated ability to inflict costly injuries upon opponents. By quantifying the match between signal magnitude and competitive ability, we can unambiguously determine an individual's relative dishonesty. Ideally, these types of empirical studies should focus on systems where one can measure (i) the benefits of honest and dishonest signaling, (ii) the costs of the honest component of signaling (handicap costs), and (iii) the costs imposed on dishonest signalers by receivers (receiver-imposed costs). Certainly such requirements are easily accommodated via analyses of arthropod behavior and performance.

REFERENCES

Adams, E.S. and Mesterton-Gibbons, M. (1995) The cost of threat displays and the stability of deceptive communication. *Journal of Theoretical Biology*, **175**, 405–421.

Arnott, G. and Elwood, R.W. (2010) Signal residuals and hermit crab displays: flaunt it if you have it!. *Animal Behaviour*, **79**, 137–143.

Angilletta, M.J. and Wilson, R.S. (2012) Cryptic asymmetry: unreliable signals mask asymmetric performance of crayfish weapons. *Biology Letters*, **8**, 551–553.

Backwell, P.R.Y., Christy, J.H., Telford, S.R., Jennions, M.D., and Passmore, N.I. (2000) Dishonest signalling in a fiddler crab. *Proceedings of the Royal Society of London, Series B: Biological Sciences*, **267**, 719–724.

Barbosa, A. and Moller, A.P. (1999a) Sexual selection and tail streamers in the barn swallow: appropriate tests of the function of size-dimorphic long tails. *Behavioral Ecology*, **10**, 112–114.

Barbosa, A. and Moller, A.P. (1999b) Aerodynamic costs of long tails in male barn swallows *Hirundo rustica* and the evolution of sexual size dimorphism. *Behavioral Ecology*, **10**, 128–135.

Basolo, A.L. and Alcaraz, G. (2003) The turn of the sword: length increases male swimming costs in swordtails. *Proceedings of the Royal Society of London, Series B: Biological Sciences*, **270**, 1631–1636.

Berglund, A., Bisazza, A., and Pilastro, A. (1996) Armaments and ornaments: an evolutionary explanation of traits of dual utility. *Biological Journal of the Linnean Society*, **58**, 385–399.

Botero, C.A., Pen, I., Komdeur, J., and Weissing, F.J. (2010) The evolution of individual variation in communication strategies. *Evolution*, **64**, 3123–3133.

Byrne, R.W. and Corp, N. (2004) Neocortex size predicts deception rate in primates. *Proceedings of the Royal Society of London, Series B: Biological Sciences*, **271**, 1693–1699.

Bywater, C.L., Angilletta, M.J., and Wilson, R.S. (2008) Weapon size is a reliable indicator of strength and social dominance in female slender crayfish (*Cherax dispar*). *Functional Ecology*, **22**, 311–316.

Bywater, C.L. and Wilson, R.S. (2012) Is honesty the best policy? Testing signal reliability in fiddler crabs when receiver-dependent costs are high. *Functional Ecology*, **26**, 804–811.

Crane, J. (1975) *Fiddler Crabs of the World (Ocypodidae: Genus* Uca*)*, Princeton University Press, Princeton.

Elwood, R.W., Pothanikat, R.M.E., and Briffa, M. (2006) Honest and dishonest displays, motivational state and subsequent decisions in hermit crab shell fights. *Animal Behaviour*, **72**, 853–859.

Enquist, M. (1985) Communication during aggressive interactions with particular reference to variation in choice of behavior. *Animal Behaviour*, **33**, 1152–1161.

Fitzgibbon, C.D. and Fanshawe, J.H. (1988) Stotting in thomson's gazelles: an honest signal of condition. *Behavioral Ecology and Sociobiology*, **23**, 69–74.

Getty, T. (1998) Handicap signalling: when fecundity and viability do not add up. *Animal Behaviour*, **56**, 127–130.

Grafen, A. (1990) Biological signals as handicaps. *Journal of Theoretical Biology*, **144**, 517–546.

Healey, M. and Olsson, M. (2009) Too big for his boots: Are social costs keeping condition-dependent status signalling honest in an Australian lizard? *Austral Ecology*, **34**, 636–640.

Hughes, M. (1996a) Size assessment via a visual signal in snapping shrimp. *Behavioral Ecology and Sociobiology*, **38**, 51–57.

Hughes, M. (1996b) The function of concurrent signals: visual and chemical communication in snapping shrimp. *Animal Behaviour*, **52**, 247–257.

Hughes, M. (2000) Deception with honest signals: signal residuals and signal function in snapping shrimp. *Behavioral Ecology*, **11**, 614–623.

Husak, J.F., Henningsen, J.P., Vanhooydonck, B., and Irschick, D.J. (2012) A performance-based approach to studying costs of reliable signals, in *Animal Signaling: a Functional Approach* (eds D.J. Irschick, M. Briffa, and J. Podos), John Wiley & Sons, USA.

Jennions, M.D. and Backwell, P.R.Y. (1996) Residency and size affect fight duration and outcome in the fiddler crab *Uca annulipes*. *Biological Journal of the Linnean Society*, **57**, 293–306.

Johnstone, R.A. (1997) The evolution of animal signals, in *Behavioural Ecology: an Evolutionary Approach* (eds J.R. Krebs and N.B. Davies), Blackwell Science Ltd., Oxford, pp. 155–178.

Jordao, J.M. and Oliveira, R.F. (2001) Major chelae make male fiddler crabs more conspicuous to visual predators: a test using human observers. *Hydrobiologia*, **449**, 241–247.

Kotiaho, J.S. (2000) Testing the assumptions of conditional handicap theory: costs and condition dependence of a sexually selected trait. *Behavioral Ecology and Sociobiology*, **48**, 188–194.

Kotiaho, J.S. (2001) Costs of sexual traits: a mismatch between theoretical considerations and empirical evidence. *Biological Reviews*, **76**, 365–376.

Krebs, J.R. and Dawkins, R. (1984) Animal signals: mind-reading and manipulation, in *Behavioural Ecology: an Evolutionary Approach*, 2nd edn (eds J.R. Krebs and N.B. Davies), Sinauer Associates, Inc. Publishers, Sunderland, MA, USA, Xi+493p. Illus. Paper, 380–402.

Lachmann, M., Szamado, S., and Bergstrom, C.T. (2001) Cost and conflict in animal signals and human language. *Proceedings of the National Academy of Sciences of the United States of America*, **98**, 13189–13194.

Lailvaux, S.P., Reaney, L.T., and Backwell, P.R.Y. (2009) Dishonest signalling of fighting ability and multiple performance traits in the fiddler crab *Uca mjoebergi*. *Functional Ecology*, **23**, 359–366.

Lappin, A.K., Brandt, Y., Husak, J.F., Macedonia, J.M., and Kemp, D.J. (2006) Gaping displays reveal and amplify a mechanically based index of weapon performance. *American Naturalist*, **168**, 100–113.

Maynard Smith, J. (1974) The theory of games and the evolution of animal conflicts. *Journal of Theoretical Biology*, **47**, 209–221.

Maynard Smith, J. (1982) *Evolution and the Theory of Games*, Cambridge University Press, Cambridge.

Maynard Smith, J. and Harper, D.G.C. (1995) Animal signals: models and terminology. *Journal of Theoretical Biology*, **177**, 305–311.

Maynard Smith, J. and Harper, D.G.C. (2003) *Animal Signals*, Oxford University Press Inc., New York.

Moller, A.P. (1989) Natural and sexual selection on a plumage signal of status and on morphology in house sparrows, *Passer domesticus*. *Journal of Evolutionary Biology*, **2**, 125–140.

Moore, J.C., Obbard, D.J., Reuter, C., West, S.A., and Cook, J.M. (2009) Male morphology and dishonest signalling in a fig wasp. *Animal Behaviour*, **78**, 147–153.

Pereira, R.A.S. and do Prado, A.P. (2005a) Recognition of competitive asymmetries reduces he severity of fighting in male Idarnes fig wasps. *Animal Behaviour*, **70**, 249–256.

Pereira, R.A.S. and do Prado, A.P. (2005b) Non-pollinating wasps distort the sex ratio of pollinating fig wasps. *Oikos*, **110**, 613–619.

Reaney, L.T., Milner, R.N.C., Detto, T., and Backwell, P.R.Y. (2008a) The effects of chela regeneration on territory ownership and mating success in the fiddler crab *Uca mjoebergi*. *Animal Behaviour*, **75**, 1473–1478.

Reaney, L.T., Milner, R.N.C., Detto, T., and Backwell, P.R.Y. (2008b) The effects of claw regeneration on territory ownership and mating success in the fiddler crab *Uca mjoebergi*. *Animal Behaviour*, **75**, 1473–1478.

Rohwer, S. and Ewald, P.W. (1981) The cost of dominance and advantage of subordination in a badge signalling system. *Evolution*, **35**, 441–454.

Searcy, W.A. and Nowicki, S. (2005) *The Evolution of Animal Communication: Reliability and Deception in Signaling Systems*, Princeton University Press, Princeton, NJ.

Senar, J.C. (1999) Plumage colouration as a signal of social status. *Proceedings of the International Ornithological Congress*, **22**, 1669–1686.

Seebacher, F. and Wilson, R.S. (2006) Fighting fit: thermal plasticity of metabolic function and fighting success in the crayfish *Cherax destructor*. *Functional Ecology*, **20**, 1045–1053.

Seebacher, F. and Wilson, R.S. (2007) Individual recognition in crayfish (Cherax dispar): the roles of strength and experience in deciding aggressive encounters. *Biology Letters*, **3**, 471–474.

Steger, R. and Caldwell, R.L. (1983) Intraspecific deception by bluffing – a defense strategy of newly molted stomatopods (arthropoda, crustacea). *Science*, **221**, 558–560.

Szalai, F. and Számadó, S. (2009) Honest and cheating strategies in a simple model of aggressive communication. *Animal Behaviour*, **78**, 949–959.

Számadó, S. (2000) Cheating as a mixed strategy in a simple model of aggressive communication. *Animal Behaviour*, **59**, 221–230.

Számadó, S. (2003) Threat displays are not handicaps. *Journal of Theoretical Biology*, **221**, 327–348.

Számadó, S. (2008) How threat displays work: species-specific fighting techniques, weaponry and proximity risk. *Animal Behaviour*, **76**, 1455–1463.

Számadó, S. (2011) The cost of honesty and the fallacy of the handicap principle. *Animal Behaviour*, **81**, 3–10.

Tibbetts, E.A. and Dale, J. (2004) A socially enforced signal of quality in a paper wasp. *Nature*, **432**, 218–222.

Tibbetts, E.A. and A. Izzo (2010) Social punishment of dishonest signalers caused by mismatch of signal and behavior. *Current Biology.* **20**, 1637–1640.

Wilson, R.S., Angilletta, M.J., James, R.S., Navas, C., and Seebacher, F. (2007) Dishonest signals of strength in male slender crayfish (*Cherax dispar*) during agonistic encounters. *American Naturalist*, **170**, 284–291.

Yachi, S. (1995) How can honest signalling evolve? The role of handicap principle. *Proceedings of the Royal Society of London, Series B: Biological Sciences*, **262**, 283–288.

Zahavi, A. (1977) The cost of honesty (further remarks on the handicap principle). *Journal of Theoretical Biology*, **67**, 603–605.

9

FUNCTIONAL APPROACH TO CONDITION

Dustin J. Wilgers and Eileen A. Hebets

School of Biological Sciences, University of Nebraska-Lincoln, Lincoln, NE, USA

INTRODUCTION

Animal signaling is commonly thought to be costly. Signaling costs can arise via a variety of avenues, including energy expenditure, predator attraction, and so on (reviews in Zuk and Kolluru, 1998; Kotiaho, 2001), and are predicted to increase with signal expression (e.g., size, amplitude, and intensity; Johnstone, 1997). Due to these costs, signaler condition, which is hypothesized to be a reflection of a signaler's genetic quality, is expected to influence the level of signal expression one can afford (Zahavi, 1975), resulting in a positive correlation between signaler condition and signal expression – that is, condition-dependent signaling (Zahavi, 1977; West-Eberhard, 1979; Andersson, 1982; Nur and Hasson, 1984; Zeh and Zeh, 1988; von Schantz *et al.*, 1999).

Despite the wealth of theoretical and empirical work on condition-dependent signaling, the term *condition* itself is somewhat enigmatic. An individual's condition is a theoretical construct associated with the acquisition and allocation of nutritional resources, and is assumed to be an integral part of an individual's health, vigor, and viability (Andersson, 1982; Nur and Hasson, 1984; Zeh and Zeh, 1988). Condition is often thought of as polygenic in nature, capturing much of the additive genetic variance responsible for viability, spanning numerous loci across the

Animal Signaling and Function: An Integrative Approach, First Edition.
Edited by Duncan J. Irschick, Mark Briffa, and Jeffrey Podos.

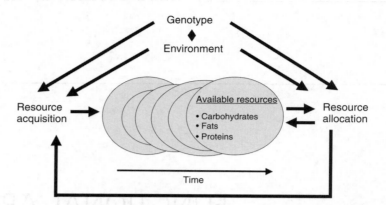

Figure 9.1. A representation of the complex interactions (indicated by diamond) between an individual's genotype and environment and their interactions with resource acquisition, available resource pool, and resource allocation. Resource acquisition and/or resource allocation can vary over the lifespan of individuals, resulting in different sizes of resource pools (i.e., condition) over time.

genome (Andersson, 1982; Rowe and Houle, 1996); however, like other quantitative traits, condition is also influenced by the environment as well as by interactions between an individual's genotype and the environment (Hunt *et al.*, 2004b).

In this chapter, we use a widely accepted working definition of condition provided by Rowe and Houle (1996) – a pool of resources acquired from the environment, which is available for allocation to various fitness-related traits (see Figure 9.1). An individual's condition sums numerous processes throughout its lifespan and is constantly fluctuating as resources are acquired and allocated to different functions (Figure 9.1). This broad definition of condition incorporates information on the resources available throughout an individual's life – it encompasses the resources used during development to create structures (including those used for resource acquisition), the resources used in the normal functioning of an individual (i.e., its physiology), and the resources currently available in an individual's energy stores (Figure 9.1). The decisions directing resource acquisition and allocation strategies (Figure 9.1) are ultimately influenced by both an individual's environment (e.g., presence/absence of predators and food abundance) and its genotype (e.g., heritable traits related to foraging ability, digestion, and learning).

PRACTICAL APPROACHES TO CONDITION

Proxies of Current Energy Reserves

Scientists interested in condition-dependent signaling typically consider an individual's current condition (or current energy stores) by utilizing proxies such as body condition (a variety of indices have been used to estimate this; see Table 9.1). Some studies use measures of overall body weight or volume, while others measure

TABLE 9.1. Examples of Proxies Used to Estimate Body Condition Across Taxa

Proxy	Measure	Group	Citation
Estimates of Reserves via Absolute Body Measures			
Body size	Wing size	Insects	Hooper et al. (1999); David et al. (2000); Blanckenhorn and Hosken (2003); Van Homrigh et al. (2007)
	Shape of abdomen	Birds	Owen (1981)
	Pectoral muscle size	Birds	Perez-Rodriguez et al. (2006)
Mass		Birds	Gonzalez et al. (1999); Hill (2000); McGraw and Hill (2000);
		Birds	Johnsen et al. (2003); Perez-Rodriguez et al. (2006)
		Insects	Kotiaho et al. (2001); Kotiaho (2002); Rantala et al. (2003); Scheuber et al. (2003a)
Volume		Spiders	Mappes et al. (1996); Rundus et al. (2011)
Growth rate		Birds	Sibly et al. (1987)
Trait assymetry		Birds	Zuk et al. (1990); Keyser and Hill (1999)
		Birds	McGraw et al. (2002)
Estimates of Reserves Controlling for Body Size			
Relative weight	Weight/volume	Fish	Neuman and Flammang (1997)
Density		Spiders	Moya-Larano et al. (2008)
Ratio	Weight/fixed body measure	Amphibians	Arntzen et al. (1999)
		Fish	Tonn et al. (1989); Greenstreet (1992); Candolin (2000)
		Lizards	van Berkum et al. (1989); van Marken Lichtenbelt et al. (1993)
		Birds	Moller (1987); Evans and McMahon (1987);
		Spiders	Shamble et al. (2009); Wilgers and Hebets (2011)
Slope adjusted ratio	Dynamic body part/static body part	Spiders	Anderson (1974); Watson (1990); Jakob (1991)
	$Weight/(body\ size)^{slope}$[a]	Fish	Kulling and Milinski (1992); Nicoletto (1993)
		Insects	Pierce et al. (1985); Juliano (1986); Baker (1989)
Residual	Weight/body size	Birds	Andersson (1992); Carranza and Hidalgo de Trucios (1993); Hamer and Furness (1993);

(continued)

TABLE 9.1. (*Continued*)

Proxy	Measure	Group	Citation
			Schluter and Gustafsson (1993); Veiga (1993); Qvarnstrom (1999); Weatherhead et al. (1999); Merila et al. (2001), Moller and Petrie (2002); Doucet and Montgomerie (2003); Sarasola et al. (2004); Ardia (2005); Bize et al. (2006)
		Crustaceans	Jennions and Backwell (1998)
		Insects	Marden and Rollins (1994); Wagner and Hoback (1999); Gray and Eckhardt (2001); Holzer et al. (2003); Scheuber et al. (2003b)
		Mammals	Dobson (1992); Dobson and Michener (1995); Woodroffe (1995); Dobson et al. (1999) Fisher (1999); Schulte-Hostedde et al. (2001); Blackwell (2002)
		Amphibians	Murphy (1994); Judge and Brooks (2001)
		Reptiles	Dunlap and Mathies (1993); Weatherhead et al. (1995); Keller et al. (1997); Cuadrado (1998); Shine et al. (2001)
		Spiders	Uetz et al. (2002); Hoefler et al. (2008); Wilder and Rypstra (2008); Lomborg and Toft (2009); Wilgers et al. (2009); Taylor et al. (2011)
ANCOVA	Weight with body size covariate	Birds	Torok et al. (2003); Parker and Garant (2004)
PCA	Factor loading with size and mass	Spiders	Lomborg and Toft (2009); Wilder and Rypstra (2008)
		Birds	Bize et al. (2006)
Scaled mass index	Body mass scaled by mass/length relationship	Mammals	Peig and Green (2009); Peig and Green (2010)
		Birds	Peig and Green (2009)
		Reptiles	Peig and Green (2009)

PCA, principal component analysis.

[a]Slope of relationship determined from entire population.

the physical size of body parts assumed to be indicative of energy reserves (e.g., shape of abdomen in birds; Owen, 1981). While potentially easy to quantify, these absolute measurements are confounded with body size, which may give little or no information about differences in current energy reserves (Piersma and Davidson, 1991), but instead could reflect larger quantities of non-energy related compounds (e.g., water, bone; Tomkins et al., 2004).

The vast majority of indices currently used attempt to control for body size by investigating the relationship between a dynamic body variable thought to represent energy reserves and a static, or less dynamic, estimate of overall body size (Jakob et al., 1996). Commonly, scientists use body weight or volume, measures that are known to change rapidly with resource acquisition, and control for body size using the length/width of skeletal (or exoskeletal in invertebrates) structures that are either fixed during certain life stages or remain effectively static over the time period of interest. Such methods vary tremendously and are the topic of much debate (e.g., Jakob et al., 1996; Garcia-Berthou, 2001; Peig and Green, 2010). Here, we simply highlight a few of the more common body condition indices and direct our readers to the relevant literature regarding the issues associated with each.

The simplest body condition index is the ratio index, which is calculated as body weight, or volume, divided by a linear measure of body size. Ratio indices provide a good descriptive index that is comparable across groups or populations and has been used consistently in the literature (Table 9.1). However, the ratio index is often not independent of body size, which limits conclusions on body condition alone (for criticisms, see Blem, 1984; Ranta et al., 1994; Jakob et al., 1996). Additional methods to quantify body condition separate the effects of energy reserves and body size by incorporating size measures as a covariate in analysis of covariance (ANCOVA) models, and analyzing variation in body weight or other dynamic measures (Garcia-Berthou, 2001). Alternatively, researchers use the residuals from a linear regression of body weight against some linear measure of body size. The use of residuals has become increasingly common over the last two decades (reviewed in Green, 2001). However, numerous studies have cautioned their use due to potential violations of statistical assumptions (e.g., linear relationships and independence; Green, 2001), alterations of trait allometric relationships (Kotiaho, 1999), or limitations in comparisons across individuals varying in size (Kotiaho, 1999) and across heterogeneous groups (e.g., populations; Jakob et al., 1996). Ultimately, the choice of body condition index is non-trivial as results and conclusions often vary widely, based on the chosen index (Bolger and Connolly, 1989; Jakob et al., 1996; Moya-Larano et al., 2008; Peig and Green, 2010).

Body condition indices have at times been found to be good predictors of specific energy reserves (Sibly et al., 1987; Schulte-Hostedde et al., 2001; Cattet et al., 2002; Ardia, 2005; Schulte-Hostedde et al., 2005); however, as mentioned previously, variation in these proxies may also simply reflect variation in other compounds (e.g., water; Schulte-Hostedde et al., 2001, 2005). Additionally, these snapshot indices do not provide information regarding specific types of energy stores, or how these energy stores vary throughout development. Given that different animals, and even different developmental stages, may have different nutritional requirements and may require different essential elements, the black box approach commonly used to estimate body condition provides little information as to the mechanisms underlying condition, and how they may differ across taxa (Lailvaux and Irschick, 2006). Condition indices also fail to illuminate variation in the ratios of specific energy reserves, which may be important as measures of different types of energy reserves do not necessarily correlate with one another and can even provide contradictory results (e.g., Blanckenhorn and Hosken, 2003). As such, a more detailed approach incorporating direct measurements of specific energy reserves for allocation to fitness-related traits may prove to be important for a detailed understanding of the evolution of condition-dependent signaling.

Direct Measures of Current Energy Reserves

Organisms convert newly ingested organic matter into three main groups of compounds: carbohydrates, fats, and proteins. Simple carbohydrates, such as glucose, are used to fuel metabolism, which may be important for energy-demanding behavioral displays. If not used immediately, simple carbohydrates are converted to energy storage compounds, such as starch and glycogen (complex carbohydrates), which can be later accessed as an energy source to fuel aerobic or anaerobic metabolism. Due to storage constraints associated with glycogen (i.e., size, hydrophilic), the majority of excess carbohydrates are broken down to synthesize compact long-term energy storage, like fats.

Fats serve both metabolic and structural functions. They are likely an important energy reserve for animals where feeding is limited for long periods of time. Importantly, fat stores can be accumulated in one life stage for use in another. Additionally, fats are useful for delivering other important resources, such as carotenoids, which are known to perform a variety of physiological functions (reviewed in Olson and Owens, 1998).

Proteins serve important structural and functional roles, and have been found to be necessary for normal development in a variety of animals (e.g., Eagle, 1959; House, 1961). The amino acids required for protein synthesis can either be consumed or be synthesized in the body. Proteins also serve as a long-term energy storage molecule; however, the efficiency and extent to which proteins can be effectively digested by certain taxa may vary (e.g., insects; Chapman, 1998). Unlike fats, there is no specialized store for proteins and thus energy must come from catabolism of both structural and functional organs (e.g., muscles, digestive organs), where excessive depletion can harm animal performance (Jenni and Jenni-Eiermann, 1998). Because of this, the relative protein contribution to the energy budget is, for example, only about 5% in migrating birds (Jenni and Jenni-Eiermann, 1998), and protein catabolism for metabolic energy is typically only utilized when other energy stores are depleted (king penguins: Robin *et al.*, 1988; green sea turtles: Jessop *et al.*, 2004).

Given differences in their accessibility, their storage, and their potential use, detailed knowledge of an organism's (i) nutritional requirements, (ii) abundance of specific energy resources (e.g., quantification of carbohydrates, fats, and/or proteins) across their lifespan, and (iii) allocation strategies will aid our understanding of targets of selection surrounding condition-dependent signaling.

Manipulating Energy Reserves

An initial first step toward documenting condition-dependent signaling often involves quantification of signal/display expression in wild caught animals and subsequent correlations with some proxy of condition (e.g., body condition index and parasite load; Hoglund *et al.*, 1992; Buchholz, 1995; Thompson *et al.*, 1997; Doucet and Montgomerie, 2003). While such an approach is informative, it cannot provide information about the relative influence of genotype versus environment in determining the witnessed variation. Thus, often in conjunction with such field-based correlative approaches, scientists employ more controlled environmental manipulations to experimentally alter current body condition and examine corresponding changes in signal expression.

Manipulating nutrition (quantity and quality), density, parasite load, temperature, and so on can lead to measurable differences among treatment groups in various proxies associated with condition (reviewed in Cotton *et al.*, 2004). Regardless of the manipulation, treatments commonly represent two extremes (e.g., high vs low nutrient levels, parasitized vs unparasitized). While using such extreme manipulations increases the power to detect a relationship between current condition and signal expression, it removes much of the potentially relevant middle of the population distribution in phenotype response. Therefore, Cotton *et al.* (2004) suggest the use of a broader set of manipulations, providing a more comprehensive assessment of the relationship between various proxies of condition and signal expression.

The majority of studies of condition-dependent signaling thus far have utilized either phenotypic manipulations across randomly sampled individuals or genetic controls (e.g., Kodric-Brown, 1989; Houde and Torio, 1992; Birkhead *et al.*, 1998; Grether, 2000). Such approaches can either swamp out or eliminate potential genetic variation among individuals and thus do not allow variation in condition to manifest solely as a result of additive genetic variation among individuals (Cotton *et al.*, 2004). Given that many hypotheses relating to the evolution of condition-dependent signal expression (e.g., various indicator mechanisms) assume phenotypic quality/viability (i.e., condition) to be heritable (e.g., Andersson, 1982; Hamilton and Zuk, 1982; Iwasa *et al.*, 1991; Iwasa and Pomiankowski, 1999), future studies are needed which focus more on the genetic basis of condition (Tomkins *et al.*, 2004).

CONDITION AND ANIMAL PERFORMANCE

Allocation of Energy Reserves

The pool of current energy reserves that an individual possesses is expected to directly relate to its ability to afford costly activities, such as the development of elaborate morphological characters and the production of behavioral displays that may enhance reproductive success, and/or the engagement in other behaviors and physiological processes that may enhance survival. As such, variation in this pool of resources should translate into variation in animal performance, where performance is defined as an organism's ability to conduct various ecologically relevant tasks related to survival (e.g., foraging, running speed, and overall endurance) and/or reproduction (e.g., fighting ability and courtship displays; see Lailvaux and Irschick, 2006; Irschick *et al.*, 2008). Here, we discuss the relationship between various measures of energy reserves and animal performance, highlighting studies in which detailed knowledge of energy stores has been crucial.

Body condition indices have been shown to correlate with various fitness measures – individuals in better body condition survive better (e.g., Naef-Daenzer *et al.*, 2001; Shine *et al.*, 2001; Murray, 2002; Morrison *et al.*, 2007) and have overall higher reproductive success (e.g., Chastel *et al.*, 1995; Dobson and Michener, 1995; Otronen, 1995; Wauters and Dhondt, 1995). Direct links between available carbohydrates and animal performance have also been documented. During hovering flight in the broadtailed hummingbird (*Selaphorus platycercus*), fasting individuals switch from primarily metabolizing fats to oxidizing mainly carbohydrates 20–60 minutes after a meal (Welch *et al.*, 2006). Similarly, up to 78% of the fuel required for hovering flight in the nectarivorous bat, *Glossophaga sorincina*, was shown to come from recently ingested carbohydrates (Welch *et al.*, 2008). For such animals that engage in energy-demanding behavioral displays, we might expect a relationship between display performance and exogenous sugar uptake, making the direct measurement of carbohydrates potentially informative. Ingested carbohydrates are also an important energy source for sex pheromone expression in the cockroach, *Nauphoeta cinerea* (South *et al.*, 2011) and when given a choice, males consumed diets with higher carbohydrate content (South *et al.*, 2011).

In addition to the immediate use of carbohydrates, glycogen stores are also known to be important in energy-demanding behavioral displays. In the mosquito, *Anopheles feeborni*, males feed on nectar during the night and store the acquired carbohydrates as glycogen for use during swarming flight the next day (Yuval *et al.*, 1994). In fiddler crabs, males wave a single large claw during courtship displays to attract females. One of the primary stores of energy in decapod crustaceans is glycogen, which is readily mobilized into blood glucose for ATP synthesis during fights and leg-waving displays. Matsumasa and Murai (2005) found variability in resting male blood glucose levels, suggesting standing variation in male quality. Additionally, they found that males who waved their legs more frequently had higher lactate levels, a byproduct of glucose catabolism, in their bloodstream, suggesting a positive correlation between glycogen energy reserves and performance. Glycogen is also a known energy source for frogs during metabolically demanding calling activity (Bevier, 1997).

Both survival and fecundity have been shown to increase with increased fat reserves across a variety of taxa (Elowe and Dodge, 1989; Atkinson and Ramsay, 1995; Vleck and Vleck, 2002) and fat stores accumulated during early developmental stages are known to influence subsequent life stages. For example, accumulation of fat stores by juveniles has been found to be crucial for terrestrial survival post-metamorphosis in amphibians (Scott et al., 2007) and for reproductive success in damselflies (Plaistow and Siva-Jothy, 1996). Studies on birds also provide evidence that body condition during juvenile stages can influence adult body condition (e.g., fat reserves), and that these fat reserves are important for survival during strenuous activities, such as migration (Merila and Svensson, 1997). Given the above-demonstrated relationships between early life fat storage and later life performance, fats provide an excellent example of the need for a more inclusive (incorporating more than simply current body condition indices) and detailed (quantifying fats vs carbohydrates or proteins) examination of an animal's energy reserves when relating it to performance.

Proteins are commonly used in the development of a variety of structures, including bird feathers, which are produced by keratins. In dark-eyed juncos, *Junco hyemalis*, birds on protein-enriched diets expressed faster feather growth rates along with larger and brighter white plumage ornaments on the tail, suggesting the intake of specific energy compounds (i.e., diet quality) can result in condition-dependent expression (McGlothlin et al., 2007). Similar results have been found in house sparrows, *Passer domesticus*, where male house sparrows on protein-enriched diets have large white wing bars (Poston et al., 2005). In addition, male house sparrows had brighter (i.e., not as black) but not bigger melanin-based black bibs when fed diets lacking melanin-precursors compared to males fed normal diets (Poston et al., 2005).

In many cases, energy-demanding activities, such as flight, do not rely solely on one energy resource type. For example, long-distance migrations require tremendous energy reserves from multiple sources (e.g., glycogen, fats, proteins). In several species, protein catabolism increases when fat reserves are near depletion (<5–10%; Schwilch et al., 2002; Bauchinger and Biebach, 2001). Nonetheless, the relative utilization of different compounds varies across birds and has been found to be a function of their diet (Gannes, 2001), suggesting a direct link between the types and amounts of resources acquired and those used during performance.

Trade-Offs in Resource Allocation

The allocation of resources from a finite pool (although the pool is rarely finite as currently available resources are frequently used for additional nutrient acquisition – e.g., to sustain foraging) is expected to result in trade-offs in resource allocation, such that allocating resources to one trait reduces the available resources for allocation to other traits (Rowe and Houle, 1996; Zera and Harshman, 2001). For individuals with fewer energy reserves, it might be necessary to allocate more resources to basic survival requirements, making investment in other traits (e.g., reproductive) relatively more costly. However, an individual's optimal resource allocation among traits aims to maximize overall fitness and thus depends not only on the amount of available resources, but also on the strengths of various selection sources (Rowe and Houle, 1996). For example, in reproductive systems exemplified by strong female choice, signalers might maximize fitness by allocating greater resources to secondary sexual traits (e.g., ornamentation and displays), or other traits under selection via females, thereby gaining increased reproductive success even at a cost to other life-history traits. Such a trade-off has been found in the field cricket, *Teleogryllus commodus*, in which high body-condition males invest so many resources to their sexual displays that they tend to die younger than lower condition males (Hunt *et al.*, 2004a). Similarly, in *Hygrolycosa rubrofasciata* wolf spiders, males induced to court at higher rates suffered greater mortality and lost more weight over the trial than did males that courted at lower rates, suggesting that allocation of energy reserves to courtship reduced their availability for allocation to survival (Mappes *et al.*, 1996). However, within the high courting group, males that maintained high courtship levels survived better, suggesting variation in an individual's ability to afford the high energetic costs associated with the display (Mappes *et al.*, 1996). This variation could be attributed to body condition, where larger resource pools may afford greater levels of reproductive trait expression while simultaneously experiencing greater viability (Jennions *et al.*, 2001).

Signaling systems, in which there is a working knowledge of the resources utilized during costly signal expression and how these resources are used in other fitness-related functions, can facilitate our understanding of the details surrounding the selection and subsequent evolution of condition-dependent signaling. For example, carotenoid-based coloration is found in numerous animal taxa (e.g., birds, fish, amphibians, reptiles, insects, mollusks, crustaceans; Matsuno, 2001; McGraw *et al.*, 2005), producing some of the most brilliant coloration in the animal kingdom (Olson and Owens, 1998). Carotenoid pigments (e.g., carotenes, xanthophylls) cannot be synthesized by animals; instead they must be acquired through consumption of carotenoid-rich food. Carotenoids also play important roles throughout the body (e.g., enhancing immune system function: antioxidants, free-radical absorbers; reviews in Lozano, 1994; Shykoff and Widmer, 1996; Lozano, 2001; and serving as photo-protectants: example in Japanese Quail; Thomson *et al.*, 2002), yet despite these important roles, carotenoids appear to be readily allocated to a signaling function. Prior research has shown that individuals in good nutritional condition (i.e., high reserves of carotenoids) and in good overall health can afford to allocate greater resources to a colorful carotenoid-based display (e.g., McGraw and Hill, 2000; McGraw and Ardia, 2003; Saks *et al.*, 2003; Mougeot *et al.*, 2007), and females favor those males with larger or brighter displays (examples in guppies: Endler, 1983; house finches: Hill, 1990; three-spined sticklebacks: Bakker and Mundwiler, 1994). The putative trade-off in carotenoid resource allocation within males and the seeming importance of this trait in female mate choice can lead us to generate testable predictions about the strength of sexual selection for condition-dependent carotenoid-based signaling. For example, we might expect females to pay more attention to carotenoid-based signals in environments with low carotenoid resource availability. Grether (2000) tested this hypothesis in the guppy,

Poecilia reticulata, across a gradient of carotenoid availabilities and found no support. Nonetheless, by understanding the physiology underlying signal expression, we gain further insights into traits that compete for limited resources within the body and the precise trade-offs that might result.

Time Scales for Resource Allocation

Resource allocation is dynamic in nature and condition-dependent signals involved in reproductive behavior can potentially reflect the pool of available resources at different time scales (Johnstone, 1995). For example, morphological traits associated with signaling can potentially provide receivers with information about the resource pool available to an individual throughout the development of a particular structure (e.g., deer antlers: Clutton-Brock *et al.*, 1982; Suttie and Kay, 1983; horns in beetles: Emlen, 1994; spider coloration: Shamble *et al.*, 2009; Rundus *et al.*, 2011; Taylor *et al.*, 2011). Alternatively, morphological traits such as carotenoid coloration may reflect more recent or even current resource pools (e.g., Grether, 2000; Rosen and Tarvin, 2006). To complicate matters further, many behavioral displays (e.g., acoustic and vibratory song, body movements) require intense motor performance, which is known to raise metabolic rates considerably, and thus require substantial immediate energy reserves. Such elaborate displays may also require complex structures (e.g., muscles) and motor skills acquired during development (Byers *et al.*, 2010), and thus reflect resource pools available at earlier life stages. Ultimately, the production of costly displays can provide information about both current energy reserves (Mappes *et al.*, 1996; Hoefler *et al.*, 2008) as well as energy reserves available during development (Nowicki *et al.*, 2002); thus, knowing the relative influence of each may be important for understanding the selection pressures associated with condition-dependent signal evolution.

CONDITION AND MATE CHOICE

The evolution of condition-dependent signaling is often attributed to sexual selection. Male–male competition provides an intuitive relationship between body condition, signaling performance, and direct fitness outcomes (e.g., access to females); and numerous examples evidence condition-dependent signals that are used in male–male competition (e.g., review in Berglund *et al.*, 1996; Veiga, 1993; Vanpé *et al.*, 2007). Female choice is also frequently touted as being involved in the evolution of condition-dependent signaling (Bondurianski, 2007), yet the connections here are less intuitive. It is the relationship between female choice and condition-dependent signaling that we will focus on subsequently.

It is hypothesized that females indirectly assess a potential mate's quality by examining courtship signal expression, ultimately resulting in the evolution of stable female preferences and corresponding exaggeration of costly male secondary sexual traits (Andersson, 1986; Heywood, 1989; Hoelzer, 1989; Grafen, 1990; Iwasa *et al.*, 1991; Iwasa and Pomiankowski, 1999). In support of this, there exists a wealth of consistent evidence demonstrating that females prefer to mate with males that exhibit more elaborate traits/displays (e.g., larger, brighter, and louder; reviews in Andersson, 1994; Johnstone, 1995; but see Griffith *et al.*, 1999; Lebas and Marshall, 2001; Shamble *et al.*, 2009 for examples of no evidence of female choice). For these choosy females, mate choice can be costly (e.g., Alatalo *et al.*, 1988; Rowe, 1994), suggesting that these choosy females are receiving some benefit from their choice of mates. Females could benefit directly through increased paternal care, increased nutritional resources, better quality territories, or reduced parasitism risk (among others). Hypothesized direct benefits such as these generate a clear prediction – a positive correlation should exist between a signaler's condition and the direct benefits conveyed to females. Indeed, condition-dependent signals have been found to correlate with paternal care (e.g., birds; Hill, 1991; Senar *et al.*, 2002), with the quantity of sperm transferred (e.g., guppies; Matthews *et al.*, 1997; crickets; Wagner and Harper, 2003) and with the production of nutritional benefits, such as spermatophores (e.g., crickets; Wagner and Harper, 2003) and nuptial gifts (e.g., fireflies; Crastley, 2004). The ability to quantify the direct benefits females receive and to then relate them to proxies of signaler condition makes direct benefits a compelling, and readily testable, hypothesis regarding female mate choice on condition-dependent signals.

Choosy females are also expected to benefit indirectly if preferred mates pass "good genes" to their offspring that increase their fitness. Evidence for underlying heritable additive genetic variance for body condition is slowly accumulating (Merila, 1996; Merila and Svensson, 1997; Sheldon *et al.*, 1997; Grether, 2000; Kotiaho *et al.*, 2001; Merila *et al.*, 2001; Blanckenhorn and Hosken, 2003), as are examples demonstrating that condition-dependent signals covary with genetic variance (i.e., genic capture; e.g., David *et al.*, 2000; Brandt and Greenfield, 2004; Parker and Garant, 2004; Missoweit *et al.*, 2008). Additionally, there is some evidence that the additive genetic variance in sexually selected traits results in viability benefits to offspring (Moller and Alatalo, 1999), although the effects may be relatively minor (Alatalo *et al.*, 1998; Moller and Alatalo, 1999). Regardless of the magnitude, offspring from highly ornamented males in good body condition have been found to experience benefits in a variety of fitness-related traits, where offspring feed at higher rates (tree frogs: Doty and Welch, 2001), survive better (guppies: Evans *et al.*, 2004), are in better body condition (collared flycatchers: Sheldon *et al.*, 1997), have increased resistance to parasites (sticklebacks: Barber *et al.*, 2001), and have better larval development (tree frogs: Welch *et al.*, 1998). However, these measurements fall short of evidencing true fitness benefits to females, which should be measured minimally by the number of offspring that each offspring produces (i.e., grandchildren; Hunt *et al.*, 2004b).

Similar to the previously highlighted variation in proxies used for estimating male condition (see Table 9.1), scientists also use a variety of different proxies for estimating female and offspring fitness, including condition itself, making it difficult to obtain an overall picture of the evidence for females receiving fitness benefits from mate choice decisions based on condition-dependent signals. Furthermore, while examples do exist that suggest benefits, both direct and indirect, to females for these choices, the evidence is surprisingly sparse (especially for indirect benefits) and the current viewpoint may be exaggerated due to publication bias toward positive results (Kotiaho and Puurtinen, 2007). We suggest that the relative scarcity of examples may reflect an insufficient approach to condition-dependent signal evolution. For example, proxies of current body condition may not accurately reflect resources relevant to female fitness, as it may be the case that the details about energy reserves that are important to females and/or reflective of energy reserves available at a different time scale (e.g., during juvenile development) are crucial. Additionally, the relative influences of the environmental component, additive genetic variance, and their interactions on body condition and corresponding offspring fitness need to be more firmly established, as signal reliability, especially in condition-dependent signaling systems with proposed indirect benefits, is hypothesized to be compromised (Greenfield and Rodriguez, 2004; Hunt *et al.*, 2004b).

SUMMARY

Evidence of condition-dependent signaling abounds. Such signals are thought to have had important effects on the evolution of animal communication systems (Iwasa *et al.*, 1991). However, the ambiguous and broad nature of condition makes it currently un-measurable, causing scientists to rely on brief snapshots of body condition to estimate an individual's current energy reserves. The variety of proxies used to estimate body condition both across and within taxa has allowed considerable progress in advancing our understanding of ultimate explanations of animal signaling; however, using readily quantifiable proxies of body condition may cloud this understanding and slow our advance. We advocate that a more proximate approach is now in order. We argue that a focus on the physiological basis of condition, on the processes underlying resource allocation, and on the relationship between these and measures of whole organism performance will provide a more complete understanding of underlying mechanisms resulting in a signal's condition-dependence (Lailvaux and Irschick, 2006). Additionally, more proximate approaches to understanding the relationship between condition-dependent signal expression and female mating decisions, and the putative associated fitness benefits, will improve our understanding of selection pressures that might influence the evolution of condition-dependent signaling in a reproductive context. Focusing on more proximate physiological underpinnings of what condition means and how it directly influences signal expression and female fitness benefits lays the foundation for future comparisons across taxa that share similar mechanisms, which may illuminate interesting broad-scale patterns.

REFERENCES

Alatalo, R.V., Carlson, A., and Lundberg, A. (1988) The search cost in mate choice of the pied flycatcher. *Animal Behavior*, **36**, 289–291.

Alatalo, R.V., Kotiaho, J.S., Mappes, J., and Parri, S. (1998) Mate choice for offspring performance: major benefits or minor costs? *Proceedings of the Royal Society of London, Series B: Biological Sciences*, **1998**, 2297–2301.

Anderson, J.F. (1974) Responses to starvation in the spiders *Lycosa lenta* Hentz and *Filistata hibernalis* (Hentz). *Ecology*, **55**, 576–585.

Andersson, M. (1982) Sexual selection, natural selection and quality advertisement. *Biological Journal of the Linnean Society*, **17**, 375–393.

Andersson, M. (1986) Evolution of condition-dependent sex ornaments and mating preferences: sexual selection based on viability differences. *Evolution*, **40**, 804–816.

Andersson, M. (1994) *Sexual Selection*, Princeton University Press, Princeton, NJ.

Andersson, S. (1992) Female preference for long tails in lekking Jackson's widowbirds: experimental evidence. *Animal Behaviour*, **43**, 379–388.

Ardia, D.R. (2005) Super size me: an experimental test of the factors affecting lipid content and the ability of residual body mass to predict lipid stores in nestling European Starlings. *Functional Ecology*, **19**, 414–420.

Arntzen, J.W., Smithson, A., and Oldham, R.S. (1999) Marking and tissue sampling effects on body condition and survival in the newt *Triturus cristatus*. *Journal of Herpetology*, **33**, 567–576.

Atkinson, S.N. and Ramsay, M.A. (1995) The effects of prolonged fasting on the body composition and reproductive success of female polar bears (*Ursus maritamus*). *Functional Ecology*, **9**, 559–567.

Baker, R.L. (1989) Condition and size of damselflies: a field study of food limitation. *Oecologia*, **81**, 111–119.

Bakker, T.C.M. and Mundwiler, B. (1994) Female mate choice and male red coloration in a natural three-spined stickleback (*Gasterosteus aculeatus*) population. *Behavioral Ecology*, **5**, 74–80.

Barber, I., Arnott, S.A., Braithwaite, V.A., Andrew, J., and Huntingford, F.A. (2001) Indirect fitness consequences of mate choice in sticklebacks: offspring of brighter males grow slowly but resist parasitic infections. *Proceedings of the Royal Society of London, Series B: Biological Sciences*, **268**, 71–76.

Bauchinger, U. and Biebach, H. (2001) Differential catabolism of muscle protein in Garden Warblers (*Sylvia borin*): flight and leg muscle act as a protein source during long-distance migration. *Journal of Comparative Physiology B*, **171**, 293–301.

Berglund, A., Bisazza, A., and Pilastro, A. (1996) Armaments and ornaments: an evolutionary explanation of traits of dual utility. *Biological Journal of the Linnean Society*, **58**, 385–399.

Bevier, C.R. (1997) Utilization of energy substrates during calling activity in tropical frogs. *Behavioral Ecology and Sociobiology*, **41**, 343–352.

Birkhead, T.R., Fletcher, F., and Pellatt, E.J. (1998) Sexual selection in the zebra finch *Taeniopygia guttata*: condition, sex traits and immune capacity. *Behavioral Ecology and Sociobiology*, **44**, 179–191.

Bize, P., Piault, R., Moureau, B., and Heeb, P. (2006) A UV signal of offspring condition mediates context-dependent parental favouritism. *Proceedings of the Royal Society of London, Series B: Biological Sciences*, **273**, 2063–2068.

Blackwell, G.L. (2002) A potential multivariate index for condition of small mammals. *New Zealand Journal of Zoology*, **29**, 195–203.

Blanckenhorn, W.U. and Hosken, D.J. (2003) Heritability of three condition surrogates in the yellow dung fly. *Behavioral Ecology*, **14**, 612–618.

Blem, C.R. (1984) Ratios in avian physiology. *Auk*, **101**, 153–155.

Bolger, T. and Connolly, P.L. (1989) The selection of suitable indices for the measurement and analysis of fish condition. *Journal of Fish Biology*, **34**, 171–182.

Bondurianski, R. (2007) The evolution of condition-dependent sexual dimorphism. *American Naturalist*, **169**, 9–19.

Brandt, L.S.E. and Greenfield, M.D. (2004) Condition-dependent traits and the capture of genetic variance in male advertisement song. *Journal of Evolutionary Biology*, **17**, 821–828.

Buchholz, R. (1995) Female choice, parasite load and male ornamentation in wild turkeys. *Animal Behaviour*, **50**, 929–943.

Byers, J., Hebets, E., and Podos, J. (2010) Female mate choice based on male motor performance. *Animal Behaviour*, **79**, 771–778.

Candolin, U. (2000) Increased signalling effort when survival prospects decrease: male–male competition ensures honesty. *Animal Behaviour*, **60**, 417–422.

Carranza, J. and Hidalgo de Trucios, S.J. (1993) Condition-dependence and sex traits in the male great bustard. *Ethology*, **94**, 187–200.

Cattet, M.R.L., Caulkett, N.A., Obbard, M.E., and Stenhouse, G.B. (2002) A body-condition index for ursids. *Canadian Journal of Zoology*, **80**, 1156–1161.

Chapman, R.F. (1998) *The Insects: Structure and Function*, 4th edn, Harvard University Press, Cambridge, Massachusetts.

Chastel, O., Weimerskirch, H., and Jouventin, P. (1995) Influence of body condition on reproductive decision and reproductive success in the blue petrel. *The Auk*, **4**, 964–972.

Clutton-Brock, T.H., Guinness, F.E., and Albon, S.D. (1982) *Red Deer. Behavior and Ecology of Two Sexes*, University of Chicago Press, Chicago, IL.

Cotton, S., Fowler, K., and Pomiankowski, A. (2004) Do sexual ornaments demonstrate heightened condition-dependent expression as predicted by the handicap hypothesis? *Proceedings of the Royal Society of London, Series B: Biological Sciences*, **271**, 771–783.

Crastley, C.K. (2004) Flash signals, nuptial gifts and female preferences in *Photinus* fireflies. *Integrative and Comparative Biology*, **44**, 238–241.

Cuadrado, M. (1998) The influence of female size on the extent and intensity of mate guarding by males in *Chamaeleo chamaeleon*. *Journal of Zoology (London)*, **246**, 351–358.

David, P., Bjorksten, T., Fowler, K., and Pomiankowski, A. (2000) Condition-dependent signalling of genetic variation in stalk-eyed flies. *Nature*, **406**, 186–188.

Dobson, F.S. (1992) Body mass, structural size, and life history patterns of the Columbian ground squirrel. *American Naturalist*, **140**, 109–125.

Dobson, F.S. and Michener, G.R. (1995) Maternal traits and reproduction in Richardson's ground squirrels. *Ecology*, **76**, 851–862.

Dobson, F.S., Risch, T.S., and Murie, J.O. (1999) Increasing returns in the life-history of Columbian ground squirrels. *Journal of Animal Ecology*, **68**, 73–86.

Doty, G.V. and Welch, A.M. (2001) Advertisement call duration indicates good genes for offspring feeding rate in gray tree frogs (*Hyla versicolor*). *Behavioral Ecology and Sociobiology*, **49**, 150–156.

Doucet, S.M. and Montgomerie, R. (2003) Multiple sexual ornaments in satin bowerbirds: ultraviolet plumage and bowers signal different aspects of male quality. *Behavioral Ecology*, **14**, 503–509.

Dunlap, K.D. and Mathies, T. (1993) Effects of nymphal ticks and their interaction with malaria on the physiology of male fence lizards. *Copeia*, **1993**, 1045–1048.

Eagle, H. (1959) Amino acid metabolism in mammalian cell cultures. *Science*, **130**, 432–437.

Elowe, K.D. and Dodge, W.E. (1989) Factors affecting black bear reproductive success and cub survival. *Journal of Wildlife Management*, **53**, 962–968.

Emlen, D.J. (1994) Environmental control of horn length dimorphism in the beetle *Onthophagus acuminatus* (Coleoptera: Scarabaeidae). *Proceedings of the Royal Society of London, Series B: Biological Sciences*, **256**, 131–136.

Endler, J.A. (1983) Natural and sexual selection on color patterns in poeciliid fishes. *Environmental Biology of Fishes*, **9**, 173–190.

Evans, J.P., Kelley, J.L., Bisazza, A., Finazzo, E., and Pilastro, A. (2004) Sire attractiveness influences offspring performance in guppies. *Proceedings of the Royal Society of London, Series B: Biological Sciences*, **271**, 2035–2042.

Evans, R.M. and McMahon, B.F. (1987) Within-brood variation in growth and condition in relation to brood reduction in the American white pelican. *Wilson Bulletin*, **99**, 190–201.

Fisher, D.O. (1999) Offspring sex-ratio variation in the brindled nailtail wallaby, *Onychogalea fraenata*. *Behavioral Ecology and Sociobiology*, **45**, 411–419.

Gannes, L.Z. (2001) Comparative fuel use of migrating passerines: effects of fat stores, migration distance, and diet. *The Auk*, **118**, 665–677.

Garcia-Berthou, E. (2001) On the misuse of residuals in ecology: testing regression residuals vs. the analysis of covariance. *Journal of Animal Ecology*, **70**, 708–711.

Gonzalez, G., Sorci, G., Moler, A.P., Ninni, P., Haussy, C., and De Lope, F. (1999) Immunocompetence and condition-dependent sexual advertisement in male house sparrows (*Passer domesticus*). *Journal of Animal Ecology*, **68**, 1225–1234.

Grafen, A. (1990) Sexual selection unhandicapped by the Fisher process. *Journal of Theoretical Biology*, **144**, 473–516.

Gray, D.A. and Eckhardt, G. (2001) Is cricket courtship song condition dependent? *Animal Behaviour*, **62**, 871–877.

Green, A.J. (2001) Mass/length residuals: measures of body condition or generators of spurious results? *Ecology*, **82**, 1473–1483.

Greenfield, M.D. and Rodriguez, R.L. (2004) Genotype–environment interaction and the reliability of mating signals. *Animal Behaviour*, **68**, 1461–1468.

Greenstreet, S.P.R. (1992) Migration of hatchery reared juvenile Atlantic salmon, *Salmo salar* L. down a release ladder. 2. Effect of fish developmental strategy on speed and pattern of movement. *Journal of Fish Biology*, **40**, 667–681.

Grether, G.F. (2000) Carotenoid limitation and mate preference evolution: a test of the indicator hypothesis in guppies (*Poecilia reticulata*). *Evolution*, **54**, 1712–1724.

Griffith, S.C., Owens, I.P.F., and Burke, T. (1999) Female choice and annual reproductive success favour less-ornamented male house sparrows. *Proceedings of the Royal Society of London, Series B: Biological Sciences*, **266**, 765–770.

Hamer, K.C. and Furness, R.W. (1993) Parental investment and brood defence by male and female great skuas *Catharacta skua*: the influence of food supply, laying date, body size and body condition. *Journal of Zoology*, **230**, 7–18.

Hamilton, W.D. and Zuk, M. (1982) Heritable true fitness and bright birds: a role for parasites? *Science*, **218**, 384–387.

Heywood, J.S. (1989) Sexual selection by the handicap mechanism. *Evolution*, **43**, 1387–1397.

Hill, G.E. (1990) Female house finches prefer colourful males: sexual selection for a condition-dependent trait. *Animal Behavior*, **40**, 563–572.

Hill, G.E. (1991) Plumage coloration is a sexually selected indicator of male quality. *Nature*, **350**, 337–339.

Hill, G.E. (2000) Energetic constraints on expression of carotenoid-based plumage coloration. *Journal of Avian Biology*, **31**, 559–566.

Hoefler, C.D., Persons, M.H., and Rypstra, A.L. (2008) Evolutionarily costly courtship displays in a wolf spider: a test of viability indicator theory. *Behavioral Ecology*, **19**, 974–979.

Hoelzer, G.A. (1989) The good parent process of sexual selection. *Animal Behaviour*, **40**, 1067–1078.

Hoglund, J., Alatalo, R.V., and Lundberg, A. (1992) The effects of parasites on male ornaments and female choice in the lek-breeding black grouse (*Tetrao tetrix*). *Behavioral Ecology and Sociobiology*, **30**, 71–76.

Holzer, B., Jacot, A., and Brinkhof, M.W.G. (2003) Condition-dependent signaling affects male sexual attractiveness in field crickets, *Gryllus campestris*. *Behavioral Ecology*, **14**, 353–359.

Hooper, R.E., Tsubaki, Y., and Siva-Jothy, M.T. (1999) Expression of a costly, plastic secondary sexual trait is correlated with age and condition in a damselfly with two male morphs. *Physiological Entomology*, **24**, 364–369.

Houde, A.E. and Torio, A.J. (1992) Effect of parasitic infection on male color pattern and female choice in guppies. *Behavioral Ecology*, **3**, 346–351.

House, H.L. (1961) Insect nutrition. *Annual Review of Entomology*, **6**, 13–26.

Hunt, J., Brooks, R., Jennions, M.D., Smith, M.J., Bentsen, C.L., and Bussiere, L.F. (2004a) High-quality male field crickets invest heavily in sexual display but die young. *Nature*, **432**, 1024–1027.

Hunt, J., Bussiere, L.F., Jennions, M.D., and Brooks, R. (2004b) What is genetic quality? *Trends in Ecology and Evolution*, **19**, 329–333.

Irschick, D.J., Meyers, J.J., Husak, J.F., and Le Galliard, J.-F. (2008) How does selection operate on whole organism functional performance capacities? A review and synthesis. *Evolutionary Ecology Research*, **10**, 177–196.

Iwasa, Y. and Pomiankowski, A. (1999) Good parent and good genes models of handicap evolution. *Journal of Theoretical Biology*, **200**, 97–109.

Iwasa, Y., Pomiankowski, A., and Nee, S. (1991) The evolution of costly mate preferences. 2. The handicap principle. *Evolution*, **45**, 1431–1442.

Jakob, E.M. (1991) Costs and benefits of group living for pholcid spiderlings: losing food, saving silk. *Animal Behaviour*, **41**, 711–722.

Jakob, E.M., Marshall, S.D., and Uetz, G.W. (1996) Estimating fitness: a comparison of body condition indices. *Oikos*, **77**, 61–67.

Jenni, L. and Jenni-Eiermann, S. (1998) Fuel supply and metabolic constraints in migrating birds. *Journal of Avian Biology*, **29**, 521–528.

Jennions, M.D. and Backwell, P.R.Y. (1998) Variation in courtship rate in the fiddler crab *Uca annulipes*: is it related to male attractiveness? *Behavioral Ecology*, **9**, 605–611.

Jennions, M.D., Moller, A.P., and Petrie, M. (2001) Sexually selected traits and adult survival: a meta-analysis. *Quarterly Review of Biology*, **76**, 3–36.

Jessop, T.S., Hamann, M., and Limpus, C.J. (2004) Body condition and physiological changes in male green turtles during breeding. *Marine Ecology Progress Series*, **276**, 281–288.

Johnsen, A., Delhey, K., Andersson, S., and Kempenaers, B. (2003) Plumage colour in nestling blue tits: sexual dichromatism, condition dependence and genetic effects. *Proceedings of the Royal Society of London, Series B: Biological Sciences*, **270**, 1263–1270.

Johnstone, R.A. (1995) Sexual selection, honest advertisement and the handicap principle – reviewing the evidence. *Biological Reviews of the Cambridge Philosophical Society*, **70**, 1–65.

Johnstone, R.A. (1997) The evolution of animal signals, in *Behavioral Ecology: An Evolutionary Approach* (eds J.R. Krebs and N.B. Davies), Wiley-Blackwell, USA.

Judge, K.A. and Brooks, R.J. (2001) Chorus participation by male bullfrogs, *Rana catesbeiana*: a test of the energetic constraint hypothesis. *Animal Behaviour*, **62**, 849–861.

Juliano, S. (1986) Food limitation of reproduction and survival for populations of *Brachinus* (Coleoptera: Carabidae). *Ecology*, **67**, 1036–1045.

Keller, C., Diaz-Paniagua, C., and Andreu, A.C. (1997) Post-emergent field activity and growth rates of hatchling spur-thighed tortoises, *Testudo graeca*. *Canadian Journal of Zoology*, **75**, 1089–1098.

Keyser, A.J. and Hill, G.E. (1999) Condition-dependent variation in the blue-ultraviolet coloration of a structurally based plumage ornament. *Proceedings of the Royal Society of London, Series B: Biological Sciences*, **266**, 771–777.

Kodric-Brown, A. (1989) Dietary carotenoids and male mating success in the guppy: an environmental component to female choice. *Behavioral Ecology and Sociobiology*, **25**, 393–401.

Kotiaho, J.S. (1999) Estimating fitness: comparison of body condition indices revisited. *Oikos*, **87**, 399–400.

Kotiaho, J.S. (2001) Costs of sexual traits: a mismatch between theoretical considerations and empirical evidence. *Biological Reviews*, **76**, 365–376.

Kotiaho, J.S. (2002) Sexual selection and condition dependence of courtship display in three species of horned dung beetles. *Behavioral Ecology*, **13**, 791–799.

Kotiaho, J.S. and Puurtinen, M. (2007) Mate choice for indirect genetic benefits: scrutiny of the current paradigm. *Functional Ecology*, **21**, 638–644.

Kotiaho, J.S., Simmons, L.W., and Tomkins, J.L. (2001) Towards a resolution of the lek paradox. *Nature*, **410**, 684–686.

Kulling, D. and Milinski, M. (1992) Size-dependent predation risk and partner quality in predator inspection of sticklebacks. *Animal Behaviour*, **44**, 949–955.

Lailvaux, S. and Irschick, D.J. (2006) A functional perspective on sexual selection: insights and future prospects. *Animal Behavior*, **72**, 263–273.

Lebas, N.R. and Marshall, N.J. (2001) No evidence of female choice for a condition-dependent trait in the agamid lizard, *Ctenophorus ornatus*. *Behaviour*, **138**, 965–980.

Lomborg, J.P. and Toft, S. (2009) Nutritional enrichment increases courtship intensity and improves mating success in male spiders. *Behavioral Ecology*, **20**, 700–708.

Lozano, G.A. (1994) Carotenoids, parasites, and sexual selection. *Oikos*, **70**, 309–311.

Lozano, G.A. (2001) Carotenoids, immunity, and sexual selection: comparing apples and oranges? *American Naturalist*, **158**, 200–203.

Mappes, J., Alatalo, R.V., Kotiaho, J.S., and Parri, S. (1996) Viability costs of condition-dependent sexual male display in a drumming wolf spider. *Proceedings of the Royal Society of London, Series B: Biological Sciences*, **263**, 785–789.

Marden, J.H. and Rollins, R.A. (1994) Assessment of energy reserves by damselflies engaged in aerial contests for mating territories. *Animal Behaviour*, **44**, 949–955.

Matsumasa, M. and Murai, M. (2005) Changes in blood glucose and lactate levels of male fiddler crabs: effects of aggression and claw waving. *Animal Behaviour*, **69**, 569–577.

Matsuno, T. (2001) Aquatic animal carotenoids. *Fisheries Science*, **67**, 771–783.

Matthews, I.M., Evans, J.P., and Magurran, A.E. (1997) Male display rate reveals ejaculate characteristics in the Trinidadian guppy *Poecilia reticulata*. *Proceedings of the Royal Society of London, Series B: Biological Sciences*, **264**, 695–700.

McGlothlin, J.W., Duffy, D.L., Henry-Freeman, J.L., and Ketterson, E.D. (2007) Diet quality affects an attractive white plumage pattern in dark-eyed juncos (*Junco hyemalis*). *Behavioral Ecology and Sociobiology*, **61**, 1391–1399.

McGraw, K.J. and Ardia, D.R. (2003) Carotenoids, immunocompetence, and the information content of sexual colors: an experimental test. *American Naturalist*, **162**, 704–712.

McGraw, K.J. and Hill, G.E. (2000) Differential effects of endoparasitism on the expression of carotenoid- and melanin-based ornamental coloration. *Proceedings of the Royal Society of London, Series B: Biological Sciences*, **267**, 1525–1531.

McGraw, K.J., Hudon, J., Hill, G.E., and Parker, R.S. (2005) A simple and inexpensive chemical test for behavioral ecologists to determine the presence of carotenoid pigments in animal tissues. *Behavioral Ecology and Sociobiology*, **57**, 391–397.

McGraw, K.J., Mackillop, E.A., Dale, J., and Hauber, M.E. (2002) Different colors reveal different information: how nutritional stress affects the expression of melanin- and structurally based ornamental plumage. *Journal of Experimental Biology*, **205**, 3747–3755.

Merila, J. (1996) Genetic variation in offspring condition: an experiment. *Functional Ecology*, **10**, 465–474.

Merila, J., Kruuk, L.E.B., and Sheldon, B.C. (2001) Natural selection on the genetical component of variance in body condition in a wild bird population. *Journal of Evolutionary Biology*, **14**, 918–926.

Merila, J. and Svensson, E. (1997) Are fat reserves in migratory birds affected by condition in early life? *Journal of Avian Biology*, **28**, 279–286.

Missoweit, M., Engqvist, L., Lubjuhn, T., and Sauer, K.P. (2008) Nuptial feeding in the scorpionfly *Panorpa vulgaris*: maintenance of genetic variance in sexual advertisement through dependence on condition influencing traits. *Evolutionary Ecology*, **22**, 689–699.

Moller, A. (1987) Variation in badge size in male house sparrows *Passer domesticus*: evidence for status signalling. *Animal Behaviour*, **35**, 1637–1644.

Moller, A.P. and Alatalo, R.V. (1999) Good-genes effects in sexual selection. *Proceedings of the Royal Society of London, Series B: Biological Sciences*, **266**, 85–91.

Moller, A.P. and Petrie, M. (2002) Condition dependence, multiple sexual signals, and immunocompetence in peacocks. *Behavioral Ecology*, **13**, 248–253.

Morrison, R.I.G., Davidson, N.C., and Wilson, J.R. (2007) Survival of the fattest: body stores on migration and survival in red knots *Calidris canutus islandica*. *Journal of Avian Biology*, **38**, 479–487.

Mougeot, F., Perez-Rodriguez, L., Martinez-Padilla, J., Leckie, F., and Redpath, M. (2007) Parasites, testosterone and honest carotenoid-based signalling of health. *Functional Ecology*, **21**, 886–898.

Moya-Larano, J., Macias-Ordonez, R., Blanckenhorn, W.U., and Fernandez-Montraveta, C. (2008) Analysing body condition: mass, volume or density? *Journal of Animal Ecology*, **77**, 1099–1108.

Murphy, C.G. (1994) Determinants of chorus tenure in barking treefrogs (*Hyla gratiosa*). *Behavioral Ecology and Sociobiology*, **34**, 285–294.

Murray, D.L. (2002) Differential body condition and vulnerability to predation in snowshoe hares. *Journal of Animal Ecology*, **71**, 614–625.

Naef-Daenzer, B., Widmer, F., and Nuber, M. (2001) Differential post-fledging survival of great and coal tits in relation to their condition and fledging date. *Journal of Animal Ecology*, **70**, 730–738.

Neuman, R.M. and Flammang, M.K. (1997) Relative weight as a body condition index for chain pickerel. *Journal of Freshwater Ecology*, **12**, 19–26.

Nicoletto, P.F. (1993) Female sexual response to condition-dependent ornaments in the guppy, *Poecilia reticulata*. *Animal Behaviour*, **46**, 441–450.

Nowicki, S., Searcy, W.A., and Peters, S. (2002) Brain development, song learning and mate choice in birds: a review and experimental test of the "nutritional stress hypothesis". *Journal of Comparative Physiology A*, **188**, 1003–1014.

Nur, N. and Hasson, O. (1984) Phenotypic plasticity and the handicap principle. *Journal of Theoretical Biology*, **110**, 275–297.

Olson, V.A. and Owens, I.P.F. (1998) Costly sexual signals: are carotenoids rare, risky or required? *Trends in Ecology and Evolution*, **13**, 510–514.

Otronen, M. (1995) Energy reserves and mating success in males of the yellow dung fly, *Scathophaga sterocoraria*. *Functional Ecology*, **9**, 683–688.

Owen, M. (1981) Abdominal profile – a condition index for wild geese in the field. *The Journal of Wildlife Management*, **45**, 227–230.

Parker, T.H. and Garant, D. (2004) Quantitative genetics of sexually dimorphic traits and capture of genetic variance by a sexually-selected condition-dependent ornament in red junglefowl (*Gallus gallus*). *Journal of Evolutionary Biology*, **17**, 1277–1285.

Peig, J. and Green, A.J. (2009) New perspectives for estimating body condition from mass/length data: the scaled mass index as an alternative method. *Oikos*, **118**, 1883–1891.

Peig, J. and Green, A.J. (2010) The paradigm of body condition: a critical reappraisal of current methods based on mass and length. *Functional Ecology*, **24**, 1323–1332.

Perez-Rodriguez, L., Blas, J., Vinuela, J., Marchant, T.A., and Bortolotti, G.R. (2006) Condition and androgen levels: are condition-dependent and testosterone-mediated traits two sides of the same coin? *Animal Behaviour*, **72**, 97–103.

Pierce, C., Crowley, P., and Johnson, D. (1985) Behavior and ecological interactions of larval Odonata. *Ecology*, **66**, 1504–1512.

Piersma, T. and Davidson, N.C. (1991) Confusion of mass and size. *Auk*, **108**, 441–444.

Plaistow, S. and Siva-Jothy, M.T. (1996) Energetic constraints and male mate-securing tactics in the damselfly *Calopteryx splendens xanthostoma* (Charpentier). *Proceedings of the Royal Society of London, Series B: Biological Sciences*, **263**, 1233–1239.

Poston, J.P., Hasselquist, D., Stewart, I.R.K., and Westneat, D.F. (2005) Dietary amino acids influence plumage traits and immune responses of male house sparrows, *Passer domesticus*, but not as expected. *Animal Behaviour*, **70**, 1171–1181.

Qvarnstrom, A. (1999) Genotype-by-environment interactions in the determination of the size of a secondary sexual character in the collared flycatcher (*Ficedula albicollis*). *Evolution*, **53**, 1564–1572.

Ranta, E., Laurila, A., and Elmberg, J. (1994) Reinvinting the wheel: analysis of sexual dimorphism in body size. *Oikos*, **70**, 313–321.

Rantala, M.J., Kortet, R., Kotiaho, J.S., Vainikka, A., and Suhonen, J. (2003) Condition dependence of pheromones and immune function in the grain beetle *Tenebrio molitor*. *Functional Ecology*, **17**, 534–540.

Robin, J.P., Frain, M., Sardet, C., Groscolas, R., and Le Maho, Y. (1988) Protein and lipid utilization during long-term fasting in emperor penguins. *American Journal of Physiology: Regulatory, Integrative and Comparative Physiology*, **254**, R61–R68.

Rosen, R.F. and Tarvin, K.A. (2006) Sexual signals of the male American goldfinch. *Ethology*, **112**, 1008–1019.

Rowe, L. (1994) The costs of mating and mate choice in water striders. *Animal Behavior*, **48**, 1049–1056.

Rowe, L. and Houle, D. (1996) The lek paradox and the capture of genetic variance by condition dependent traits. *Proceedings of the Royal Society of London, Series B: Biological Sciences*, **263**, 1415–1421.

Rundus, A.S., Sullivan-Beckers, L., Wilgers, D.J., and Hebets, E.A. (2011) Females are choosier in the dark: environment-dependent reliance on courtship components and its impact on fitness. *Evolution*, **65**, 268–282.

Saks, L., Ots, I., and Horak, P. (2003) Carotenoid-based plumage coloration of male greenfinches reflects health and immunocompetence. *Oecologia*, **134**, 301–307.

Sarasola, J.H., Negro, J.J., and Travaini, A. (2004) Nutritional condition and serum biochemistry for free-living Swainson's hawks wintering in central Argentina. *Comparative Biochemistry and Physiology A*, **137**, 697–701.

Scheuber, H., Jacot, A., and Brinkhof, M.W.G. (2003a) Condition dependence of a multicomponent sexual signal in the field cricket *Gryllus campestris*. *Animal Behaviour*, **65**, 721–727.

Scheuber, H., Jacot, A., and Brinkhof, M.W.G. (2003b) The effect of past condition on a multicomponent sexual signal. *Proceedings of the Royal Society of London, Series B: Biological Sciences*, **270**, 1779–1784.

Schluter, D. and Gustafsson, L. (1993) Maternal inheritance of condition and clutch size in the collared flycatcher. *Evolution*, **47**, 658–667.

Schulte-Hostedde, A.I., Millar, J.S., and Hickling, G.J. (2001) Evaluating body condition in small mammals. *Canadian Journal of Zoology*, **79**, 1021–1029.

Schulte-Hostedde, A.I., Zinner, B., Millar, J.S., and Hickling, G.J. (2005) Restitution of mass-size residuals: validating body condition indices. *Ecology*, **86**, 155–163.

Schwilch, R., Grattarola, A., Spina, F., and Jenni, L. (2002) Protein loss during long-distance migratory flight in passerine birds: adaptation and constraint. *Journal of Experimental Biology*, **205**, 687–695.

Scott, D.E., Casey, E.D., Donovan, M.F., and Lynch, T.K. (2007) Amphibian lipid levels at metamorphosis correlate to post-metamorphic terrestrial survival. *Oecologia*, **153**, 521–532.

Senar, J.C., Figuerola, J., and Pascual, J. (2002) Brighter yellow blue tits make better parents. *Proceedings of the Royal Society of London, Series B: Biological Sciences*, **269**, 257–261.

Shamble, P.S., Wilgers, D.J., Swoboda, K.A., and Hebets, E.A. (2009) Courtship effort is a better predictor of mating success than ornamentation for male wolf spiders. *Behavioral Ecology*, **20**, 1242–1251.

Sheldon, B.C., Merila, J., Qvarnstrom, A., Gustafsson, L., and Ellegren, H. (1997) Paternal genetic contribution to offspring condition predicted by size of male secondary sexual character. *Proceedings of the Royal Society of London, Series B: Biological Sciences*, **264**, 297–302.

Shine, R., LeMaster, M.P., Moore, I.T., Olsson, M.M., and Mason, R.T. (2001) Bumpus in the snake den: effects of sex, size, and body condition on mortality of red-sided garter snakes. *Evolution*, **55**, 598–604.

Shykoff, J.A. and Widmer, A. (1996) Parasites and carotenoid-based signal intensity: how general should the relationship be? *Naturwissenschaften*, **83**, 113–121.

Sibly, R.M., Jones, P.T., and Houston, D.C. (1987) The use of body dimensions of lesser black-backed gulls *Larus fuscus* to indicate size and estimate body reserves. *Functional Ecology*, **1**, 275–279.

South, S.H., House, C.M., Moore, A.J., Simpson, S.J., and Hunt, J. (2011) Male cockroaches prefer a high carbohydrate diet that makes them more attractive to females: implications for the study of condition-dependence. *Evolution*, **65**, 1594–1606.

Suttie, J.M. and Kay, R.N.B. (1983) The influence of nutrition and photoperiod on the growth of antlers of young red deer, in *Antler Development in Cervidae* (ed R.D. Brown), Ceasar Kleberg Wildlife Reserve Institute, Kingsville, TX.

Taylor, L.A., Clark, D.L., and McGraw, K.J. (2011) Condition dependence of male display coloration in a jumping spider (*Habronattus pyrrithrix*). *Behavioral Ecology and Sociobiology*, **65**, 1133–1146.

Thompson, C.W., Hillgarth, N., Leu, M., and McClure, H.E. (1997) High parasite load in house finches (*Carpodacus mexicanus*) is correlated with reduced expression of a sexually selected trait. *American Naturalist*, **149**, 270–294.

Thomson, L.R., Toyoda, Y., Langner, A., Delori, K.M., Garnett, K.M., Craft, N., Nichols, C.R., Cheng, K.M., and Dorey, C.K. (2002) Elevated retinal zeaxanthin and prevention of light-induced photoreceptor cell death in quail. *Investigative Ophthalmology and Visual Science*, **43**, 3538–3549.

Tomkins, J.L., Radwan, J., Kotiaho, J.S., and Tregenza, T. (2004) Genic capture and resolving the lek paradox. *Trends in Ecology and Evolution*, **19**, 323–328.

Tonn, W.M., Paszkowski, C.A., and Holopainen, I.J. (1989) Responses of crucian carp populations to different predation pressure in a manipulated pond. *Canadian Journal of Zoology*, **67**, 2841–2849.

Torok, J., Hegyi, G., and Garamszegi, L.Z. (2003) Depigmented wing patch size is a condition-dependent indicator of viability in male collared flycatchers. *Behavioral Ecology*, **14**, 382–388.

Uetz, G.W., Papke, R., and Kilinc, B. (2002) Influence of feeding regime on body size, body condition and a male secondary sexual character in *Schizocosa ocreata* wolf spiders (Araneae, Lycosidae): condition-dependence in a visual signaling trait. *Journal of Arachnology*, **30**, 461–469.

van Berkum, F., Huey, R., Tsuji, J., and Garland, T. Jr. (1989) Repeatability of individual differences in locomotor performance and body size during early ontogeny of the lizard *Sceloporus occidentalis* (Baird and Girard). *Functional Ecology*, **3**, 97–105.

van Homrigh, A., Higgie, M., McGulgan, K., and Blows, M.W. (2007) The depletion of genetic variance by sexual selection. *Current Biology*, **17**, 528–532.

van Marken Lichtenbelt, W.D., Wesselingh, R.A., Vogel, J.T., and Albers, K.B.M. (1993) Energy budgets in free living iguanas in a seasonal environment. *Ecology*, **74**, 1157–1172.

Vanpé, C., Gaillard, J.M., Kjellander, P., Mysterud, A., Magnien, P., Delorme, D., van Laere, G., Klein, F., Liberg, O., and Hewison, A.J.M. (2007) Antler size provides an honest signal of male phenotypic quality in roe deer. *The American Naturalist*, **169**, 481–493.

Veiga, J.P. (1993) Badge size, phenotypic quality, and reproductive success in the house sparrow: a study on honest advertisement. *Evolution*, **47**, 1161–1170.

Vleck, C.M. and Vleck, D. (2002) Physiological condition and reproductive consequences in Adelie penguins. *Integrative and Comparative Biology*, **42**, 76–83.

von Schantz, T., Bensch, S., Grahn, M., Hasselquist, D., and Wittzell, H. (1999) Good genes, oxidative stress and condition-dependent sexual signals. *Proceedings of the Royal Society of London, Series B: Biological Sciences*, **266**, 1–12.

Wagner, W.E. and Harper, C.J. (2003) Female life span and fertility are increased by the ejaculates of preferred males. *Evolution*, **57**, 2054–2066.

Wagner, W.E. and Hoback, W.W. (1999) Nutritional effects on male calling behaviour in the variable field cricket. *Animal Behaviour*, **57**, 89–95.

Watson, P.J. (1990) Female-enhanced male competition determines the first mate and principal sire in the spider *spurthighed litigiosa* (Linyphiidae). *Behavioral Ecology and Sociobiology*, **26**, 77–90.

Wauters, L.A. and Dhondt, A.A. (1995) Lifetime reproductive success and its correlates in female Eurasian red squirrels. *Oikos*, **72**, 402–410.

Weatherhead, P.J., Barry, F.E., Brown, G.P., and Forbes, M.R.L. (1995) Sex ratios, mating behavior and sexual size dimorphism of the northern water snake, *Nerodia sipedon*. *Behavioral Ecology and Sociobiology*, **36**, 301–311.

Weatherhead, P.J., Dufour, K.W., Lougheed, S.C., and Eckert, C.G. (1999) A test of the good-genes-as-heterozygosity hypothesis using red-winged blackbirds. *Behavioral Ecology*, **10**, 619–625.

Welch, A.M., Semlitsch, R.D., and Gerhardt, H.C. (1998) Call duration as an indicator of genetic quality in male gray tree frogs. *Science*, **280**, 1928–1930.

Welch, K.C. Jr., Bakken, B.H., del Rio, C.M., and Suarez, R.K. (2006) Hummingbirds fuel hovering flight with newly ingested sugar. *Physiological and Biochemical Zoology*, **79**, 1082–1087.

Welch, K.C. Jr., Herrera, G.M., and Suarez, R.K. (2008) Dietary sugar as a direct fuel for flight in the nectarivorous bat *Glossophaga soricina*. *Journal of Experimental Biology*, **211**, 310–316.

West-Eberhard, M.J. (1979) Sexual selection, social competition, and evolution. *Proceedings of the American Philosophical Society*, **123**, 222–234.

Wilder, S.M. and Rypstra, A.L. (2008) Diet quality affects mating behaviour and egg production in a wolf spider. *Animal Behaviour*, **76**, 439–445.

Wilgers, D.J. and Hebets, E.A. (2011) Complex courtship displays facilitate male reproductive success and plasticity in signalling across variable environments. *Current Zoology*, **57**, 175–186.

Wilgers, D.J., Nicholas, A.C., Reed, D.H., Stratton, G.E., and Hebets, E.A. (2009) Condition-dependent alternative mating tactics in a sexually cannibalistic wolf spider. *Behavioral Ecology*, **20**, 891–900.

Woodroffe, R. (1995) Body condition affects implantation date in the European badger, *Meles meles*. *Journal of Zoology (London)*, **236**, 183–188.

Yuval, B., Holliday-Hanson, M.L., and Washing, R.K. (1994) Energy budget of swarming male mosquitos. *Ecological Entomology*, **19**, 74–78.

Zahavi, A. (1975) Mate selection – selection for a handicap. *Journal of Theoretical Biology*, **53**, 205–214.

Zahavi, A. (1977) The cost of honesty (further remarks on the handicap principle). *Journal of Theoretical Biology*, **67**, 603–605.

Zeh, D.W. and Zeh, J.A. (1988) Condition-dependent sex ornaments and field tests of sexual selection theory. *American Naturalist*, **132**, 454–459.

Zera, A.J. and Harshman, L.G. (2001) The physiology of life history trade-offs in animals. *Annual Review of Ecological Systems*, **32**, 95–126.

Zuk, M. and Kolluru, G.R. (1998) Exploitation of sexual signals by predators and parasitoids. *Quarterly Review of Biology*, **73**, 415–438.

Zuk, M., Thornhill, R., Ligon, J.D., and Johnson, K. (1990) Parasites and mate choice in red jungle fowl. *American Zoologist*, **30**, 235–244.

Animal Signaling and Function: An Integrative Approach, First Edition.
Edited by Duncan J. Irschick, Mark Briffa, and Jeffrey Podos.
© 2015 John Wiley & Sons, Inc. Published 2015 by John Wiley & Sons, Inc.